Barron's Regents Exams and Answers

Biology
The Living Environment

GABRIELLE I. EDWARDS
Former Science Consultant
Board of Education
City of New York
Former Assistant Principal Supervision
Science Department
Franklin D. Roosevelt High School
Brooklyn, New York

MARION CIMMINO
Former Teacher, Biology and Laboratory Techniques
Franklin D. Roosevelt High School
Brooklyn, New York

FRANK J. FODER
Teacher, Advanced Placement Biology
Franklin D. Roosevelt High School
Brooklyn, New York

G. SCOTT HUNTER
Former Teacher, Biology and Advanced Biology
Former Consultant, State Education Department Bureaus of Science and Testing
Former School Business Administrator, Schodack, NY
Former Superintendent of Schools
Mexico and Chatham, New York

BARRON'S

Barron's Educational Series, Inc.

Published by Kaplan, Inc., d/b/a Barron's Educational Series
750 Third Avenue
New York, NY 10017
www.barronseduc.com

ISBN: 978-0-8120-3197-3
ISSN 1069-2940

Printed in Canada

9 8 7 6 5 4 3 2

Contents

How to Use This Book

ORGANIZATION OF THE BOOK

Study Questions

Section 1 of the book consists of 173 questions taken from past Regents examinations in biology. Most of the questions require that you select a correct response from four choices given. A few questions provide you with a list of words or phrases from which to select the one that best matches a given description. Others are constructed-response, graphical analysis, or reading comprehension questions. You should become familiar with the formats of the questions that appear on the Living Environment Regents examination.

Each question in Section 1 of this book has a well-developed answer. Each answer provides the number of the correct response, the reason why the response is correct, and explanations of why the other choices are incorrect.

A useful feature of Section 1 is the student Self-Appraisal Guide. This device allows you to determine where your learning strengths and weaknesses lie in the major topics of each unit. For specific topics within the units, the numbers of related questions are given. As you attempt to answer each question in Section 1, you may wish to circle on the appraisal form the numbers of the questions that you are unable to answer. The circled items then help you to identify at a glance subject matter areas in which you need additional study.

New York State Learning Standards

COMMENCEMENT STANDARDS

There are several commencement standards required of students in New York State public schools regarding their performance in math, science, and technology. The Core Curriculum for The Living Environment addresses two of these standards:

***Standard 1*: Students will use mathematical analysis, scientific inquiry, and engineering design, as appropriate, to pose questions, seek answers, and develop solutions.**

***Standard 4*: Students will understand and apply scientific concepts, principles, and theories pertaining to the physical setting and living environment and recognize the historical development of ideas in science.**

The Core Curriculum for The Living Environment was built from these two commencement standards. It is important to recognize that the Core Curriculum is not a syllabus. It does not prescribe what will be taught and learned in any particular classroom. Instead, it defines the skills and understandings that you must master in order to achieve the commencement standards for life science.

Instead of memorizing a large number of details at the commencement level, then, you are expected to develop the skills needed to deal with science on the investigatory level, generating new knowledge from experimentation and sharpening your abilities in data analysis. You are also expected to read and understand scientific literature, taking from it the facts and concepts necessary for a real understanding of issues in science. You are required to pull many facts together from different sources to develop your own opinions about the moral and ethical problems facing modern society concerning technological advances. These are thinking skills that do not respond to simple memorization of facts and scientific vocabulary.

KEY IDEAS, PERFORMANCE INDICATORS, AND MAJOR UNDERSTANDINGS

Each commencement standard is subdivided into a number of Key Ideas. Key Ideas are broad, unifying statements about what you need to know. Within Standard 1, three Key Ideas are concerned with laboratory investigation and data analysis. Together, these unifying principles develop your ability to deal with data and understand how professional science is carried out in biology. Within Standard 4, seven Key Ideas present a set of concepts that are central to the science of biology. These Key Ideas develop your understanding of the essential characteristics of living things that allow them to be successful in diverse habitats.

Within each Key Idea, Performance Indicators are presented that indicate which skills you should be able to demonstrate through your mastery of the Key Idea. These Performance Indicators give guidance to both you and your teacher about what is expected of you as a student of biology.

Performance Indicators are further subdivided into Major Understandings. Major Understandings give specific concepts that you must master in order to achieve each Performance Indicator. It is from these Major Understandings that the Regents assessment material will be drawn.

LABORATORY COMPONENT

A meaningful laboratory experience is essential to the success of this or any other science course. You are expected to develop a good sense of how scientific inquiry is carried out by the professional scientist and how these same techniques can assist in the full understanding of concepts in science. The Regents requirement of 1,200 minutes of successful laboratory experience, coupled with satisfactory written reports of your findings, should be considered a minimum.

Students are required to complete four laboratory experiences required by the New York State Education Department and tested on the Regents Examination. See Barron's *Let's Review: Biology—The Living Environment* for a complete treatment of this requirement.

Regents Examinations

Section 2 of the book consists of actual Biology Regents examinations and answers. These Regents examinations are based on the New York State Core Curriculum for the Living Environment.

Assessments: Format and Scoring

The format of the Regents assessment for The Living Environment is as follows, based on actual Regents examinations.

Part A: Variable number of multiple-choice questions that test the student's knowledge of specific factual information. **All** questions must be answered on Part A.

Part B: Variable number of questions, representing a mixture of multiple-choice and constructed-response items. Questions may be based on the student's direct knowledge of biology, interpretation of experimental data, analysis of readings in science, and ability to deal with representations of biological phenomena. **All** questions must be answered on Part B.

Part C: Variable number of multiple-choice and constructed-response questions. Questions may be based on the student's direct knowledge of biology, interpretation of experimental data, analysis of readings in science, and ability to deal with representations of biological phenomena. **All** questions must be answered on Part C.

Part D—Laboratory component: This component of the examination aims to assess student knowledge of and skills on any of four required laboratory experiences supplied to schools by the New York State Education Department. Questions in Part D can be a combination of multiple-choice and constructed-response questions similar to those found in Parts A, B, and C of the Living Environment Regents Examination. The content of these questions will reflect specific laboratory experiences. You are strongly encouraged to include review of these laboratory experiences as part of your year-end Regents preparation activity.

The following chart summarizes the current laboratory requirement for New York State public schools:

LABORATORY REQUIREMENTS

Laboratory Title	Description
The Beaks of Finches	Explores the adaptive advantages of beaks with different physical characteristics
Relationships and Biodiversity	Explores the relationship between DNA structure and the biochemistry of inheritance
Making Connections	Explores the effects of physical activity on human metabolic activities
Diffusion Through a Membrane	Explores the nature of cross-membrane transport in living cells
Adaptations for Reproductive Success in Flowering Plants*	Not yet available
DNA Technology*	Not yet available
Environmental Conditions and Seed Germination*	Not yet available

*Not yet available as of this writing.

Studying questions from past Regents examinations is an invaluable aid in developing a mind set that will enable you to approach questions with understanding. Although exact questions are not repeated, question types are repeated. If you practice questions that require interpretation, problem solving, and graph construction, you will do well on the entire exam. During the school year, the 30 required laboratory lessons teach you certain manipulative skills. Questions involving identification, measurement, and other laboratory procedures are based on the laboratory exercises. Review of past materials gives you insight as to the types of questions that you may be asked to answer. Study the questions in the Regents exams in this book diligently.

HOW TO STUDY

General Suggestions for Study

You've spent all year learning many different facts and concepts about biology—far more than you could ever hope to remember the "first time around." Your teacher has drilled you on these facts and concepts; you've done homework, taken quizzes and tests, performed laboratory experiments, and reviewed the material at intervals throughout the year.

Now it's time to put everything together. The Regents exam is only a few weeks away. If you and your teacher have planned properly, you will have finished all new information about 3 weeks before the exam. Now you have to make efficient use of the days and weeks ahead to review all that you've learned in order to score high on the Regents.

The task ahead probably seems impossible, but it doesn't have to be! You've actually retained much more of the year's material than you realize! The review process should be one that helps you to recall the many facts and concepts you've stored away in your memory. Your Barron's resources, including Barron's *Let's Review: The Living Environment,* will help you to review this material efficiently.

You also have to get yourself into the right frame of mind. It won't help to be nervous and stressed during the review process. The best way to avoid being stressed during any exam is to be well rested, prepared, and confident. We're here to help you prepare and to build your confidence. So let's get started on the road to a successful exam experience!

To begin, carefully read and follow the steps outlined below.

1. Start your review early; don't wait until the last minute. Allow at least 2 weeks to prepare for the Regents exam. Set aside an hour or two a day over the next few weeks for your review. Less than an hour a day is insufficient time for you to concentrate on the material meaningfully; more than 2 hours daily will yield diminishing returns on your investment of time.
2. Find a quiet, comfortable place to study. You should seat yourself at a well-lighted work surface, free of clutter, in a room without distractions of any kind. You may enjoy watching TV or listening to music curled up in a soft chair, but these and other diversions should be avoided when doing intense studying.
3. Make sure you have the tools you need, including this book, a pen and pencil, and some scratch paper for taking notes and doing calculations. Keep your class notebook at hand for looking up information between test-taking sessions. It will also help to have a good review text, such as

Barron's *Let's Review: The Living Environment,* available for reviewing important concepts quickly and efficiently.

4. Concentrate on the material in the "Study Questions and Answers" section of this book. Read carefully and thoughtfully. Think about the questions that you review, and try to make sense out of them. Choose the answers carefully. (See the following section, "Using This Book for Study," for additional tips on question-answering techniques.)
5. Use available resources, including a dictionary and the glossary in this book, to look up the meanings of unfamiliar words in the practice questions. Remember that the same terms can appear on the Regents exam you will take, so take the opportunity to learn them now.
6. Remember: study requires time and effort. Your investment in study now will pay off when you take the Regents exam.

Using This Book for Study

This book is an invaluable tool if used properly. Read carefully and try to answer *all* the questions in Section 1 and on the practice exams. The more you study and practice, the more you will increase your knowledge of biology and the likelihood that you will earn a high grade on the Regents exam. To maximize your chances, use this book in the following way:

1. Answer all of the questions in the section entitiled "Study Questions and Answers." Check your responses by using "Answers to Topic Questions," including "Wrong Choices Explained," following each question set. Record the number of correct responses on each topic in the "Self-Appraisal Guide" at the end of the section to identify your areas of strength and weakness. Use a good review text, such as Barron's *Let's Review: The Living Environment,* to study each area on which you did poorly. Finally, go back to the questions you missed on the first round and be sure that you fully understand what each question asks and why the correct answer is what it is.
2. When you have completed the questions in "Study Questions and Answers," go on to the examination section. Select the first complete examination and take it under test conditions.
3. Interpret the term *test conditions* as follows:
   - Be well rested; get a good night's sleep before attempting *any* exam.
   - Find a quiet, comfortable room in which to work.
   - Allow no distractions of any kind.
   - Select a well-lighted work surface free of clutter.
   - Have your copy of this book with you.
   - Bring to the room a pen, a pencil, some scratch paper, and a watch or alarm clock set for the 3-hour exam limit.

4. Take a deep breath, close your eyes for a moment, and RELAX! Tell yourself, "I know this stuff!" You have lots of time to take the Regents exam; use it to your advantage by reducing your stress level. Forget about your plans for later. For the present, your number 1 priority is to do your best, whether you're taking a practice exam in this book, or the real thing.
5. Read all test directions carefully. Note how many questions you must answer to complete each part of the exam. If test questions relate to a reading passage, diagram, chart, or graph, be sure you fully understand the given information before you attempt to answer the questions that relate to it.
6. When answering multiple-choice questions on the Regents exam, TAKE YOUR TIME! Pay careful attention to the "stem" of the question; read it over several times. These questions are painstakingly written by the test preparers, and every word is chosen to convey a specific meaning. If you read the question carelessly, you may answer a question that was never asked! Then read each of the four multiple-choice answers carefully, using a *pencil* to mark in the test booklet the answer you think is correct.
7. Remember that three of the multiple-choice answers are *incorrect;* these incorrect choices are called "distracters" because they seem like plausible answers to poorly prepared or careless students. To avoid being fooled by these distracters, you must think clearly, using everything you have learned about biology since the beginning of the year. This elimination process is just as important to your success on the Living Environment Regents exam as knowing the correct answer! If more than one answer seems to be correct, reread the question to find the words that will help you to distinguish between the correct answer and the distracters. When you have made your best judgment about the correct answer, circle the number in *pencil* in your test booklet.
8. Constructed-response questions appear in a number of different forms. You may be asked to select a term from a list, write the term on the answer sheet, and define the term. You may be asked to describe some biological phenomenon or state a biological fact using a complete sentence. You may be asked to read a value from a diagram of a measuring instrument and write that value in a blank on the answer paper. When answering this type of question, care should be exercised to follow directions precisely. If a complete sentence is called for, it must contain a subject and a verb, must be punctuated, and must be written understandably in addition to answering the scientific part of the question accurately. Values must be written clearly and accurately and include a unit of measure, if appropriate. Failure to follow the directions for a question may result in a loss of credit for that question.

9. A special type of constructed-response question is the essay or paragraph question. Typically, essay or paragraph questions provide an opportunity to earn multiple credits for answering the question correctly. As in the constructed-response questions described above, it is important that you follow the directions given if you hope to earn the maximum number of credits for the question. Typically, the question outlines exactly what must be included in your essay to gain full credit. Follow these directions step by step, double-checking to be certain that all question components are addressed in your answer. In addition, your essay or paragraph should follow the rules of good grammar and good communication so that it is readable and understandable. And, of course, it should contain correct information that answers all the parts of the question asked.
10. Graphs and charts are a special type of question that requires you to organize and represent data in graphical format. Typically for such questions, you are expected to place unorganized data in ascending order in a data chart or table. You may also be asked to plot organized data on a graph grid, connect the plotted points, and label the graph axes appropriately. Finally, questions regarding data trends and extrapolated projections may be asked, requiring you to analyze the data in the graph and draw inferences from it. As with all examination questions, always follow all directions for the question. Credit can be granted only for correctly following directions and accurately interpreting the data.
11. When you have completed the exam, relax for a moment. Check your time; have you used the entire 3 hours? Probably not. Resist the urge to quit. Go back to the beginning of the exam, and, in the time remaining, *retake the exam in its entirety*. Try to ignore the penciled notations you made the first time. If you come up with a different answer the second time through, read over the question with extreme care before deciding which response is correct. Once you have decided on the correct answer, mark your choice in ink in the answer booklet.
12. Score the exam using the Answer Key at the end of the exam. Review the "Answers Explained" section for each question to aid your understanding of the exam and the material. Remember that it's just as important to understand why the incorrect responses are incorrect as it is to understand why the correct responses are correct!
13. Finally, focus your between-exam study on your areas of weakness in order to improve your performance on the next practice exam. Complete all the practice exams in this book using the techniques outlined above.
14. Be sure to sign the declaration on your answer sheet. Unless this declaration is signed, your paper cannot be scored.

Test-Taking Tips—A Summary

The following pages contain seven tips to help you achieve a good grade on the Living Environment Regents exam.

TIP 1

Be confident and prepared.

SUGGESTIONS

- Review previous tests.
- Use a clock or watch, and take previous exams at home under examination conditions (i.e., don't have the radio or television on).
- Get a review book. (The preferred book is Barron's *Let's Review: The Living Environment*.)
- Talk over the answers to questions on these tests with someone else, such as another student in your class or someone at home.
- Finish all your homework assignments.
- Look over classroom exams that your teacher gave during the term.
- Take class notes carefully.
- Practice good study habits.
- Know that there are answers for every question.
- Be aware that the people who made up the Regents exam want you to pass.
- Remember that thousands of students over the last few years have taken and passed a Biology Regents. You can pass, too!
- Complete your study and review at least one day before taking the examination. Last-minute "cramming" may hurt, rather than enhance, your performance on the exam.

- Visit *www.barronseduc.com* or *www.barronsregents.com* for the latest information on the Regents exams.
- Use Barron's website to communicate directly with subject specialists.
- Be well rested when you enter the exam room. A good night's sleep is essential preparation for any examination.
- On the night before the exam day: lay out all the things you will need, such as clothing, pens, and admission cards.
- Bring with you two pens, two pencils, an eraser, and, if your school requires it, an identification card. Decide before you enter the room that you will remain for the entire 3-hour examination period, and either bring a wristwatch or sit where you can see a clock.
- Once you are in the exam room, arrange things, get comfortable, and attend to personal needs (the bathroom).
- Before beginning the exam, take a deep breath, close your eyes for a moment, and RELAX! Repeat this technique any time you feel yourself "tensing up" during the exam.
- Keep your eyes on your own paper; do not let them wander over to anyone else's paper.
- Be polite in making any reasonable demands of the exam room proctor, such as changing your seat or having window shades raised or lowered.

TIP 2

Read test instructions carefully.

SUGGESTIONS

- Be familiar with the format of the examination.
- Know how the test will be graded.
- Read all directions carefully. Be sure you fully understand supplemental information (reading passages, charts, diagrams, graphs) before you attempt to answer the questions that relate to it.
- Underline important words and phrases.
- Ask for assistance from the exam room proctor if you do not understand the directions.

TIP 3

Read each question carefully and read each choice before recording your answer.

SUGGESTIONS

- When answering the questions, TAKE YOUR TIME! Be sure to read the "stem" of the question and each of the four multiple-choice answers very carefully.
- If you are momentarily "stumped" by a question, put a check mark next to it and go on; come back to the question later if you have time.
- Remember that three of the multiple-choice answers (known as "distracters") are incorrect. If more than one answer seems to be correct, reread the question to find the words that will help you to distinguish between the correct answer and the distractors.
- When you have made your best judgment about the correct answer, circle the appropriate number in pencil on your answer sheet.

TIP 4

Budget your test time (3 hours).

SUGGESTIONS

- Bring a watch or clock to the test.
- The Regents examination is designed to be completed in 1½ to 2 hours.
- If you are absolutely uncertain of the answer to a question, mark your question booklet and move on to the next question.
- If you persist in trying to answer every difficult question *immediately,* you may find yourself rushing or unable to finish the remainder of the examination.
- When you have completed the exam, relax for a moment. Then go back to the beginning, and, in the time remaining, *retake the exam in its entirety*. Pay particular attention to questions you skipped the first time. Once you have decided on a correct response for multiple-choice questions, mark an "X" in ink through the penciled circle on the answer sheet.

- Plan to stay in the room for the entire three hours. If you finish early, read over your work—there may be some things that you omitted or that you may wish to add. You also may wish to refine your grammar, spelling, and penmanship.

TIP 5
Use your reasoning skills.

SUGGESTIONS

- Answer *all* questions.
- Relate (connect) the question to anything that you studied, wrote in your notebook, or heard your teacher say in class.
- Relate (connect) the question to any film, demonstration, or experiment you saw in class, any project you did, or to anything you may have learned from newspapers, magazines, or television.
- Look over the entire test to see whether one part of it can help you answer another part.

TIP 6
Don't be afraid to guess.

SUGGESTIONS

- In general, go with your first answer choice.
- Eliminate obvious incorrect choices.
- If still unsure of an answer, make an educated guess.
- There is no penalty for guessing; therefore, answer ALL questions. An omitted answer gets no credit.

TIP 7

Sign the Declaration.

SUGGESTIONS

- Be sure to sign the declaration on your answer sheet.
- Unless this declaration is signed, your paper cannot be scored.

New York State Regents Biology—The Living Environment: Core Curriculum and Assessment Map

The chart on pages 17–22 lists the Key Ideas and Performance Indicators from the New York State Regents Commencement Standard 1 and Standard 4 for *The Living Environment* (1999). For each Regents examination (assessment) listed across the top of the chart, item (question) numbers have been matched to the Performance Indicator(s) most closely associated with their content.

The information compiled in this chart represents a "map" showing how closely these assessments have mirrored the curriculum required by the state of New York for this course of study. It provides guidance to the teacher and student concerning the emphasis placed on each Performance Indicator by the New York State Education Department assessment development process on four recent Regents examinations.

By following this map, you will be sure to hit every major topic that is required by the state of New York for achievement of the Standards for *The Living Environment*. At the same time, you will be better able to focus your efforts on areas of needed study that will maximize your performance on the Regents examination without wasting valuable time on areas of the curriculum that are less likely to be represented on the examination.

Although some areas of the Core Curriculum have been less-well emphasized than others on Regents examinations to date, both teachers and students are cautioned that *all* parts of the Core Curriculum are required and are subject to testing. The Regents examination may be changed at any time and without notice to include items testing Performance Indicators that may not have been well emphasized in the past.

Students, you can find in Barron's *Let's Review: Biology—The Living Environment* (Hunter) an appropriate body of content, easy-to-read explanation, and practice exercises (with Answers Explained) developed especially for the New York State high school biology student. This resource will ease your way through *The Living Environment* course of study from start to finish and will prepare you well for the culminating Regents examination.

Teachers, you are encouraged to develop Standards-Based Learning Units (SBLUs) and Essential Questions that will provide appropriate local content to illustrate these required Performance Indicators. Barron's *Let's Review: Biology—The Living Environment* (Hunter) provides an excellent source of material for the development of SBLUs and Essential Questions that is consistent with the New York State Core Curriculum for this course. In addition, it includes many concepts that have traditionally been taught in the New York State Regents biology program since 1989 and relates this content to the new Regents Core Curriculum for *The Living Environment*.

LIVING ENVIRONMENT CURRICULUM MAP AND QUESTION INDEX

STANDARD 1, KEY IDEA 1: THE CENTRAL PURPOSE OF SCIENTIFIC INQUIRY IS TO DEVELOP EXPLANATIONS OF NATURAL PHENOMENA IN A CONTINUING AND CREATIVE PROCESS.

Performance Indicator	Description	August 2016	June 2017	August 2017	June 2018
1.1	The student should be able to elaborate on basic scientific and personal explanations of natural phenomena and develop extended visual models and mathematical formulations to represent one's thinking.		47	33	
1.2	The student should be able to hone ideas through reasoning, library research, and discussion with others, including experts.	39			
1.3	The student should be able to work toward reconciling competing explanations and clarify points of agreement and disagreement.		72	52	31, 32
1.4	The student should be able to coordinate explanations at different levels of scale, points of focus, and degrees of complexity and specificity, and recognize the need for such alternative representations of the natural world.	47	54, 55		

NOTE TO TEACHERS: Periodically, errors are made in assigning Regents exam questions to the appropriate New York State Curriculum Standards (a.k.a.: attribution). Every effort has been made to assure that each exam question is attributed to the appropriate Standard.

STANDARD 1, KEY IDEA 2: BEYOND THE USE OF REASONING AND CONSENSUS, SCIENTIFIC INQUIRY INVOLVES THE TESTING OF PROPOSED EXPLANATIONS INVOLVING THE USE OF CONVENTIONAL TECHNIQUES AND PROCEDURES AND USUALLY REQUIRING CONSIDERABLE INGENUITY.

Performance Indicator	Description	August 2016	June 2017	August 2017	June 2018
2.1	The student should be able to devise ways of making observations to test proposed explanations.			31, 37	33
2.2	The student should be able to refine research ideas through library investigations, including electronic information retrieval and reviews of literature, and through peer feedback obtained from review and discussion.				
2.3	The student should be able to develop and present proposals including formal hypotheses to test explanations (i.e., predict what should be observed under specific conditions if the experiment is true).			70	59
2.4	The student should be able to carry out research for testing explanations, including selecting and developing techniques, acquiring and building apparatus, and recording observations as necessary.	32, 56, 57	40, 49		

STANDARD 1, KEY IDEA 3: THE OBSERVATIONS MADE WHILE TESTING PROPOSED EXPLANATIONS, WHEN ANALYZED USING CONVENTIONAL AND INVENTED METHODS, PROVIDE NEW INSIGHTS INTO NATURAL PHENOMENA.

Performance Indicator	Description	August 2016	June 2017	August 2017	June 2018
3.1	The student should be able to use various methods of representing and organizing observations (e.g., diagrams, tables, charts, graphs, equations, matrices) and insightfully interpret the organized data.	44, 45	44, 45, 46	36, 44, 45, 47	44, 45
3.2	The student should be able to apply statistical analysis techniques when appropriate to test if chance alone explains the results.				
3.3	The student should be able to assess correspondence between the predicted result contained in the hypothesis and the actual result, and reach a conclusion as to whether the explanation on which the prediction was based is supported.		61		37, 60
3.4	The student should be able to, based on the results of the test and through public discussion, revise the explanation and contemplate additional research.	33, 36	63	66	
3.5	The student should be able to develop a written report for public scrutiny that describes the proposed explanation, including a literature review, the research carried out, its result, and suggestions for further research.				

STANDARD 4, KEY IDEA 1: LIVING THINGS ARE BOTH SIMILAR TO AND DIFFERENT FROM EACH OTHER AND FROM NONLIVING THINGS.

Performance Indicator	Description	August 2016	June 2017	August 2017	June 2018
1.1	The student should be able to explain how diversity of populations within ecosystems relates to the stability of ecosystems.	14, 52, 53, 54, 55, 69	11, 24, 30*, 43, 53	1, 6, 22, 29	11, 14, 24
1.2	The student should be able to describe and explain the structures and functions of the human body at different organizational levels (e.g., systems, tissues, cells, organelles).	1, 31, 40, 48	18, 28, 29, 37	5, 9, 54, 55, 60, 63	7, 8, 29, 35, 42, 43, 63
1.3	The student should be able to explain how a one-celled organism is able to function despite lacking the levels of organization present in more complex organisms.	38, 46	27, 37, 39	24	

STANDARD 4, KEY IDEA 2: ORGANISMS INHERIT GENETIC INFORMATION IN A VARIETY OF WAYS THAT RESULT IN CONTINUITY OF STRUCTURE AND FUNCTION BETWEEN PARENTS AND OFFSPRING.

Performance Indicator	Description	August 2016	June 2017	August 2017	June 2018
2.1	The student should be able to explain how the structure and replication of genetic material result in offspring that resemble their parents.	5, 11, 15, 35	34, 38, 41	7, 23, 34	2, 15, 17, 66, 67, 68
2.2	The student should be able to describe and explain how the technology of genetic engineering allows humans to alter genetic makeup of organisms.	4, 8, 49, 50, 51	20, 36, 42, 64, 66	4, 16, 25, 48, 50	4, 22

*Question 30 was eliminated from the June 2017 exam.

STANDARD 4, KEY IDEA 3: INDIVIDUAL ORGANISMS AND SPECIES CHANGE OVER TIME.

Performance Indicator	Description	August 2016	June 2017	August 2017	June 2018
3.1	The student should be able to explain the major patterns of evolution.	9, 13, 16, 17, 18, 37	8, 9, 22, 23, 25, 26, 31, 48	17, 26, 40, 41, 49, 71, 72	34, 48, 49, 55, 61, 62

STANDARD 4, KEY IDEA 4: THE CONTINUITY OF LIFE IS SUSTAINED THROUGH REPRODUCTION AND DEVELOPMENT.

Performance Indicator	Description	August 2016	June 2017	August 2017	June 2018
4.1	The student should be able to explain how organisms, including humans, reproduce their own kind.	12, 19, 20, 21, 22	19, 52, 56, 57, 60	18, 27, 28, 32, 51, 65	9, 13, 16, 18, 21, 23, 69, 70

STANDARD 4, KEY IDEA 5: ORGANISMS MAINTAIN A DYNAMIC EQUILIBRIUM THAT SUSTAINS LIFE.

Performance Indicator	Description	August 2016	June 2017	August 2017	June 2018
5.1	The student should be able to explain the basic biochemical processes in living organisms and their importance in maintaining dynamic equilibrium.	2, 23	15, 17, 33, 50, 51	3, 12, 42, 43, 61, 62	27, 39, 41, 56
5.2	The student should be able to explain disease as a failure of homeostasis.	24, 42, 63, 64	13, 16, 58, 59, 62	35, 53	12, 40, 71, 72
5.3	The student should be able to relate processes at the system level to the cellular level in order to explain dynamic equilibrium.	7, 10, 25, 41, 43	14	2, 38, 39, 46	3, 38, 47, 64, 65

STANDARD 4, KEY IDEA 6: PLANTS AND ANIMALS DEPEND ON EACH OTHER AND THEIR PHYSICAL ENVIRONMENT.

Performance Indicator	Description	August 2016	June 2017	August 2017	June 2018
6.1	The student should be able to explain factors that limit growth of individuals and populations.	65, 66, 67, 70		56, 57, 68	52
6.2	The student should be able to explain the importance of preserving diversity of species and habitats.	26, 27, 71, 72	2, 10, 12, 21, 67, 71	10, 58, 67, 69	1, 6, 25, 30, 50, 51
6.3	The student should be able to explain how the living and nonliving environments change over time and respond to disturbances.	28	68, 69, 70	11	53, 54

STANDARD 4, KEY IDEA 7: HUMAN DECISIONS AND ACTIVITIES HAVE A PROFOUND IMPACT ON THE PHYSICAL AND LIVING ENVIRONMENT.

Performance Indicator	Description	August 2016	June 2017	August 2017	June 2018
7.1	The student should be able to describe the range of interrelationships of humans with the living and nonliving environment.		7	19	57
7.2	The student should be able to explain the impact of technological development and growth in the human population on the living and nonliving environment.	3, 29, 34, 58, 60, 61	4, 5, 6, 32, 35	15, 30	5, 10, 19, 20, 36, 46
7.3	The student should be able to explain how individual choices and societal actions can contribute to improving the environment.	6, 30, 59, 68	1, 3, 65	8, 13, 14, 20, 21, 59	26, 28, 58

Tips for Teachers

CLASSROOM USE

All teachers will be able to use this book with their students as a companion to their regular textbooks and will find that their students gain considerable self-confidence and ability in test taking through its consistent use.

The Living Environment Core Curriculum (1999) defines the skills and abilities students should have at the point of commencement at the upper-secondary level. It is assumed that science concepts have been taught and assessed at an age-appropriate level throughout their career, so that little additional detail needs to be presented at the upper-secondary level.

An excellent companion to this book (and any comprehensive biology text) is Barron's *Let's Review: Biology—The Living Environment* (Hunter). The factual material and organization of this book lend themselves well to the development of Standards-Based Learning Units (SBLUs) and Essential Questions. The level of detail is consistent with what students really have to know in order to do well on the New York State Regents Examination on the Living Environment.

APPLICATION-BASED CURRICULUM

The curriculum focus can be characterized as application-based—one that is less concerned about content and more concerned about thinking. It is less about *how much* students know and more about *what they can do* with what they know. The latter, after all, is what real learning is all about; these are the abilities that will last a lifetime, not facts and scientific terminology.

This being said, it is acknowledged that students will have a difficult time expressing their views and making moral and ethical judgments about science if they lack a working knowledge of scientific principles and do not have at least a passing understanding of the terms used by biologists. For this reason, teachers and administrators will need to develop local curricula that complement the Core Curriculum. It is up to the teacher or administrator to decide what examples and factual knowledge best illustrate the concepts presented in the Core Curriculum, what concepts need to be reinforced and enhanced, what experiences will add measurably to students' understanding of science, and what examples of local interest should be included.

The teacher will immediately recognize the need to go beyond this level in the classroom, with examples, specific content, and laboratory experiences that complement and illuminate these Major Understandings. It is at this level that the locally developed curriculum is essential. Each school system is challenged to develop an articulated K-12 curriculum in mathematics, science, and technology that will position students to achieve a passing standard at the elementary and intermediate levels, such that success is maximized at the commencement level.

The addition of factual content must be accomplished without contradicting the central philosophy of the learning standards. If local curricula merely revert to the fact-filled syllabi of the past, then little will have been accomplished in the standards movement other than to add yet another layer of content and requirements on the heads of students. A balance must be struck between the desire to build students' ability to think and analyze and the desire to add to the content they are expected to master.

LABORATORY EXPERIENCE

The reduction of factual detail in the Core Curriculum (1982–1999) should allow a more in-depth treatment of laboratory investigations to be planned and carried out than was possible under the previous syllabus. Laboratory experiences should be designed to address Standard 1 (inquiry techniques) but should also take into account Standards 2 (information systems), 6 (interconnectedness of content), and 7 (problem-solving approaches). They should also address the laboratory skills listed in Appendix A of the Core Curriculum.

Part D of the examination assesses student knowledge of and skills on any of four required laboratory experiences supplied to schools by the New York State Education Department. The specific laboratory experiences required in any year will vary according to a preset schedule (see chart on next page).

Questions on this section can be a combination of multiple-choice and constructed-response questions similar to those found in Parts A, B, and C of the Living Environment Regents Examination. The content of these questions reflect the four specific laboratory experiences required for a particular year. Teachers are strongly encouraged to include review of these laboratory experiences as part of their year-end Regents preparation activity.

The following chart summarizes the current laboratory requirement for New York State public schools:

Laboratory Title	**Description**
The Beaks of Finches	Explores the adaptive advantages of beaks with different physical characteristics
Relationships and Biodiversity	Explores the relationship between DNA structure and the biochemistry of inheritance
Making Connections	Explores the effects of physical activity on human metabolic activities
Diffusion Through a Membrane	Explores the nature of cross-membrane transport in living cells
Adaptations for Reproductive Success in Flowering Plants*	Not yet available
DNA Technology*	Not yet available
Environmental Conditions and Seed Germination*	Not yet available

*Not yet available as of this writing

Study Questions and Answers

QUESTIONS ON STANDARD 1—Scientific Inquiry: Students will use mathematical analysis, scientific inquiry, and engineering design, as appropriate, to pose questions, seek answers, and develop solutions.

Key Idea 1—Purpose of Scientific Inquiry: The central purpose of scientific inquiry is to develop explanations of natural phenomena in a continuing and creative process.

Performance Indicator	Description
1.1	The student should be able to elaborate on basic scientific and personal explanations of natural phenomena and develop extended visual models and mathematical formulations to represent one's thinking.
1.2	The student should be able to hone ideas through reasoning, library research, and discussion with others, including experts.
1.3	The student should be able to work toward reconciling competing explanations and clarify points of agreement and disagreement.
1.4	The student should be able to coordinate explanations at different levels of scale, points of focus, and degrees of complexity and specificity, and recognize the need for such alternative representations of the natural world.

Base your answers to questions 1 through 4 on the passage below and on your knowledge of biology.

To Tan or Not to Tan

Around 1870, scientists discovered that sunshine could kill bacteria. In 1903, Niels Finsen, an Icelandic researcher, won the Nobel Prize for his use of sunlight therapy against infectious diseases. Sunbathing then came into wide use as a treatment for tuberculosis, Hodgkin's disease (a form of cancer), and common wounds. The discovery of vitamin D, the "sunshine vitamin," reinforced the healthful image of the Sun. People learned that it was better to live in a sun-filled home than a dark dwelling. At that time, the relationship between skin cancer and exposure to the Sun was not known.

In the early twentieth century, many light-skinned people believed that a deep tan was a sign of good health. However, in the 1940s, the rate of skin cancer began to increase and reached significant proportions by the 1970s. At this time, scientists began to realize how damaging deep tans could really be.

Tanning occurs when ultraviolet radiation is absorbed by the skin, causing an increase in the activity of melanocytes, cells that produce the pigment melanin. As melanin is produced, it is absorbed by cells in the upper region of the skin, resulting in the formation of a tan. In reality, the skin is building up protection against damage caused by the ultraviolet radiation. It is interesting to note that people with naturally dark skin also produce additional melanin when their skin is exposed to sunlight.

Exposure to more sunlight means more damage to the cells of the skin. Research has shown that, although people usually do not get skin cancer as children, each time a child is exposed to the Sun without protection, the chance of that child getting skin cancer as an adult increases.

Knowledge connecting the Sun to skin cancer has greatly increased since the late 1800s. Currently, it is estimated that ultraviolet radiation is responsible for more than 90% of skin cancers. Yet, even with this knowledge, about 2 million Americans use tanning parlors that expose patrons to high doses of ultraviolet radiation. A recent survey showed that at least 10% of these people would continue to do so even if they knew for certain that it would give them skin cancer.

Many of the deaths due to this type of cancer can be prevented. The cure rate for skin cancer is almost 100% when it is treated early. Reducing exposure to harmful ultraviolet radiation helps to prevent it. During the past 15 years, scientists have tried to undo the tanning myth. If the word "healthy" is separated from the word "tan," maybe the occurrence of skin cancer will be reduced.

1. State *one* known benefit of daily exposure to the Sun. [1]

___

2. Explain what is meant by the phrase "the tanning myth." [1]

___

3. Which statement concerning tanning is correct?
(1) Tanning causes a decrease in the ability of the skin to regulate body temperature.
(2) Radiation from the Sun is the only radiation that causes tanning.
(3) The production of melanin, which causes tanning, increases when skin cells are exposed to the Sun.
(4) Melanocytes decrease their activity as exposure to the Sun increases, causing a protective coloration on the skin. 3____

4. Which statement concerning ultraviolet radiation is *not* correct?
(1) It may damage the skin.
(2) It stimulates the skin to produce antibodies.
(3) It is absorbed by the skin.
(4) It may stimulate the skin to produce excess pigment. 4____

5. Current knowledge concerning cells is a result of the investigations and observations of many scientists. The work of these scientists forms a well-accepted body of knowledge about cells. This body of knowledge is an example of a
(1) hypothesis (3) theory
(2) controlled experiment (4) research plan 5____

6. In his theory of evolution, Lamarck suggested that organisms will develop and pass on to offspring variations that they need in order to survive in a particular environment. In a later theory of evolution, Darwin proposed that changing environmental conditions favor certain variations that promote the survival of organisms. Which statement is best illustrated by this information?
(1) Scientific theories that have been changed are the only ones supported by scientists.
(2) All scientific theories are subject to change and improvement.
(3) Most scientific theories are the outcome of a single hypothesis.
(4) Scientific theories are not subject to change. 6____

Base your answers to questions 7 and 8 on the passage below and on your knowledge of biology.

The number in the parentheses ( ) at the end of a sentence is used to identify that sentence.

They Sure Do Look Like Dinosaurs

When making movies about dinosaurs, film producers often use ordinary lizards and enlarge their images many times (1). We all know, however, that although they look like dinosaurs and are related to dinosaurs, lizards are not actually dinosaurs (2).

Recently, some scientists have developed a hypothesis that challenges this view (3). These scientists believe that some dinosaurs were actually the same species as some modern lizard that had grown to unbelievable sizes (4). They think that such growth might be due to a special type of DNA called repetitive DNA, often referred to as "junk" DNA because scientists do not understand its functions (5). These scientists studied pumpkins that can reach sizes of nearly 1,000 pounds and found them to contain large amounts of repetitive DNA (6). Other pumpkins that grow to only a few ounces in weight have very little of this kind of DNA (7). In addition, cells that reproduce uncontrollably have almost always been found to contain large amounts of this DNA (8).

7. State *one* reason why scientists formerly thought of repetitive DNA as "junk." [1]

____

____

8. Write the number of the sentence that provides evidence supporting the hypothesis that increasing amounts of repetitive DNA are responsible for increased sizes of organisms. [1]

____

ANSWERS EXPLAINED Questions 1–8

Standard 1, Key Idea 1, Performance Indicators 1–4

1. One response is required. Acceptable responses include:

- *Kills bacteria*
- *Produces vitamin D*
- *Treats diseases and/or wounds*

2. One response is required. Acceptable responses include:
- *The "tanning myth" involves people believing that a tan is a sign of good health.*
- *The "tanning myth" says that a good tan is good for people.*

3. **3** *The production of melanin, which causes tanning, increases when skin cells are exposed to the Sun* is the correct statement concerning tanning. Melanin is a dark pigment that is produced in specialized skin cells in response to ultraviolet radiation in sunlight or an artificial source. This information is found in the third paragraph of the passage.

WRONG CHOICES EXPLAINED:

(1) *Tanning causes a decrease in the ability of the skin to regulate body temperature* is not a correct statement concerning tanning. There is no known relationship between tanning and body temperature regulation.

(2) *Radiation from the Sun is the only radiation that causes tanning* is not a correct statement concerning tanning. Tanning can also occur when the skin is exposed to artificial sources of ultraviolet radiation. A reference is made in the fourth paragraph of the passage to "tanning parlors" where people can be exposed to artificial doses of ultraviolet radiation.

(4) *Melanocytes decrease their activity as exposure to the Sun increases, causing a protective coloration on the skin* is not a correct statement concerning tanning. Melanin is produced as a protective pigment that helps prevent deep penetration of ultraviolet radiation into the deep layers of the skin. When ultraviolet radiation is absorbed by melanocytes, their activity increases, not decreases.

4. **2** *It stimulates the skin to produce antibodies* is not a correct statement concerning ultraviolet radiation. There is no information in the passage relating to the production of antibodies as a result of absorption of ultraviolet radiation, and no known research indicates this type of relationship.

WRONG CHOICES EXPLAINED:

(1), (3), (4) *It may damage the skin, it is absorbed by the skin,* and *it may stimulate the skin to produce excess pigment* are all correct statements concerning ultraviolet radiation. Ultraviolet radiation is an invisible but extremely powerful form of electromagnetic radiation. It can penetrate unshielded living tissues and alter the genetic makeup of the cells it encounters. In humans, this radiation can cause the production of melanin from melanocytes; in extreme cases, it can stimulate the growth of skin cancer.

5. **3** The body of knowledge described in this question is an example of a *theory*. When scientists begin to study a phenomenon in nature, their first step is normally to investigate it through repeated observation and experimentation. As a result of the analysis of the large quantity of data gathered during this process, the scientists then formulate a theory ("well-accepted body of knowledge") that describes the phenomenon in a way that is consistent with the data.

WRONG CHOICES EXPLAINED:

(1) A *hypothesis* is not the body of knowledge described in this question. Scientists develop a hypothesis ("educated guess") around their preliminary observations concerning a natural phenomenon. The hypothesis may be proven accurate or inaccurate as a result of the experimentation used to test it. For this reason, a hypothesis cannot be considered a "well-accepted body of knowledge."

(2) A *controlled experiment* is not the body of knowledge described in this question. A controlled experiment is a scientific method used to test an experimental hypothesis. The data that results from a controlled experiment can be used to support the development of a "well-accepted body of knowledge," but it does not constitute that body of knowledge.

(4) A *research plan* is not the body of knowledge described in this question. A research plan may be a series of controlled experiments designed to test various aspects of a natural phenomenon. The data that results from the research plan can be used to support the development of a "well-accepted body of knowledge," but it does not constitute that body of knowledge.

6. **2** *All scientific theories are subject to change and improvement* is the statement best illustrated by the information given. Both Lamarck and Darwin developed their theories of evolution based on observations made and inferences drawn before there was a good understanding of the genetic basis of variation. Lamarck's earlier theory of "use and disuse" was disproven by later experiments of other scientists. Darwin's later theory of "natural selection," though much closer to the currently accepted scientific theory of evolution, has been modified and improved on by the work of later scientists who have had the benefit of modern-day research in genetics, paleontology, and other sciences.

WRONG CHOICES EXPLAINED:

(1) *Scientific theories that have been changed are the only ones supported by scientists* is not the statement best illustrated by the information given. Scientists generally support theories that have stood the test of good scientific research. A theory that has not changed, as long as it is still supported by such research, is generally supported by most scientists.

(3) *Most scientific theories are the outcome of a single hypothesis* is not the statement best illustrated by the information given. In fact, scientific theories are based on the results of many experiments that each contain their own independent hypotheses.

(4) *Scientific theories are not subject to change* is not the statement best illustrated by the information given. Scientists are constantly questioning and reevaluating scientific theories. It is likely that a vast majority of all scientific theories undergo at least some modification.

7. One response is required that indicates a reason why scientists formerly thought of repetitive DNA as "junk." Acceptable responses include: [1]

- *Scientists did not understand the function of repetitive DNA.*
- *They didn't know what it did, and so they thought it was junk.*

8. One credit is allowed for indicating that either sentence 6 or sentence 7 provides evidence supporting the hypothesis that increased amounts of repetitive DNA are responsible for increased sizes of organisms. These sentences give information about the results of scientific investigations that measured the amount of repetitive DNA in the cells of giant pumpkins and miniature pumpkins and found that giant pumpkins contain more of this kind of DNA than miniature pumpkins.

Key Idea 2—Methods of Scientific Inquiry: Beyond the use of reasoning and consensus, scientific inquiry involves the testing of proposed explanations involving the use of conventional techniques and procedures and usually requiring considerable ingenuity.

Performance Indicator	Description
2.1	The student should be able to devise ways of making observations to test proposed explanations.
2.2	The student should be able to refine research ideas through library investigations, including electronic information retrieval and reviews of literature, and through peer feedback obtained from review and discussion.
2.3	The student should be able to develop and present proposals including formal hypotheses to test explanations (i.e., predict what should be observed under specific conditions if the experiment is true).
2.4	The student should be able to carry out research for testing explanations, including selecting and developing techniques, acquiring and building apparatus, and recording observations as necessary.

Base your answers to questions 9 and 10 on the diagram below of the field of view of a light compound microscope and on your knowledge of microscopes.

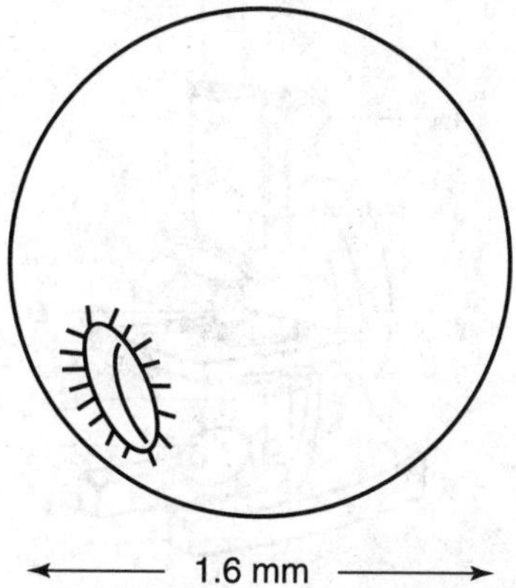

9. In order to center the organism in the field of view, the slide should be moved

(1) down and to the right
(2) down and to the left
(3) up and to the right
(4) up and to the left

9____

10. The approximate length of the organism is

(1) 500 µm
(2) 1,600 µm
(3) 50 µm
(4) 1.6 µm

10____

11. After viewing an organism under low power, a student switches to high power. The student should first

(1) adjust the mirror
(2) center the organism
(3) raise the objective and switch to high power
(4) close the diaphragm

11____

12. Using one or more complete sentences, explain why a specimen viewed under the high-power objective of a microscope appears darker than when it is viewed under low power.

____

____

Base your answers to questions 13 and 14 on the diagram below of a compound light microscope.

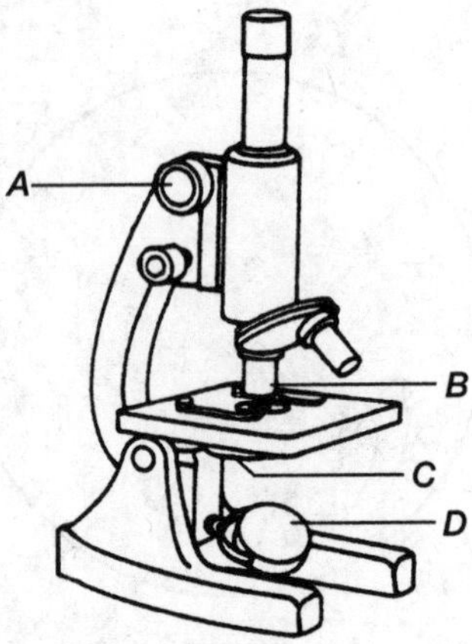

13. The letter *C* represents
(1) the mirror
(2) the diaphragm
(3) the eyepiece
(4) the high-power objective
13_____

14. Select and name *one* of the labeled parts, and in one or more complete sentences describe its function.

____

____

15. The letter "p" as it normally appears in print is placed on the stage of a compound light microscope. Which best represents the image observed when a student looks through the microscope?
(1) p (2) q (3) b (4) d
15_____

16. To separate the parts of a cell by differences in density, a biologist would probably use
(1) a microdissection instrument
(2) an ultracentrifuge
(3) a compound light microscope
(4) an electron microscope
16_____

17. The diagram below represents the field of view of a microscope. What is the approximate diameter, in micrometers, of the cell shown in the field?

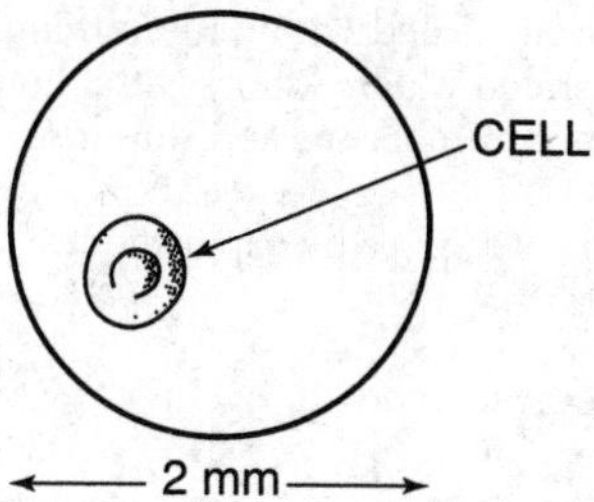

(1) 50 μm
(2) 500 μm
(3) 1,000 μm
(4) 2,000 μm

17_____

18. Base your answer to the question on the diagram below.

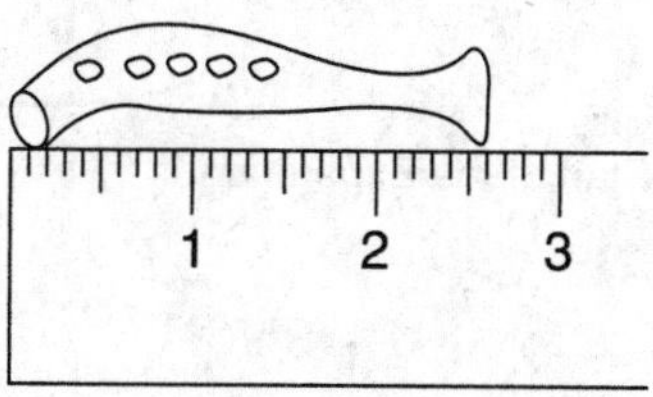

How many millimeters long is the organism resting on the metric ruler?

____

Questions 19 through 21 are based on the experiment described below.

A test tube was filled with a molasses solution, sealed with a membrane, and inverted into a beaker containing 200 mL of distilled water. A second test tube was filled with a starch solution, sealed with a membrane, and inverted into a beaker containing 200 mL of distilled water. After several hours, the water in each beaker was tested for the presence of molasses and starch.

The diagrams show the setup of the experiment.

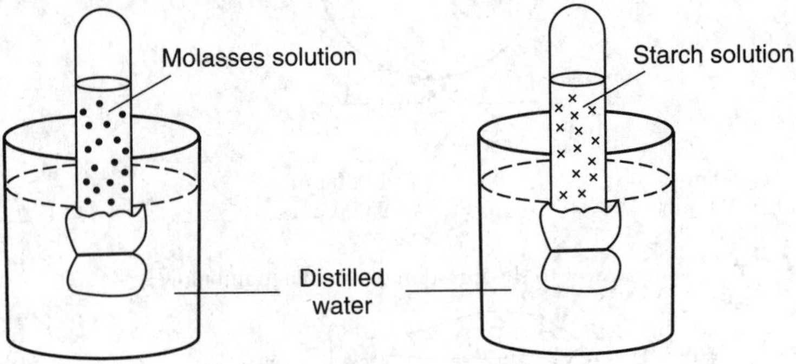

At the Start of the Experiment

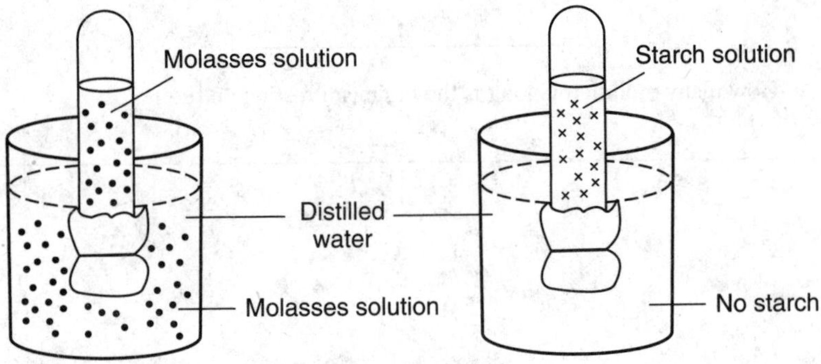

After Several Hours

Answer each question related to the experiment in one or more complete sentences.

19. What principle was being tested in the experiment?

20. What reagents were used in the experiment to test for the presence of molasses and starch?

___

21. Draw one conclusion from this experiment.

___

Questions 22 and 23 are based on the experiment described below.

An opaque disk was placed on several leaves of a geranium plant. The remaining leaves of the plant were untreated. After the plant had been exposed to sunlight, a leaf on which a disk had been placed was removed and tested as shown in parts *B* and *C* of the diagram below.

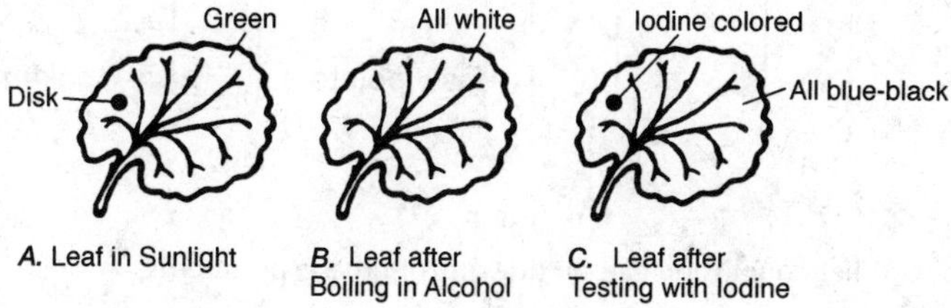

Answer each question related to the experiment in one or more complete sentences.

22. What conclusion can be drawn from the result of the experiment?

___

23. What process was being investigated by the experiment?

___

Questions 24 and 25 are based on the experiment described below.

A student added 15 mL of water to each of three test tubes, labeled *A*, *B*, and *C*. A 1-cc piece of raw potato was added to tube *B*. A 1-cc piece of cooked potato was added to tube *C*. Five drops of hydrogen peroxide (H_2O_2) were added to each test tube. The results are shown in the following diagram.

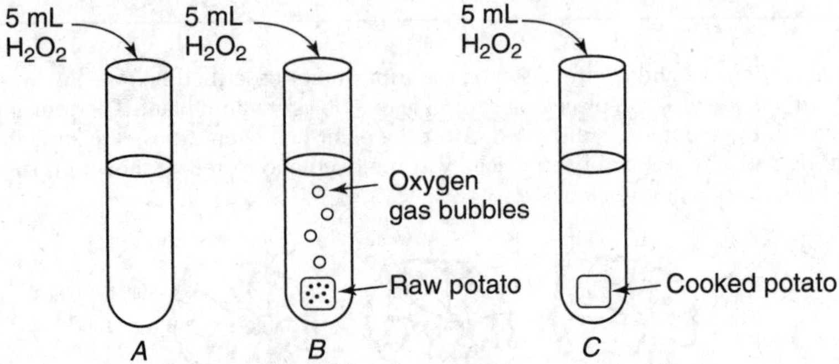

24. What conclusion can be drawn from the experiment?

---

25. Which test tube is the control? Explain the reason for your choice.

Base your answers to questions 26 through 28 on the diagram of the measuring device shown below.

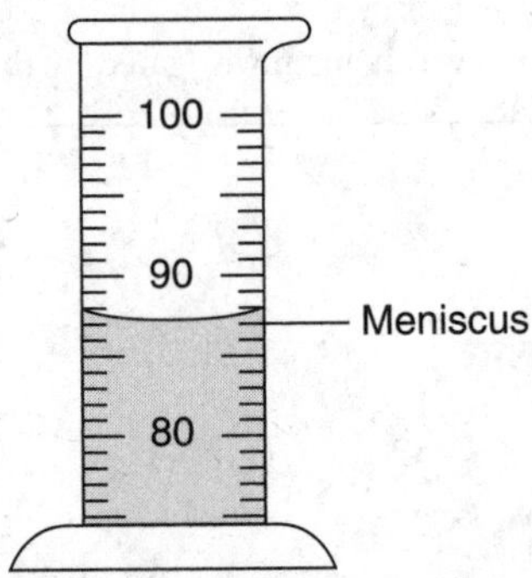

26. What is the name of this measuring device?

---

27. In one complete sentence describe the procedure that you would follow to read the meniscus.

____

28. What must a student do to obtain a volume of 85 milliliters of liquid in this measuring device?

(1) Add 2.0 mL.
(2) Remove 2.0 mL.
(3) Add 2.5 mL.
(4) Remove 8.7 mL.

28_____

ANSWERS EXPLAINED Questions 9–28

Standard 1, Key Idea 2, Performance Indicators 1–4

9. **2** Specimens viewed under the microscope appear upside-down, backward, and reversed.

WRONG CHOICES EXPLAINED:

(1), (3), (4) With any of these choices, the specimen would be moved out of the field of view.

10. **1** The field of view is given as 1.6 mm. 1 mm = 1000 µm. 1.6 mm × 1000 µm = 1600 µm. The diagram shows that three specimens would fit across the field of view. One-third of 1600 µm = 533 µm. Of the choices given, *500 µm* (choice 1) is closest to this value.

WRONG CHOICES EXPLAINED:

(2), (3), (4) Each of these choices is mathematically incorrect.

11. **2** The student should first *center the organism*. The field of view is smaller under high power; therefore, less of a specimen can be seen. If the organism is not centered, it may fall out of the field of view under high power.

WRONG CHOICES EXPLAINED:

(1) The *mirror is adjusted* for maximum light under low power. Because the diameter of the high-power objective is very small, it is impossible to adjust the light under high power.

(3) A compound light microscope is parfocal; that is, it is not necessary to *lift the high-power objective* to focus under high power. The specimen remains in focus when switching from low power to high power.

(4) *Closing the diaphragm* reduces the amount of light entering the objective. Therefore, the specimen would appear very dark and would be difficult to see.

12. *The diameter of the high-power objective is smaller than the diameter of the low-power objective. Less light enters through the high-power objective, and therefore the specimen appears darker.*

13. **2** The letter *C* represents *the diaphragm.*

WRONG CHOICES EXPLAINED:

(1) *The mirror* is represented by *D.*

(3), (4) *The eyepiece* and *the high-power objective* are not labeled on the diagram.

14. *Coarse adjustment (A)—used to focus a specimen under the low-power objective.*

or

Low-power objective (B)—along with the standard eyepiece, magnifies a specimen 100×.

or

Diaphragm (C)—regulates the amount of light entering the objectives.

or

Mirror (D)—provides a source of light that illuminates the specimen.

15. **4** The image of a specimen as seen under a microscope is upside-down (*d*). The right side is on the left side, and the top is on the bottom.

WRONG CHOICES EXPLAINED:

(1) In this choice (*p*) there is no change in the way the image of the letter appears.

(2) In this choice (*q*) the image of the letter is reversed in only one direction: The right and left sides are reversed.

(3) In this choice (*b*) the image of the letter is reversed in only one direction: The top and bottom are reversed.

16. **2** The *ultracentrifuge* is a machine that spins at a very high speed. A test tube of a liquid containing the parts of ruptured cells is placed in the machine. Each cell part has its own density (mass per unit volume). When the machine rotates, the cell parts fall to different levels in the test tube depending on their density.

WRONG CHOICES EXPLAINED:

(1) *A microdissection instrument* enables a biologist to remove a cell part from a single living cell. A micromanipulator is an example of such an instrument.

(3) A cell is transparent under a light microscope. Its structures cannot be seen unless the cell is stained. *A compound light microscope* can be used to view, but not to separate, cell parts.

(4) *An electron microscope* uses beams of electrons to view freeze-dried specimens; it cannot be used to separate cell parts for study.

17. **2** Study the information given in the diagram. Notice that the diameter of the circle is 2 mm. Since 1 mm is equal to 1,000 μm, 2 mm are equal to 2,000 μm. In relation to the entire circle, how large is the cell? Is the cell one-half as large or one-fourth as large? Dividing the circle into four parts shows us that the diameter of the cell is about one-quarter the diameter of the circle. Dividing 4 into 2,000 results in *500 μm*.

WRONG CHOICES EXPLAINED:

(1) *50 μm* is too small. The cell is ten times larger than 50.
(3) *1,000 μm* is too large. The cell is not one-half the diameter.
(4) *2,000 μm* is the diameter of the circle. The cell is only one-fourth as large.

18. The organism is *26 millimeters* long.

19. The principle of *diffusion* was being tested in the experiment.

20. *Benedict's solution* was used to test for the presence of molasses in the beaker. *Iodine* was used to test for the presence of starch in the beaker.

21. *Molasses can diffuse through a membrane.*

or

Starch cannot diffuse through a membrane.

22. *No starch was produced in the area covered by the disk.*

23. *The process of photosynthesis was being investigated.*

24. *Raw potato contains an enzyme that breaks down hydrogen peroxide.*

or

Cooking a potato destroys the enzyme that breaks down hydrogen peroxide.

25. *Test tube A is the control*. A control is the part of the experiment that provides the basis of comparison for the variable being tested.

26. The device is known as a *graduated cylinder*.

27. *The meniscus should be read at eye level.*

28. **2** To obtain a volume of 85 mL, *2.0 mL* must be removed. The graduated cylinder contains 87 mL of liquid.

Key Idea 3—Analysis in Scientific Inquiry: The observations made while testing proposed explanations, when analyzed using conventional and invented methods, provide new insights into natural phenomena.

Performance Indicator	Description
3.1	The student should be able to use various methods of representing and organizing observations (e.g., diagrams, tables, charts, graphs, equations, matrices) and insightfully interpret the organized data.
3.2	The student should be able to apply statistical analysis techniques, when appropriate, to test if chance alone explains the results.
3.3	The student should be able to assess correspondence between the predicted result contained in the hypothesis and actual result, and reach a conclusion as to whether the explanation on which the prediction was based is supported.
3.4	The student should be able to, based on the results of the test and through public discussion, revise the explanation and contemplate additional research.
3.5	The student should be able to develop a written report for public scrutiny that describes the proposed explanation, including a literature review, the research carried out, its result, and suggestions for further research.

29. If curve *A* in the diagram represents a population of hawks in a community, what would most likely be represented by curve *B*?

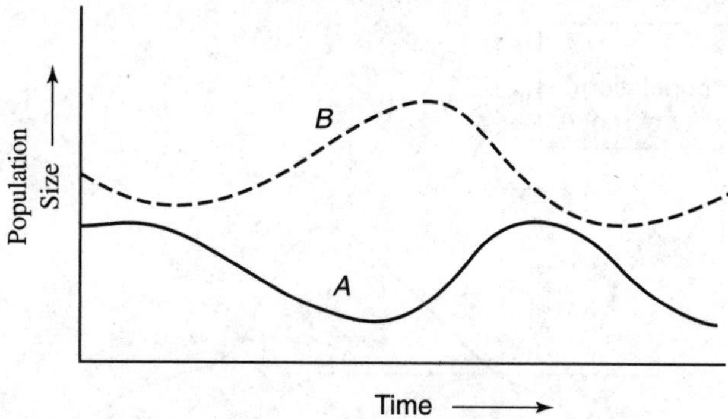

(1) the dominant trees in that community
(2) a population with which the hawks have a mutualistic relationship
(3) variations in the numbers of producers in that community
(4) a population on which the hawks prey 29_____

Base your answers to questions 30 and 31 on this graph and on your knowledge of biology. The graph below depicts changes in the population growth rate of Kaibab deer.

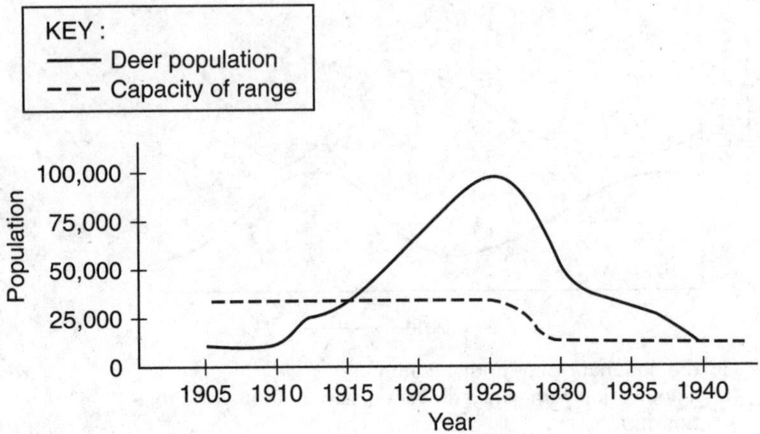

30. About how many deer could the range have supported in 1930 without some of them starving to death?
(1) 12,000 (3) 50,000
(2) 35,000 (4) 100,000 30____

31. In which year were the natural predators of the deer most likely being killed off faster than they could reproduce?
(1) 1905 (2) 1920 (3) 1930 (4) 1940 31____

32. Which process is illustrated by the diagram?

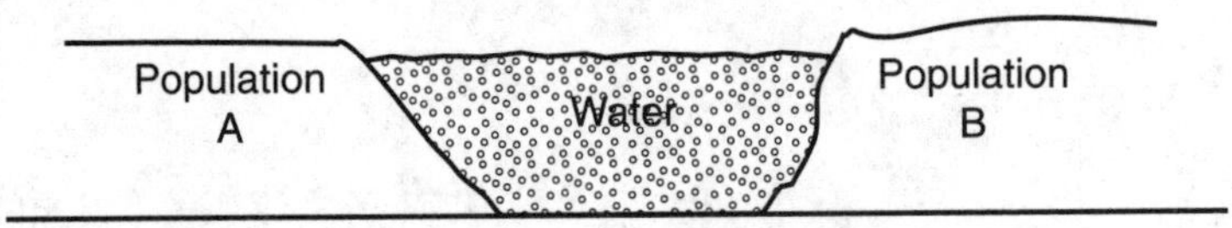

(1) migration (3) speciation
(2) adaptive radiation (4) isolation 32____

Base your answers to questions 33 through 36 on the information and data table below and on your knowledge of biology. The table shows the average systolic and diastolic blood pressure measured in millimeters of mercury (mm Hg) for humans between the ages of 2 and 14 years.

Data Table

	Average Blood Pressure (mm Hg)	
Age	Systolic	Diastolic
2	100	60
6	101	64
10	110	72
14	119	76

DIRECTIONS **(33–36):** Using the information in the data table, construct a line graph on the grid provided, following the directions below.

33. Mark an appropriate scale on each labeled axis.

34. Plot the data for systolic blood pressure on your graph. Surround each point with a small triangle and connect the points.

35. Plot the data for diastolic blood pressure on your graph. Surround each point with a small circle and connect the points.

Example: 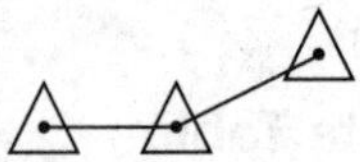

Example: 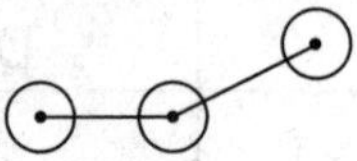

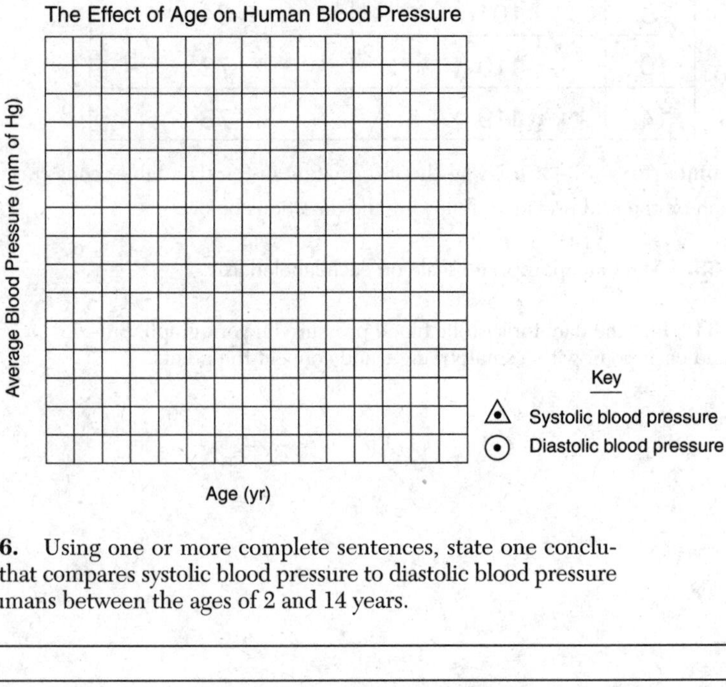

36. Using one or more complete sentences, state one conclusion that compares systolic blood pressure to diastolic blood pressure in humans between the ages of 2 and 14 years.

____

____

37. The graph below shows the results of an experiment.

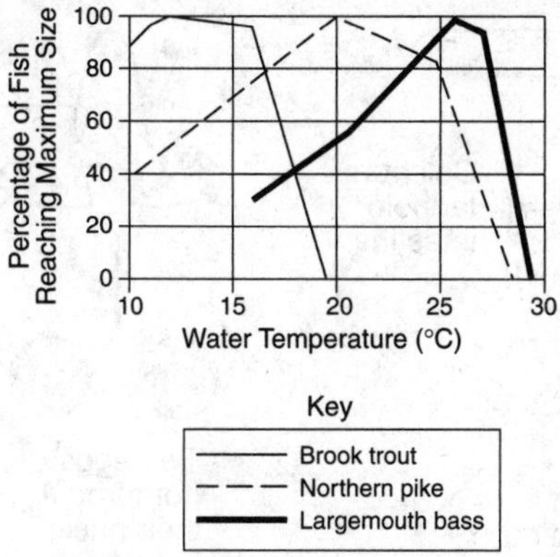

At 16°C, what percentage of the brook trout reached maximum size?

(1) 30% (3) 75%
(2) 55% (4) 95% 37____

38. An experiment is represented in the diagram below.

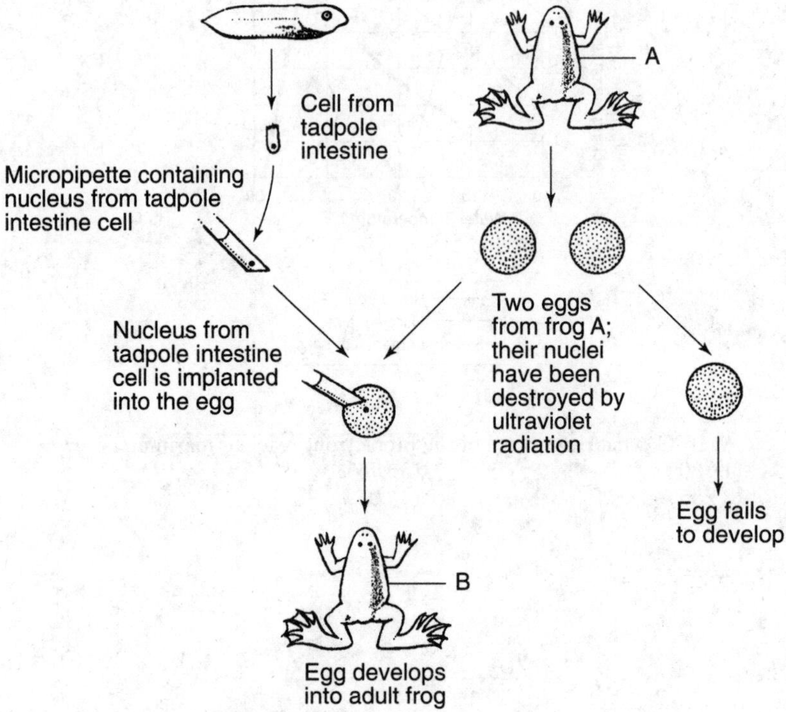

An inference that can be made from this experiment is that
(1) adult frog B will have the same genetic traits as the tadpole
(2) adult frog A can develop only from an egg and a sperm
(3) fertilization must occur in order for frog eggs to develop into adult frogs
(4) the nucleus of a body cell fails to function when transferred to other cell types

38_____

39. The charts below show the relationship of recommended weight to height in men and women age 25–29.

Height-Weight Charts

MEN Age 25–29 Weight (lb)				
Height Feet	Height Inches	Small Frame	Medium Frame	Large Frame
5	2	128–134	131–141	138–150
5	3	130–136	133–143	140–153
5	4	132–138	135–145	142–156
5	5	134–140	137–148	144–160
5	6	136–142	139–151	146–164
5	7	138–145	142–154	149–168
5	8	140–148	145–157	152–172
5	9	142–151	148–160	155–176
5	10	144–154	151–163	158–180
5	11	146–157	154–166	161–184
6	0	149–160	157–170	164–188
6	1	152–164	160–174	168–192
6	2	155–168	164–178	172–197
6	3	158–172	167–182	176–202
6	4	162–176	171–187	181–207

WOMEN Age 25–29 Weight (lb)				
Height Feet	Height Inches	Small Frame	Medium Frame	Large Frame
4	10	102–111	109–121	118–131
4	11	103–113	111–123	120–134
5	0	104–115	113–126	122–137
5	1	106–118	115–129	125–140
5	2	108–121	118–132	128–143
5	3	111–124	121–135	131–147
5	4	114–127	124–138	134–151
5	5	117–130	127–141	137–155
5	6	120–133	130–144	140–159
5	7	123–136	133–147	143–163
5	8	126–139	136–150	146–167
5	9	129–142	139–153	149–170
5	10	132–145	142–156	152–173
5	11	135–148	145–159	155–176
6	0	138–151	148–162	158–179

The recommended weight for a 6'0" tall man with a small frame is closest to that of a

(1) 5'10" man with a medium frame
(2) 5'9" woman with a large frame
(3) 6'0" man with a medium frame
(4) 6'0" woman with a medium frame 39____

Base your answers to questions 40 through 43 on the information below and on your knowledge of biology.

A group of biology students extracted the photosynthetic pigments from spinach leaves using the solvent acetone. A spectrophotometer was used to measure the percent absorption of six different wavelengths of light by the extracted pigments. The wavelengths of light were measured in units known as nanometers (nm). One nanometer is equal to one-billionth of a meter. The following data were collected:

yellow light (585 nm)—25.8% absorption
blue light (457 nm)—49.8% absorption
orange light (616 nm)—32.1% absorption
violet light (412 nm)—49.8% absorption
red light (674 nm)—41.0% absorption
green light (533 nm)—17.8% absorption

40. Complete all three columns in the data table below so that the wavelength of light either increases or decreases from the top to the bottom of the data table.

Color of Light	Wavelength of Light (nm)	Percent Absorption by Spinach Extract

DIRECTIONS **(41–42):** Using the information in the data table, construct a line graph on the grid provided, following the directions below.

41. Mark an appropriate scale on the axis labeled "Percent Absorption."

42. Plot the data from the data table on your graph. Surround each point with a small circle and connect the points.

Example: 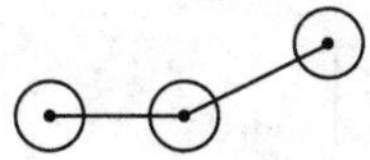

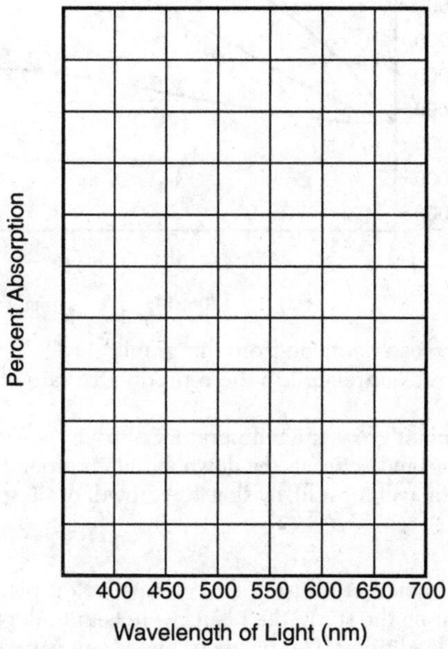

43. Which statement is a valid conclusion that can be drawn from the data obtained in this investigation?

(1) Photosynthetic pigments in spinach plants absorb blue light and violet light more efficiently than red light.
(2) The data would be the same for all pigments in spinach plants.
(3) Green light and yellow light are not absorbed by spinach plants.
(4) All plants are efficient at absorbing violet light and red light. 43____

44. The graph below represents the results of an investigation of the growth of three identical bacterial cultures incubated at different temperatures.

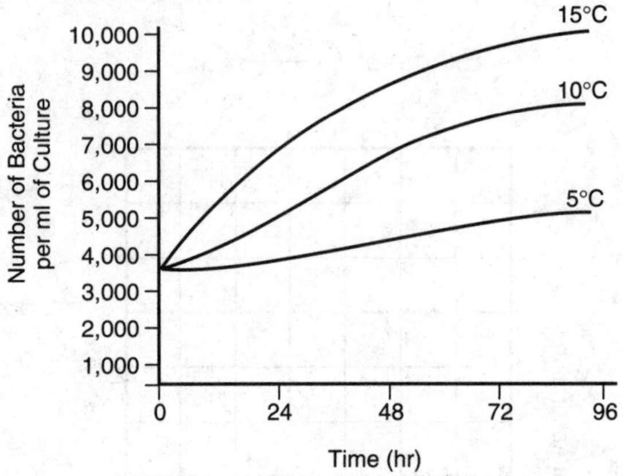

Which inference can be made from this graph?
(1) Temperature is unrelated to the reproductive rate of bacteria.
(2) Bacteria cannot grow at a temperature of 5°C.
(3) Life activities in bacteria slow down at high temperatures.
(4) Refrigeration will most likely slow the growth of these bacteria. 44____

45. A study was conducted using two groups of ten plants of the same species. During the study, the plants were kept under identical environmental conditions. The plants in one group were given a growth solution every 3 days. The heights of the plants in both groups were recorded at the beginning of the study and at the end of a 3-week period. The data showed that the plants given the growth solution grew faster than those not given the solution.

When other researchers conduct this study to test the accuracy of the results, they should
(1) give growth solution to both groups
(2) make sure that the conditions are identical to those in the first study
(3) give an increased amount of light to both groups of plants
(4) double the amount of growth solution given to the first group 45____

46. Worker bees acting as scouts are able to communicate the distance of a food supply from the hive by performing a "waggle dance." The graph below shows the relationship between the distance of a food supply from the hive and the number of turns in the waggle dance every 15 seconds.

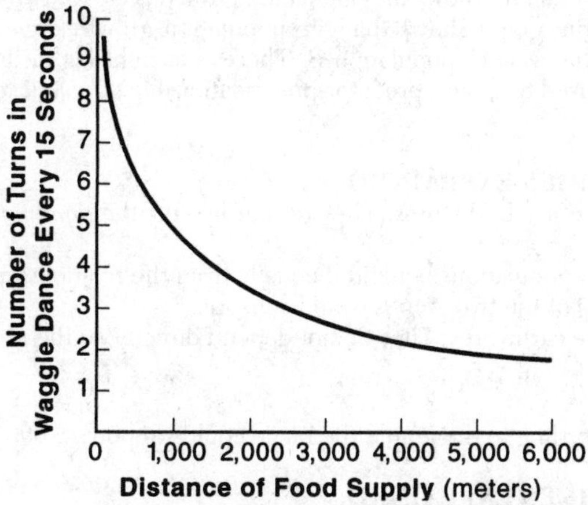

Using one or more complete sentences, state the relationship between the distance of the food supply from the hive and the number of turns a bee performs in the waggle dance every 15 seconds.

---

47. Based on experimental results, a biologist in a laboratory reports a new discovery. If the experimental results are valid, biologists in other laboratories should be able to perform

(1) an experiment with a different variable and obtain the same results
(2) the same experiment and obtain different results
(3) the same experiment and obtain the same results
(4) an experiment under different conditions and obtain the same results 47____

ANSWERS EXPLAINED Questions 29–47

Standard 1, Key Idea 3, Performance Indicators 1–5

29. **4** The diagram shows the population growth cycle for two organisms. An examination of the graph shows that the population growth cycle of the hawks closely follows the cycle of population *B*. There is a slight lag in the cycles. This type of graph is used to show a predator-prey relationship. *The hawks prey on population B.*

WRONG CHOICES EXPLAINED:

(1) Hawks are *not* herbivores. They do not live off the *dominant trees* in the community.

(2) If the two populations benefited equally from the relationship (were *mutualistic*), the peaks of the two graphs would coincide.

(3) Hawks are carnivores. They do not depend directly on the *producers* in the community.

30. **1** According to the graph, the range could support *12,000* deer in 1930.

WRONG CHOICES EXPLAINED:

(2) The carrying capacity of the range was *35,000*. The *carrying capacity* is the maximum number of individuals that can be supported by the area. The number is usually constant unless severe environmental changes occur.

(3) The actual number of deer occupying the range in 1930 was *50,000*.

(4) A population of *100,000* deer was reached in 1925.

31. **2** In nature, a predator-prey relationship keeps the prey population in check. In *1920*, the population of deer increased. It was at this time that the predators were removed by human hunting.

WRONG CHOICES EXPLAINED:

(1) The deer probably entered the region in *1905*. It takes time for an organism to adjust to a new environment.

(3) In *1930*, the deer population, which had exceeded the carrying capacity, was declining. The decline was caused by starvation. The deer had consumed almost all of the available vegetation in the area.

(4) By *1940*, the deer population reached the new carrying capacity of the range. The carrying capacity had been greatly reduced by overgrazing by the previously unchecked deer population.

32. **4** As the result of *isolation*, the members of populations *A* and *B* are separated and are prevented from interbreeding. Variations that occur in one area are not transmitted to the individuals in the other area. Consequently, over a long period of time, the genetic differences become accentuated, and the new variations are maintained.

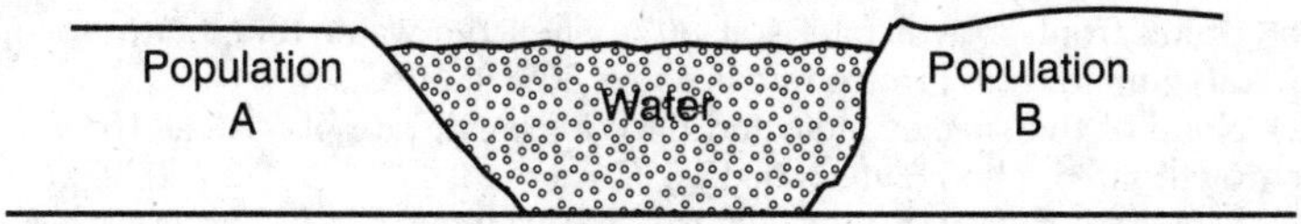

WRONG CHOICES EXPLAINED:

(1) *Migration* is not possible when populations are separated by a geographical barrier such as water.

(2) *Adaptive radiation* refers to a branching evolution and is not depicted in the diagram.

(3) *Speciation* indicates formation of new species from a parent population. This is not indicated in the diagram.

33–35.

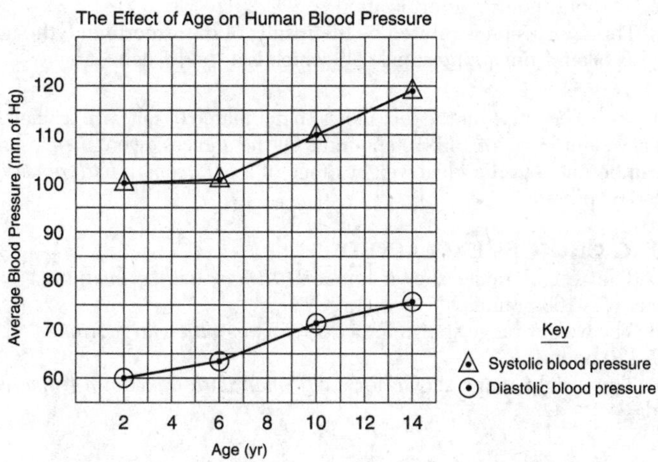

36. *Systolic pressure is higher than diastolic pressure.* or *Both systolic pressure and diastolic pressure increase between the ages of 2 and 14.*

[*Note:* Other correct complete-sentence responses are acceptable.]

37. **4** Brook trout growth is represented by the lighter solid line on the graph. Tracing along the horizontal axis to 16°C and then tracing up to the point at which the brook trout line is encountered, we see that the growth rate is between 80% and 100% but closer to 100%; we estimate it is about 95%.

WRONG CHOICES EXPLAINED:

(1) Brook trout growth rate is at *30%* when the water temperature is about 18°C. Largemouth bass growth rate is about 30% at 16°C.

(2) None of the species indicated have a growth rate of *55%* at 16°C. Brook trout growth rate is at 55% at about 17.5°C.

(3) Brook trout growth rate falls to *75%* at about 17°C. The growth rate of northern pike is at 75% at about 16°C.

38. **1** The diagram represents an experiment in cloning. Because the nucleus (which contains the genetic material) used in this experiment comes from a tadpole, the egg will produce new cells that have *genetic characteristics identical to those of the tadpole.*

WRONG CHOICES EXPLAINED:

(2) It is unclear from this diagram how frog *A* came into being.

(3) This conclusion is refuted by the results of this experiment; a new frog was created without the use of fertilization.

(4) This conclusion is refuted by the results of this experiment; the tadpole nucleus was taken from an intestinal cell, which is a "body" cell.

39. **4** The chart on the left shows that a man 6'0" tall with a small frame has an ideal weight range of 149–160 pounds. Of the choices given, the closest comparison can be made to the ideal weight range of a *6'0" woman with a medium frame* at 148–162 pounds.

WRONG CHOICES EXPLAINED:

(1) The weight range shown for a *5'10" man with a medium frame* is 151–163 pounds.

(2) The weight range shown for a *5'9" woman with a large frame* is 149–170 pounds.

(3) The weight range shown for a *6'0" man with a medium frame* is 157–170 pounds.

40.

Color of Light	Wavelength of Light (nm)	Percent Absorption by Spinach Extract
violet	412	49.8
blue	457	49.8
green	533	17.8
yellow	585	25.8
orange	616	32.1
red	674	41.0

41–42.

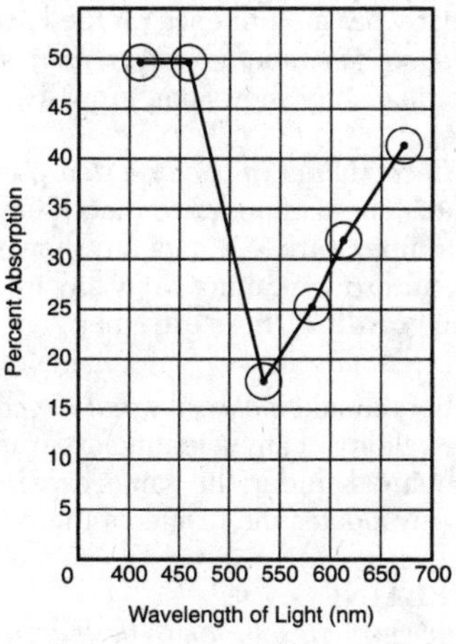

43. **1** The statement *photosynthetic pigments in spinach plants absorb blue light and violet light more efficiently than red light* is a valid conclusion that can be drawn from the data. The high point of the chart/graph data is clearly shown to be above the blue and violet wavelengths of light.

WRONG CHOICES EXPLAINED:

(2) The statement *the data would be the same for all pigments in spinach plants* is not supported by the results of this experiment. The chart/graph data show considerable variation in the experimental results as the wavelength of light varies.

(3) The statement *green light and yellow light are not absorbed by spinach plants* is not supported by the results of this experiment. Although the chart/graph data show a lower absorption rate at these wavelengths, there is still some absorption in this range.

(4) The statement *all plants are efficient at absorbing violet light and red light* is not supported by the results of this experiment. The experimental data are limited to the absorption of light by pigments found in one type of plant. These data cannot be extended to all plants unless all other types of plants are tested under the same experimental conditions and the results are found to be similar.

44. **4** Of those given, *refrigeration will most likely slow the growth of these bacteria* is the most reasonable inference that can be made from the graph data. The graph clearly shows a slower rate of growth (reproduction) at 5°C than at 10°C or 15°C.

WRONG CHOICES EXPLAINED:

(1) The inference that *temperature is unrelated to the reproductive rate of bacteria* is not supported by the data. Temperature is the independent (experimental) variable in this study. It clearly has an influence on the bacterial reproductive rate.

(2) The inference that *bacteria cannot grow at a temperature of 5°C* is not supported by the data. The graph clearly shows that growth at this temperature, while slow, occurs at a steady pace.

(3) The inference that *life activities in bacteria slow down at high temperatures* is not supported by the data. The data indicate that, if anything, bacterial activity increases with increasing temperature. No data are shown for bacterial growth at temperatures above 15°C, and so we cannot draw any inference about what happens to the rate of bacterial growth at these extremes.

45. **2** These researchers should *make sure that the conditions are identical to those in the first study.* The validity of any scientific experiment can be verified only if the same results are obtained under the same experimental conditions. Any change in these conditions invalidates the results of the verification study.

WRONG CHOICES EXPLAINED:

(1) If the researchers *give growth solution to both groups,* there will be no control group against which to compare the experimental group. The results of the verification study will be invalid because the experimental conditions will have been changed.

(3) If the researchers *give an increased amount of light to both groups of plants,* the original experimental method will be altered. The results of the verification study will be invalid because the experimental conditions will have been changed.

(4) If the researchers *double the amount of growth solution given to the first group,* the original experimental method will not be followed. The results of the verification study will be invalid because the experimental conditions will have been changed.

46. *The number of turns in the waggle dance decreases as the distance of the food supply from the hive increases.* Or *The closer to the hive the food source is located, the more turns there are in the waggle dance.* [*Note:* Any correct, complete-sentence answer is acceptable.]

47. **3** Other biologists in other laboratories should be able to perform *the same experiment and obtain the same results* if the experimental results are valid. Any experimental results obtained by one scientist must be validated through independent research by other scientists following the same procedures.

WRONG CHOICES EXPLAINED:

(1) If different scientists perform *an experiment with a different variable and obtain the same results,* the original experimental results will be invalidated.

(2), (4) If different scientists perform *the same experiment and obtain different results,* or *an experiment under different conditions and obtain the same results,* they will neither validate nor invalidate the results of the original experiment. All variables and conditions must be kept the same if the experimental results are to be properly tested.

QUESTIONS ON STANDARD 4—Biological Concepts: Students will understand and apply scientific concepts, principles, and theories pertaining to the physical setting and living environment and recognize the historical development of ideas in science.

Key Idea 1—Application of Scientific Principles: Living things are both similar to and different from each other and from nonliving things.

Performance Indicator	Description
1.1	The student should be able to explain how diversity of populations within ecosystems relates to the stability of ecosystems.
1.2	The student should be able to describe and explain the structures and functions of the human body at different organizational levels (e.g., systems, tissues, cells, organelles).
1.3	The student should be able to explain how a one-celled organism is able to function despite lacking the levels of organization present in more complex organisms.

48. In which life function is the potential energy of organic compounds converted to a form of stored energy that can be used by the cell?

(1) transport (3) excretion
(2) respiration (4) regulation 48____

49. Which life activity is *not* required for the survival of an individual organism?

(1) nutrition (3) reproduction
(2) respiration (4) synthesis 49____

50. Which function of human blood includes the other three?

(1) transporting nutrients (3) maintaining homeostasis
(2) transporting oxygen (4) collecting wastes 50____

51. In the human body, the blood with the greatest concentration of oxygen is found in the
(1) left atrium of the heart (3) nephrons of the kidney
(2) cerebrum of the brain (4) lining of the intestine 51____

52. Which type of vessel normally contains valves that prevent the backward flow of materials?
(1) artery (3) capillary
(2) arteriole (4) vein 52____

DIRECTIONS (53–55): For each of questions 53 through 55, select the excretory structure, chosen from the list below, that best answers the question. Then record its number in the space provided at the right.

Excretory Structures
(1) Alveolus
(2) Nephron
(3) Sweat gland
(4) Liver

53. Which structure forms urine from water, urea, and salts?

____

54. Which structure removes carbon dioxide and water from the blood?

____

55. Which structure is involved in the breakdown of red blood cells?

____

56. The bones of the lower arm are connected to the muscles of the upper arm by
(1) ligaments (3) cartilage
(2) tendons (4) skin 56____

57. The diagram below shows the same type of molecules in area *A* and area *B*. With the passage of time, some molecules move from area *A* to area *B*.

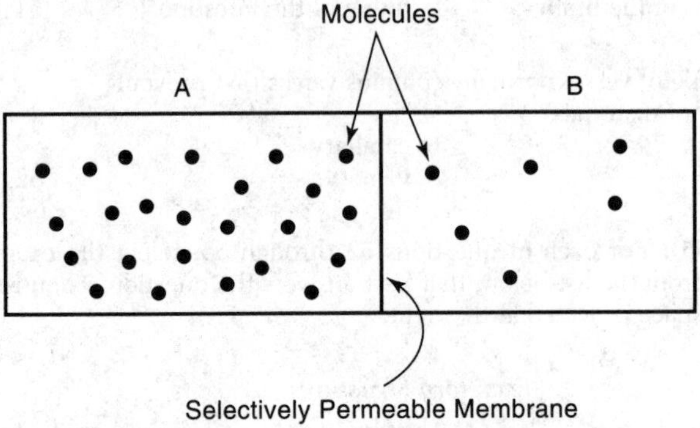

The movement is the result of the process of

(1) phagocytosis (3) diffusion
(2) pinocytosis (4) cyclosis 57____

58. Which is the principal inorganic compound found in cytoplasm?

(1) lipid (3) water
(2) carbohydrate (4) nucleic acid 58____

59. A specific organic compound contains only the elements carbon, hydrogen, and oxygen in a ratio of 1:2:1. This compound is most probably a

(1) nucleic acid (3) protein
(2) carbohydrate (4) lipid 59____

60. Compared to ingested food molecules, end-product molecules of digestion are usually

(1) smaller and more soluble (3) smaller and less soluble
(2) larger and more soluble (4) larger and less soluble 60____

61. The cellular function of the endoplasmic reticulum is to

(1) provide channels for the transport of materials
(2) convert urea to a form usable by the cell
(3) regulate all cell activities
(4) change light energy into chemical bond energy 61____

62. In which organelles are polypeptide chains synthesized?
(1) nuclei (3) ribosomes
(2) vacuoles (4) cilia 62____

63. Which organelle contains hereditary factors and controls most cell activities?
(1) nucleus (3) vacuole
(2) cell membrane (4) endoplasmic reticulum 63____

64. Centrioles are cell structures involved primarily in
(1) cell division (3) enzyme production
(2) storage of fats (4) cellular respiration 64____

65. Which cell structure contains respiratory enzymes?
(1) cell wall (3) mitochondrion
(2) nucleolus (4) vacuole 65____

66. Which process is represented below?

$$\text{simple organic molecules} \xrightarrow{\text{enzymes}} \text{complex organic molecules} + H_2O$$

(1) hydrolysis (3) digestion
(2) synthesis (4) respiration 66____

67. Amino acids derived from the digestion of a piece of meat are transported to living cells of an animal. In the cell they are
(1) converted to cellulose
(2) used to attack invading bacteria
(3) synthesized into specific proteins
(4) incorporated into glycogen molecules 67____

68. Which of the following variables has the *least* direct effect on the rate of a hydrolytic reaction regulated by enzymes?
(1) temperature
(2) pH
(3) carbon dioxide concentration
(4) enzyme concentration 68____

69. Which term refers to the chemical substance that aids in the transmission of the impulse through the area indicated by X?

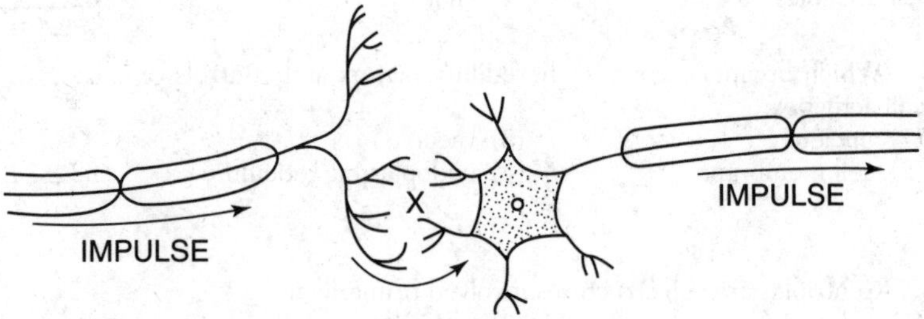

(1) neurotransmitter (3) neuron
(2) synapse (4) nerve 69____

70. Which lists human nervous-system structures in order of increasing size?
(1) neuron, nerve, ganglion, receptor
(2) nerve, ganglion, neuron, receptor
(3) neuron, receptor, ganglion, nerve
(4) ganglion, receptor, nerve, neuron 70____

71. Glands located within the digestive tube include
(1) gastric glands and thyroid glands
(2) gastric glands and intestinal glands
(3) thyroid glands and intestinal glands
(4) adrenal glands and intestinal glands 71____

72. In humans, which substance is directly responsible for controlling the calcium levels of the blood?
(1) adrenalin (3) parathormone
(2) insulin (4) thyroxin 72____

Base your answer to question 73 on the word equation below.

glucose → 2 pyruvic acid → 2 ethyl alcohol + 2 carbon dioxide + energy

73. The process represented by the word equation is known as
(1) aerobic respiration (3) chemosynthesis
(2) fermentation (4) dehydration synthesis 73____

74. The excretory organelles of some unicellular organisms are contractile vacuoles and
(1) cell membranes (3) ribosomes
(2) cell walls (4) centrioles 74____

75. Which is a type of asexual reproduction that commonly occurs in many species of unicellular protists?

(1) external fertilization (3) binary fission
(2) tissue regeneration (4) vegetative propagation 75____

ANSWERS EXPLAINED Questions 48–75

Standard 4, Key Idea 1, Performance Indicators 1–3

48. **2** *Respiration* is the life function by which ATP is made available to cells. Carbohydrate molecules are organic compounds. The breakdown of the carbohydrate molecules releases the energy stored in the bonds of the compounds. Potential energy is stored energy. The released potential energy is used to produce ATP.

WRONG CHOICES EXPLAINED:

(1) *Transport* is the life function by which materials are distributed throughout an organism.

(3) *Excretion* is the life function by which the wastes of metabolism are removed from an organism. Carbon dioxide, water, ammonia, and urea are metabolic wastes.

(4) *Regulation* is the life activity by which an organism responds to changes in its environment. The responses are controlled by the nervous system and the endocrine system.

49. **3** *Reproduction*, the life function through which a parent organism gives rise to offspring, is not necessary for the survival of the parent. Although reproduction is not required for the survival of an individual, it is necessary for the survival of a species. If a given species loses its potential for reproduction, it will become extinct.

WRONG CHOICES EXPLAINED:

(1) *Nutrition* is a collective term that refers to the biochemical processes by which cells extract nutrient molecules from food substances. The nutrients are used to build tissues, provide energy, and regulate the many biochemical activities that occur in cells. Without nutrition, cells die and, consequently, so do organisms. Each organism is dependent on adequate nutrition for survival.

(2) *Respiration* refers to the series of chemical changes that fuel molecules undergo to release chemical energy for cells. Respiration is necessary for the survival of the individual. Tissue cells cannot live without a means of obtaining chemical energy to power cellular activities such as active transport and metabolism. Of course, death of tissue cells means death of the individual.

(4) *Synthesis* occurs when small molecules are joined chemically to form large molecules. Enzymes, hormones, and body tissues are the results of syntheses, without which an individual organism cannot survive. Synthesis is a building-up process in which molecules vital to the life of the organism are produced.

50. **3** *Maintaining homeostasis* is the function of human blood that includes the other three. By transporting nutrients, oxygen, wastes, and other materials around the body, the blood helps to make essential materials available to every living body cell while removing potentially harmful materials from these tissues. Equal distribution of these materials helps to promote a steady state in the tissues essential to homeostatic balance.

WRONG CHOICES EXPLAINED:

(1), (2), (4) *Transporting nutrients, transporting oxygen,* and *collecting wastes* are all functions of the blood that are involved in maintaining homeostasis. Nutrients provide cells with dissolved food molecules. Oxygen is used by cells in the release of energy from these food molecules. Wastes such as urea and carbon dioxide are carried away from the cells for excretion into the environment.

51. **1** Blood that has just returned from the lungs has the greatest concentration of oxygen. The *left atrium of the heart* receives blood directly from the lungs.

WRONG CHOICES EXPLAINED:

(2) Brain tissue is one of the largest consumers of oxygen. The blood circulating in the *cerebellum* gives up most of its oxygen to the nerve cells.

(3) The largest concentration of metabolic wastes is found in the *nephrons*. The nephrons are filtering units in the kidney.

(4) The largest concentration of digested nutrients is found in the *lining of the intestine*. Absorption of nutrients occurs through the villi in the small intestine.

52. **4** *Veins* are blood vessels that carry blood to the heart. They contain valves that prevent the backflow of blood. The blood in veins is usually deoxygenated; the exception is the pulmonary vein in which the blood is rich in oxygen.

WRONG CHOICES EXPLAINED:

(1) *Arteries* are blood vessels that transport blood away from the heart. Arteries are rather thick-walled and pump blood in rhythm with the heart. They have no valves.

(2) Small arteries are *arterioles.* This type of blood vessel functions similarly to arteries. Arterioles lead into capillaries.

(3) *Capillaries* are the smallest blood vessels. They are one cell thick and permit diffusion of water, nutrients, gases, and other substances into and out of the bloodstream. Capillaries have no valves; they are the connecting vessels between arterioles and venules.

53. **2** The *nephron* is the unit of structure of the kidney. Each nephron has a glomerulus, Bowman's capsule, and kidney tubules. The kidney tubules filter out excess water, salts, and the wastes from protein metabolism. Urea and salts dissolved in water form urine.

54. **1** The *alveolus* is an air sac in the lung. It not only permits the diffusion of oxygen from the lungs into the bloodstream but also aids in the diffusion of carbon dioxide and water vapor out of the blood into the lungs.

55. **4** The *liver* is the largest gland in the body. One of its functions is to destroy old red blood cells and change the waste products into bile. The liver also synthesizes the anticoagulant known as heparin.

56. **2** *Tendons* are tough connective tissues made strong by fibers. Tendons connect muscles to bones. The movable joints function when muscles pull on tendons.

WRONG CHOICES EXPLAINED:

(1) *Ligaments* are strong connective tissues that contain elastic muscle fibers. Ligaments connect bone to bone.

(3) *Cartilage* is a supporting tissue that provides strength to body structures without rigidity. Cartilage supports structures such as the ears and nose and covers the ends of bones that form joints. The ground substance, or matrix, of cartilage is made of protein.

(4) *Skin* is composed of epithelial tissue. Skin serves as a body covering and has no function in the movement of bones or muscles.

57. **3** *Diffusion* is the process that results in the movement of molecules from a region of higher concentration (area A) to a region of lower concentration (area B). This net movement occurs until the concentrations of molecules have reached equilibrium between area A and area B.

WRONG CHOICES EXPLAINED:

(1), (2) *Phagocytosis* and *pinocytosis* are processes by which certain protists engulf their food and enclose it within a vacuole for digestion.

(4) *Cyclosis* refers to the streaming of cytoplasm in the cell, a simple form of intracellular transport.

58. **3** Inorganic compounds are compounds that do not contain carbon atoms. *Water,* the universal solvent, is the principal inorganic compound of cytoplasm. Water is the medium through which all chemical reactions take place in the cell.

WRONG CHOICES EXPLAINED:

(1), (2), (4) *Lipids, carbohydrates,* and *nucleic acids* are organic compounds. Organic compounds are carbon-containing compounds.

59. **2** Glucose is the building block of *carbohydrate* molecules. The ratio of carbon to hydrogen to oxygen is 1:2:1 in glucose and all reducing sugars. By dehydration synthesis, many glucose molecules form complex carbohydrates. However, the 1:2:1 ratio holds.

WRONG CHOICES EXPLAINED:

(1) A *nucleic acid* is composed of a phosphate group, a protein base, and a five-carbon sugar. DNA and RNA are nucleic acids. The CHO 1:2:1 ratio is not applicable.

(3) *Proteins* are built from amino acids, which, in addition to carbon, hydrogen, and oxygen, contain nitrogen. Some protein molecules also contain sulfur. The 1:2:1 ratio of elements does not apply to proteins because proteins are tissue builders whereas carbohydrates are fuel molecules.

(4) *Lipids* are fats and are composed of three fatty acid molecules and one glycerol molecule. The 1:2:1 ratio of carbon to hydrogen to oxygen does not apply to fats.

60. **1** The end products of digestion are usually *smaller and more soluble* than the ingested food molecules. Digestion makes available nutrient molecules that can diffuse across cell membranes and enter the cytoplasm of cells. Carbohydrates are broken down into glucose molecules. Fats are hydrolyzed into fatty acids and glycerol. Proteins are digested into their component amino acid molecules. Each of these end products of digestion is able to diffuse across cell membranes and enter into the biochemical activities of cells.

WRONG CHOICES EXPLAINED:

(2) Synthesis produces *larger molecules*. Larger molecules are more complex and are usually less soluble than smaller, simpler ones. Digestion results in smaller nutrient molecules.

(3) *Smaller molecules* are usually more soluble than larger ones. Digestion produces molecules that are more soluble than the complex nutrient molecules that were ingested.

(4) *Molecules derived from digestion* of ingested food are not larger than the molecules from which they came. Molecules produced by the digestion of complex carbohydrates, proteins, and fats are more soluble and are able to dissolve in water. Thus, these molecules can cross cell membranes.

61. **1** The endoplasmic reticulum is a network of membranes that extends throughout the cell. The membranes form channels that *provide for the movement of materials through the cell.*

WRONG CHOICES EXPLAINED:

(2) *Urea* is a metabolic waste. It is a poisonous nitrogen compound. Urea must be removed from the cells if an organism is to survive.

(3) The *nucleus* is the organelle in the cell that regulates all cellular activities.

(4) The *chloroplasts* are organelles in plant cells that contain the green pigment chlorophyll. Chloroplasts are necessary for the process of photosynthesis.

62. **3** Proteins are polypeptide chains. Proteins are synthesized in the *ribosomes.*

WRONG CHOICES EXPLAINED:

(1) The *nuclei* contain the genetic material carried in the chromosomes.

(2) *Vacuoles* are saclike organelles in the cytoplasm. Food vacuoles and contractile vacuoles are two common types of vacuoles.

(4) *Cilia* are microscopic hairs used for locomotion by some protozoans.

63. **1** The *nucleus* contains the hereditary factors. Nuclei of plant and animal cells house the chromosomes, which are composed of deoxyribonucleic acid. Molecules of DNA function as genes. Points on the chromosomes are genes. Genetic information is passed from parent to offspring by way of the genes. Chromosomes are part of the fine structure of the nucleus. DNA molecules contribute to the chemical structure. Genes are sites or points that dot the length of the chromosome. Genes, DNA molecules, and chromosomes function in passing along hereditary factors.

WRONG CHOICES EXPLAINED:

(2) The *cell membrane* encloses the contents of the cell and directs the flow of materials into and out of the cell. The cell membrane does not contribute to the passing of genetic material from one generation to the next. The function of the membrane is to control cellular transport.

(3) A *vacuole* is a fluid-filled space in the cytoplasm. Vacuoles help to regulate the internal pressure of the cell. The vacuoles in fat cells are filled with oil.

(4) The membranes that line the cytoplasmic canals within cells are known collectively as the *endoplasmic reticulum.* This cytoplasmic fine structure aids in the transport of molecules from the cell membrane to various sites within the cell. Neither the endoplasmic reticulum nor the vacuoles of the cell membrane contain hereditary structures.

64. **1** Centrioles are cell structures involved primarily in *cell division.* Centrioles are organelles that lie in the cytoplasm outside the nucleus; they are also found near the base of each flagellum and cilium. The centrioles of nonflagellated animal cells move to the spindle poles during cell division and seem to send out spindle fibers. The spindle fibers are attached to chromosomes and appear to pull the chromosomes from the center of the cell to the spindle poles.

WRONG CHOICES EXPLAINED:

(2) *Fats are stored* in cells. Fat in which the energy is channeled into heat production is stored in brown fat cells of hibernating mammals. At times, fat can be stored in arteries or accumulate around the heart. Fat cells are not involved in cell division.

(3) *Enzyme production* is controlled by the ribosomes that dot the membranes of the endoplasmic reticulum. Molecules of tRNA and mRNA regulate enzyme production.

(4) *Cellular respiration* is the process by which energy is released from glucose molecules. This process takes place in the mitochondria where oxygen is used as the final hydrogen carrier.

65. **3** Cellular respiration occurs in the mitochondria (plural of *mitochondrion*). Each step in the process of cellular respiration is regulated by enzymes. Respiratory enzymes are located in the mitochondria.

WRONG CHOICES EXPLAINED:

(1) The *cell wall* is composed of cellulose. Cell walls give shape and protection to plant cells.

(2) The *nucleolus* contains the materials needed for the synthesis of RNA. It is located in the nucleus of the cell.

(4) A *vacuole* is a rounded sac that serves as a storage place for food and waste products. Some vacuoles, such as contractile vacuoles, maintain a stable internal environment.

66. **2** *Synthesis* is the formation of complex molecules by combining simpler molecules. Water is removed from the simple molecules in this process.

WRONG CHOICES EXPLAINED:

(1) *Hydrolysis* is the addition of water to split complex molecules into simpler molecules. It is the opposite of synthesis.

(3) *Digestion* is another name for hydrolysis.

(4) *Respiration* is the process by which cells obtain energy. Glucose is converted to smaller molecules.

67. **3** Amino acids are the building blocks of proteins. The dehydration synthesis of amino acids *produces protein molecules*.

WRONG CHOICES EXPLAINED:

(1) *Cellulose* is a polysaccharide composed of hundreds of simple sugar molecules. The sugars were joined together by dehydration synthesis.

(2) Antibodies attack *invading bacteria*. Antibodies are protein molecules produced by special white blood cells.

(4) *Glycogen*, a polysaccharide, is a product of the dehydration synthesis of many glucose units.

68. **3** A *hydrolytic reaction* is a reaction in which a molecule is split. Enzymes are needed to speed up such a reaction. Any factor that affects the operation of the enzyme affects the speed at which the reaction takes place. The *concentration of carbon dioxide* has the least effect on enzyme activity.

WRONG CHOICES EXPLAINED:

(1) As the *temperature* is increased up to a point, the rate of the reaction increases. The increase in temperature increases the speed at which the enzyme and the substrate make contact with each other. The substrate is the molecule on which the enzyme acts. A very high temperature destroys the enzyme, and the reaction stops.

(2) Every enzyme works best at a particular *pH*. The enzymes in the stomach work in an acid environment, whereas the enzymes in the intestine work best in a basic medium.

(4) One molecule of an enzyme reacts with one molecule of a substrate. Increasing the *concentration of an enzyme* means that more substrate molecules will be acted on. The rate of the reaction will increase.

69. **1** A *neurotransmitter* is a chemical substance that is released by an impulse arriving at the terminal end of a neuron. The neurotransmitter diffuses across the synapse and stimulates the second nerve cell. Acetylcholine is an example of a neurotransmitter.

WRONG CHOICES EXPLAINED:

(2) A *synapse* is the space between the terminal end of one nerve cell and the dendrites of a second nerve cell. The area marked by an X in the diagram is a synapse.

(3) A *neuron* is a nerve cell that is specially adapted for the conduction of impulses.

(4) A *nerve* is made up of many neurons.

70. **1** A *neuron* is a single microscopic nerve cell. A *nerve* is composed of many nerve cells. A *ganglion* is a large mass of cell bodies of nerve cells; a ganglion functions as a coordinating center for impulses. A *receptor* is an organ specialized to receive environmental stimuli. The eye is an example of a receptor.

WRONG CHOICES EXPLAINED:

(2), (3), (4) In these three choices, either one or several structures are not arranged according to increasing size.

71. **2** *Gastric glands* are embedded in the walls of the stomach. They are duct glands that secrete gastric juice, a mixture of water, hydrochloric acid, rennin, and pepsin. Gastric juice begins the digestion of protein in the stomach. *Intestinal glands* are duct glands that line the walls of the small intestine. They secrete intestinal juice, a mixture of water, proteases, amylases, and lipases. Both types of glands lie within the digestive tube.

WRONG CHOICES EXPLAINED:

(1) Gastric glands are described above. *Thyroid glands* lie outside the digestive tract at the base of the neck, straddled across the larynx. The thyroid is a ductless gland that secretes the hormone known as thyroxin. Thyroid glands do not function in the biochemical process of digestion.

(3) *Thyroid glands* and intestinal glands are described above. Thyroxin controls the metabolism of cells. The explanation above shows why this choice is wrong.

(4) *Adrenal glands* are dual endocrine glands that lie on top of each kidney. They are not within the digestive tract. The adrenal medulla, the inner gland, secretes the hormone adrenaline, also known as epinephrine. This hormone enables the body to function in emergencies. The adrenal cortex secretes about six active hormones, including cortisone, the antiarthritis hormone.

72. **3** *Parathormone* is the hormone secreted by the parathyroid glands. The parathyroids are buried in the thyroids. Parathormone controls the level of calcium in the blood. Lack of blood calcium causes muscles to go into tetany. Tetany, or cramping, of the heart muscle causes death.

WRONG CHOICES EXPLAINED:

(1) *Adrenalin* is the hormone of the adrenal medulla, a ductless gland called the "gland of combat." Adrenaline stimulates the heart to beat faster, increases the rate of breathing, and controls the constriction and dilation of the arteriole walls.

(2) *Insulin* is secreted by the beta cells of the islets of Langerhans, which lie in the pancreas. Insulin controls sugar metabolism; specifically, it makes cell walls permeable to glucose and encourages the phosphorylation of fructose.

(4) *Thyroxin* is released by the thyroid gland. The rate of cellular metabolism is controlled by thyroxin. Iodine is used in the synthesis of thyroxin. People whose thyroid glands fail to develop become cretins; they are mentally retarded and physically undersized.

73. **2** Another name for anaerobic respiration is *fermentation*. In the process of fermentation, glucose is converted to energy, alcohol, and carbon dioxide.

WRONG CHOICES EXPLAINED:

(1) *Aerobic respiration* is another name for cellular respiration. This process requires oxygen. The following is an equation for aerobic respiration.

$$\text{glucose} + \text{oxygen} \rightarrow \text{pyruvic acid} \rightarrow \text{carbon dioxide} + \text{water} + \text{energy}$$

(3) *Chemosynthesis* is the synthesis of carbohydrates from inorganic compounds without the use of sunlight as a source of energy. Chemosynthesis is a form of autotrophic nutrition. It is carried out only by certain species of bacteria such as nitrifying bacteria.

(4) *Dehydration synthesis* is the method by which simple molecules are converted to complex molecules.

74. **1** *Cell membranes* and contractile vacuoles are excretory organelles of some unicellular organisms. The cell membrane is a selectively permeable membrane. It permits the diffusion of carbon dioxide and ammonia, two metabolic waste gases.

WRONG CHOICES EXPLAINED:

(2) *Cell walls* are composed of nonliving materials. Many canals penetrate through these walls, allowing the unrestricted passage of molecules.

(3) Proteins are synthesized in *ribosomes*.

(4) *Centrioles* are rodlike particles found in the centrosome. They function during the processes of mitosis and meiosis. Centrioles are found only in animal cells.

75. **3** A unicellular protist (e.g., an ameba) is composed of a single cell. When this cell divides by mitosis, the process is known as *binary fission*.

WRONG CHOICES EXPLAINED:

(1) *External fertilization* is an element of sexual reproduction in many aquatic multicellular species. Both the sexual nature of this process and the fact that it is carried out by multicellular animals eliminate this as a correct choice.

(2) *Tissue regeneration* implies a process that occurs in multicellular organisms.

(4) *Vegetative propagation* is a form of asexual reproduction common to certain species of multicellular plants; it cannot be carried out by unicellular protists.

Key Idea 2—Genetic Continuity: Organisms inherit genetic information in a variety of ways that result in continuity of structure and function between parents and offspring.

Performance Indicator	Description
2.1	The student should be able to explain how the structure and replication of genetic material result in offspring that resemble their parents.
2.2	The student should be able to describe and explain how the technology of genetic engineering allows humans to alter genetic makeup of organisms.

76. Corn plants grown in the dark will be white and usually much taller than genetically identical corn plants grown in light, which will be green and shorter. The most probable explanation for this is that the

(1) corn plants grown in the dark were all mutants for color and height
(2) expression of a gene may be dependent on the environment
(3) plants grown in the dark will always be genetically albino
(4) phenotype of a plant is independent of its genotype 76____

77. In order for a substance to act as a carrier of hereditary information, it must be

(1) easily destroyed by enzyme action
(2) exactly the same in all organisms
(3) present only in the nuclei of cells
(4) copied during the process of mitosis 77____

78. During synapsis in meiosis, portions of one chromosome may be exchanged for corresponding portions of its homologous chromosome. This process is known as

(1) nondisjunction (3) crossing-over
(2) polyploidy (4) hybridization 78____

79. A DNA nucleotide is composed of three parts. These three parts may be

(1) phosphate, adenine, and thymine
(2) phosphate, deoxyribose, and thymine
(3) phosphate, glucose, and cytosine
(4) adenine, thymine, and cytosine 79____

80. A double-stranded DNA molecule replicates as it unwinds and "unzips" along weak

(1) hydrogen bonds (3) phosphate groups
(2) carbon bonds (4) ribose groups 80____

Base your answers to questions 81 through 84 on your knowledge of biology and the diagrams below. The diagram on the left represents a portion of a double-stranded DNA molecule. The diagrams at the right represent specific combinations of nitrogenous bases found in compounds transporting specific amino acids.

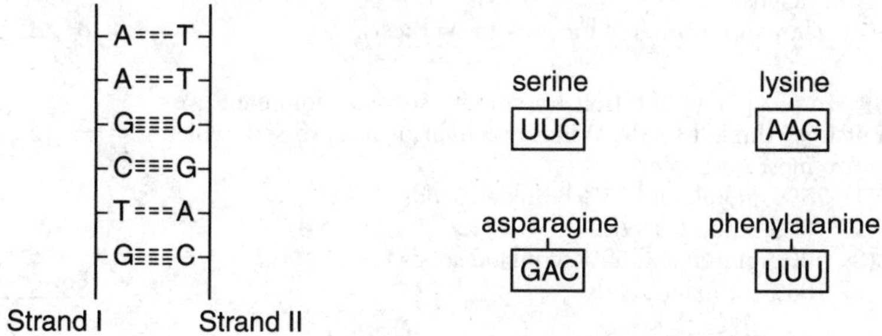

81. The amino acid whose genetic code is present in strand I is
(1) lysine
(2) serine
(3) asparagine
(4) phenylalanine
81_____

82. The thymine (*T*) of strand I is accidentally replaced by adenine (*A*). This occurrence is called
(1) segregation
(2) disjunction
(3) cytoplasmic inheritance
(4) gene mutation
82_____

83. The number of different amino acids coded by strand I is
(1) 1 (2) 2 (3) 8 (4) 12
83_____

84. Which represents the sequence of nitrogenous bases in the molecule of messenger RNA synthesized by strand I?
(1) *-T-T-C-G-U-C-*
(2) *-A-A-C-G-T-C-*
(3) *-U-U-C-G-A-C-*
(4) *-A-A-G-C-U-G-*
84_____

85. Molecules that transport amino acids to ribosomes are known as
(1) protein molecules
(2) RNA molecules
(3) mitochondria
(4) chromosomes
85_____

86. A similarity between DNA molecules and RNA molecules is that they
(1) are built from nucleotides
(2) are double-stranded
(3) contain deoxyribose sugar
(4) contain uracil
86_____

87. What is the function of DNA molecules in the synthesis of proteins?
(1) They catalyze the formation of peptide bonds.
(2) They determine the sequence of amino acids in a protein.
(3) They transfer amino acids from the cytoplasm to the nucleus.
(4) They supply energy for protein synthesis. 87____

88. In pea plants, the trait for smooth seeds is dominant over the trait for wrinkled seeds. When two hybrids are crossed, which results are most probable?
(1) 75% smooth and 25% wrinkled seeds
(2) 100% smooth seeds
(3) 50% smooth and 50% wrinkled seeds
(4) 100% wrinkled seeds 88____

89. A person who is homozygous for blood type *A* has a genotype that can be represented as
(1) I^aI^b (2) I^aI^a (3) I^ai (4) ii 89____

90. Animal breeders often cross breed members of the same litter in order to maintain desirable traits. This procedure is known as
(1) hybridization
(2) inbreeding
(3) natural selection
(4) vegetative propagation 90____

ANSWERS EXPLAINED Questions 76–90

Standard 4, Key Idea 2, Performance Indicators 1–2

76. **2** Corn plants grown in the dark will be white. The most probable explanation for this is that expression of the gene for color may *depend on the environment*. The plants have the genetic information for chlorophyll production. This can be assumed because they are genetically identical to the plants grown in the light. Light is needed to activate the chlorophyll gene.

WRONG CHOICES EXPLAINED:

(1) *Mutations* are sudden changes in the genetic material. Mutations are inherited. Because the plants grown in the dark were genetically identical to those grown in the light, neither group lacked the genetic information for chlorophyll production.

(3) *Albinism* is a condition resulting from the absence of a normal gene for color. Both groups of plants had the normal gene for color.

(4) The *phenotype* is the physical appearance of the organism. The phenotype depends on the genotype.

77. **4** The hereditary information is contained in the chromosomes. During mitosis, the *chromosomes duplicate*. The duplication of chromosomes ensures the equal distribution of identical genetic material to the new cells.

WRONG CHOICES EXPLAINED:

(1) If the hereditary information is *destroyed*, the cells cannot function. The chromosomes contain the information necessary for carrying out all cellular activities.

(2) *No two organisms are exactly alike*. No two organisms have the same hereditary material. Identical twins are the only exception to these statements.

(3) Plasmagenes are *genes located outside the nucleus*. Drug resistance in some bacteria is transmitted through plasmagenes.

78. **3** *Crossing-over* is the exchange of chromosomal material between homologous pairs of chromosomes. This process occurs during synapsis in meiosis.

WRONG CHOICES EXPLAINED:

(1) *Nondisjunction* is the failure of homologous chromosomes to separate from each other during meiosis. Cells with extra chromosomes and cells with too few chromosomes result from nondisjunction.

(2) *Polyploidy* is a condition in which the cells have extra sets of chromosomes beyond the normal 2*n* number.

(4) *Hybridization* is the crossing of two organisms that are distinctly different from each other. The purpose is to bring together new combinations of genes. Usually the individual with the new gene combinations is more sturdy than either parent. A tangelo is a cross between a tangerine and a grapefruit.

79. **2** A DNA nucleotide is composed of a *deoxyribose sugar molecule, a phosphoric acid molecule, and a nitrogen base*.

WRONG CHOICES EXPLAINED:

(1) *Adenine and thymine* are bases.

(3) *Glucose* is not the sugar molecule in DNA.

(4) *Adenine, thymine, and cytosine* are bases.

80. **1** The two strands of DNA are held together by *hydrogen bonds*. The hydrogen bonds form weak links between the base pairs of each strand.

WRONG CHOICES EXPLAINED:

(2) The sugars and bases of the nucleotides of DNA are organic compounds. Each individual compound is made up of *carbon bonds*.

(3) The nucleotides in each strand are joined together by *phosphate groups*.

(4) There are no *ribose groups* in DNA.

81. **1** *Lysine* is the amino acid whose genetic code is present on DNA strand I. Strand I of the DNA molecule contains two triplet codons (a triplet codon is a three-base sequence): AAG-CTG. Each triplet codon represents a specific amino acid. The amino acids are carried by tRNA molecules. The tRNA codon matches the DNA codon (except that U replaces T). There are two possible tRNA codons that could match the DNA strand I sequence; they are AAG and CUG. Of these, only the AAG tRNA appears in the diagram. The AAG tRNA carries the amino acid known as lysine.

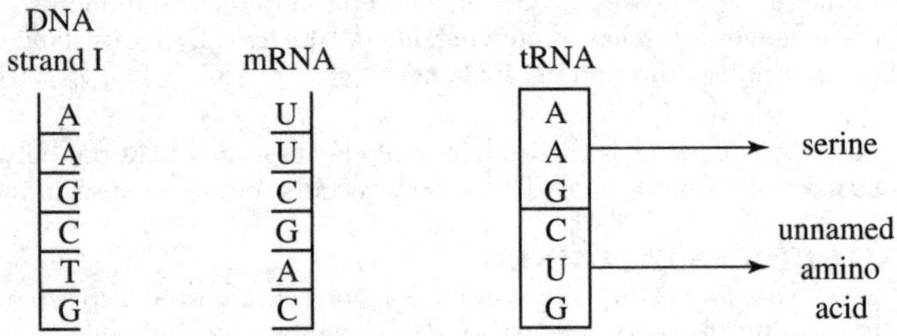

WRONG CHOICES EXPLAINED:

(2) The DNA triplet code for *serine* is TTC. This codon does not appear on strand I of the DNA molecule in the diagram.

(3) The DNA triplet code for *asparagine* is GAC. This codon does not appear on strand I of the DNA molecule in the diagram.

(4) The DNA triplet code for *phenylalenine* is TTT. This codon does not appear on strand I of the DNA molecule in the diagram.

82. **4** A gene controls the production of a protein. The substitution of one base for another changes the triplet code. One amino acid will be substituted for another. The result is a *mutation*. The replacement of glutamic acid by valine in a hemoglobin molecule causes sickle cell anemia.

WRONG CHOICES EXPLAINED:

(1) *Segregation* is the separation of alleles from each other during the formation of gametes.

(2) *Disjunction* is the separation of homologous chromosomes during the process of meiosis.

(3) *Cytoplasmic inheritance* is the inheritance of genes located in the cytoplasm, not in the nucleus. The cytoplasmic genes are called plasmagenes.

83. **2** *Two* different amino acids are coded by strand I. Strand I has two triplet codes, six bases.

WRONG CHOICES EXPLAINED:

(1) Only three bases would have to be shown in the diagram to code for *1* amino acid.

(3) 24 bases are needed for *8* amino acids.

(4) 48 bases are needed for *12* amino acids.

84. **3** *UUCGAC* is the correct sequence. Base pairing is an important concept in DNA duplication and RNA synthesis. Adenine pairs with thymine; cytosine pairs with guanine. There is no thymine in RNA. Uracil takes its place.

WRONG CHOICES EXPLAINED:

(1) *TTCGUC* is not correct. Because thymine is present in the base sequences, the molecule cannot be RNA.

(2) *AACGTC* is not correct. The base sequences are not complementary to either strand I or strand II.

(4) *AAGCUG* is not correct. The base sequences are complementary to strand II in the diagram not strand I.

85. **2** Amino acids are transported to the ribosomes by *RNA molecules* known as transfer RNA, tRNA.

WRONG CHOICES EXPLAINED:

(1) *Protein molecules* are synthesized in the ribosomes. The code for the synthesis is contained in mRNA.

(3) Cellular respiration takes place in the *mitochondria*.

(4) *Chromosomes* are structures found in the nucleus. They are composed of DNA and protein. The genes are located on the chromosomes.

86. **1** Both DNA and RNA *are built from nucleotides*.

WRONG CHOICES EXPLAINED:

(2) Only DNA is *double-stranded*. RNA is single-stranded.

(3) *Deoxyribose* is the sugar in the DNA nucleotides. Ribose is the sugar in the RNA nucleotides.

(4) The base thymine is replaced by *uracil* in RNA nucleotides.

87. **2** The *sequence of amino acids* in a protein is determined by the triplet codes in DNA. The codes are carried to the ribosomes when mRNA is synthesized. A DNA strand is the template in mRNA synthesis.

WRONG CHOICES EXPLAINED:

(1) Enzymes *catalyze the formation of peptide bonds*. A peptide bond is a C–N bond formed by the dehydration synthesis of amino acids.

(3) *Amino acids are transferred* from the cytoplasm to the ribosomes, not to the nucleus. Transfer RNA is the carrier molecule.

(4) ATP molecules *supply the energy for protein synthesis*.

88. **1** A *hybrid* is an individual that has two different alleles for a particular trait. The hybrid may be represented by the symbols *Ss*. The hybrid is smooth; the smooth trait is dominant over the wrinkled trait. When two hybrids are crossed, *75% of the offspring will have the smooth trait and 25% of the offspring will have the wrinkled trait*.

	S	*s*
S	*SS*	*Ss*
s	*Ss*	*ss*

WRONG CHOICES EXPLAINED:

(2), (3), (4) These choices are incorrect based on the information provided by the Punnett square.

89. **2** The term *homozygous* means *pure for the trait*. A homozygous individual has two identical alleles for a gene. There are three alleles for blood type: I^a, I^b, and *i*. The allele I^a produces a protein for blood type A; the allele I^b produces a protein for blood type B; the allele *i* does not produce either protein. The type of blood is determined by the combination of alleles. The genotype refers to the allelic combination. Because the person in the question is homozygous for type A blood, his genotype is I^aI^a.

WRONG CHOICES EXPLAINED:

(1) The I^a and I^b alleles are both dominant over the *i* allele. When both I^a and I^b alleles occur in the same person, the person has type AB blood.

(3) The type of blood represented by the genotype I^ai is type A.

(4) The genotype of a person with type O blood is *ii*.

90. **2** The mating of members of the same litter to maintain desirable traits is known as *inbreeding*. Because mating pairs come from the same litter, they are genetically similar to each other. Inbreeding is used to maintain pure breeds.

WRONG CHOICES EXPLAINED:

(1) *Hybridization*, or outbreeding, is the mating of organisms with contrasting traits. It is the opposite of inbreeding.

(3) Factors in the environment select the organisms that are best adapted to survive in the environment. This principle is known as *natural selection*. It is an essential feature in the theory of evolution.

(4) *Vegetative propagation* is asexual reproduction in plants.

Key Idea 3—Organic Evolution: Individual organisms and species change over time.

Performance Indicator	Description
3.1	The student should be able to explain the major patterns of evolution.

91. In modern classification, protozoa and algae are known as molds, and bacteria are known as
(1) bryophytes (3) protists
(2) plants (4) animals 91____

92. Most modern biologists agree that an ideal classification system should reflect
(1) nutritional similarities among organisms
(2) habitat requirements of like groups
(3) distinctions between organisms based on size
(4) evolutionary relationships among species 92____

93. Which term includes the other three?
(1) genus (3) kingdom
(2) species (4) phylum 93____

94. In one modern classification system, organisms are grouped into three
(1) kingdoms (3) genera
(2) phyla (4) species 94____

95. Which is one basic assumption of the heterotroph hypothesis?
(1) More complex organisms appeared before less complex organisms.
(2) Living organisms did not appear until there was oxygen in the atmosphere.
(3) Large autotrophic organisms appeared before small photosynthesizing organisms.
(4) Autotrophic activity added molecular oxygen to the environment. 95____

96. According to the heterotroph hypothesis, scientists believe that life arose in

(1) a desert environment (3) a vacuum
(2) a forest environment (4) an ocean environment 96____

97. From an evolutionary standpoint, the greatest advantage of sexual reproduction is the

(1) variety of organisms produced
(2) appearance of similar traits generation after generation
(3) continuity within a species
(4) small number of offspring produced 97____

98. According to modern theories of evolution, which of the following factors would be *least* effective in bringing about species changes?

(1) geographic isolation (3) genetic recombination
(2) changing environments (4) asexual reproduction 98____

99. A factor that tends to cause species to change is a

(1) stable environment (3) recombination of genes
(2) lack of migration (4) decrease in mutations 99____

100. If a fossil mammoth were discovered frozen in ice, its cells could be analyzed to determine whether its proteins were similar to those of the modern elephant. This type of investigation is known as comparative

(1) anatomy (3) biochemistry
(2) embryology (4) ecology 100____

101. If members of the same species have been geographically isolated from each other for an extended period of time, which will they most likely exhibit?

(1) mutations identical to each other
(2) random recombination occurring in the same manner
(3) evolution of traits of high adaptive value for their particular environments
(4) evolution into two new species which will have no problem interbreeding 101____

102. Skeletal similarities between two animals of different species are probably due to the fact that both species

(1) live in the same environment
(2) perform the same functions
(3) are genetically related to a common ancestor
(4) have survived until the present time 102____

103. The best means of discovering if there is a close evolutionary relationship between animals is to compare
(1) blood proteins
(2) use of forelimbs
(3) foods consumed
(4) habitats occupied
103____

104. In the process of evolution, the effect of the environment is to
(1) prevent the occurrence of mutations
(2) act as a selective force on variations in species
(3) provide conditions favorable for the formation of fossils
(4) provide stable conditions favorable to the survival of all species
104____

105. In a stable population in which the gene frequencies have been constant for a long time, the rate of evolution
(1) increases
(2) decreases
(3) remains the same
(4) increases, then decreases
105____

106. Certain strains of bacteria that were susceptible to penicillin in the past have now become resistant. The probable explanation for this is that
(1) the mutation rate must have increased naturally
(2) the strains have become resistant because they needed to do so for survival
(3) a mutation was retained and passed on to succeeding generations because it had high survival value
(4) the principal forces influencing the pattern of survival in a population are isolation and mating
106____

107. The frequency of traits that presently offer high adaptive value to a population may *decrease* markedly in future generations if
(1) conditions remain stable
(2) the environment changes
(3) all organisms with these traits survive
(4) mating remains random
107____

108. Since the publication of Darwin's theory, evolutionists have developed the concept that
(1) a species produces more offspring than can possibly survive
(2) the individuals that survive are those best fitted to the environment
(3) through time, favorable variations are retained in a species
(4) mutations are partially responsible for the variations within a species
108____

109. One factor that Darwin was unable to explain satisfactorily in his theory of evolution was
(1) natural selection (3) survival of the fittest
(2) overproduction (4) the source of variations 109____

ANSWERS EXPLAINED Questions 91–109

Standard 4, Key Idea 3, Performance Indicator 1

91. **3** The *protists* include all unicellular organisms and organisms that have both plant and animal features within one cell. Protozoa and algae are protists.

WRONG CHOICES EXPLAINED:

(1) *Bryophytes* are multicellular green plants that do not have vascular tissue. Mosses are examples of bryophytes.

(2) Multicellular photosynthetic organisms make up the *plant* kingdom, which includes both vascular and nonvascular plants.

(4) The *animal* kingdom is composed of multicellular organisms that cannot manufacture their own food. The organisms within this kingdom lack cell walls, and most are capable of some type of locomotion.

92. **4** A classification system should reflect *evolutionary relationships among species*. Evolutionary relationships are determined on the basis of the similarities in the anatomy, embryology, and biochemistry among organisms.

WRONG CHOICES EXPLAINED:

(1) All animals from protozoans to humans utilize the same nutrients in a similar manner. The process of photosynthesis is the same in tree cells and unicellular algae. *Nutritional similarities among organisms* are *not* useful in a system of classification.

(2) Both the whale and the fish live in an ocean environment. However, the whale is a mammal. Other than *sharing the same habitat*, the whale has no fish characteristics.

(3) Algae and protozoans are both microscopic organisms. However, algae have plant characteristics and protozoans have animal characteristics. *Size cannot be used* as the basis for a system of classification.

93. **3** According to the classification system, the largest grouping of organisms is the *kingdom*. Following this, the other groups, in order, are phylum, class, order, family, genus, species. Depending on its chief characteristics, an organism is placed in one of five kingdoms, Monera, Protist, Fungi, Animal, or Plant.

94. **1** In one modern classification system, organisms are grouped into three *kingdoms:* Animal, Plant, and Protist. Organisms that are not typical plants or animals are classified as protists. Examples are protozoa, slime molds, and bacteria.

WRONG CHOICES EXPLAINED:

(2) A *phylum* (plural phyla) is a large grouping that consists of classes, orders, families, genera, and species.

(3) A *genus* (plural genera) is a classification group composed of species. Members of a genus are more closely related than groups belonging to a given phylum.

(4) The *species* is the unit of classification. All members of a species are so closely related that they can mate and produce viable offspring.

95. **4** A heterotroph is an organism that must get its food from a source outside its own body cells; it cannot synthesize its food from inorganic materials. The heterotroph hypothesis proposes that the first living things on Earth were heterotrophs that obtained their food from the organic materials in the primitive seas. Autotrophs are organisms, such as green plants, that can synthesize their own food. At some stage in Earth's history, *autotrophic activity* used up the carbon dioxide in the air and, as a consequence of photosynthesis, *added molecular oxygen to the atmosphere.*

WRONG CHOICES EXPLAINED:

(1) Coacervates and then relatively *simple cells developed before more complex organisms*.

(2) Living heterotrophs *appeared before molecular oxygen was added to the atmosphere*.

(3) *Small cells that carried on photosynthesis appeared before the more complex vascular and seed plants.* The course of evolution is from the simple to the complex.

96. **4** According to the heterotroph hypothesis, life on earth evolved through a sequence of stages. The gases of the primitive atmosphere, such as methane, ammonia, and hydrogen, were washed down by heavy rains into the *early oceans*. They were acted on by ultraviolet radiation, cosmic rays, the earth's heat, and radioactivity. The bonding together of the molecules resulted in the formation of larger organic molecules.

WRONG CHOICES EXPLAINED:

(1) A *desert* environment could not support "first" life. The intense heat and the rapid evaporation of water are conditions that do not allow for the movement or maintenance of molecules in a fluid medium.

(2) A *forest* environment does not provide the pools of warm water on a continuous basis necessary for aggregate molecules to form.

(3) A *vacuum,* a place without air, cannot support life.

97. **1** Sexual reproduction helps to maximize the number of different allelic combinations that occur in offspring, leading to a greater *variety of organisms produced* within the species' population as a whole. Individuals displaying favorable traits in a changing environment are more likely to survive and to pass these traits on to their offspring, a fact that helps to promote evolutionary change.

WRONG CHOICES EXPLAINED:

(2), (3) *Appearance of similar traits generation after generation* and *continuity within a species* describe conditions that promote stability and uniformity within species, both of which are maximized during asexual reproduction.

(4) Evolutionarily speaking, the larger the number of offspring produced during reproduction, the more successful a particular species variety tends to be in competing with other varieties of the same species. *Small number of offspring produced,* therefore, is not an evolutionary advantage of sexual or any other type of reproduction.

98. **4** Variations among organisms are necessary for speciation. *Asexual reproduction* is least effective in bringing about changes in species. There are no variations among organisms that are reproduced asexually. These organisms are genetically like their parents.

WRONG CHOICES EXPLAINED:

(1) *Geographic isolation* increases the chance that a group of organisms will develop a new gene pool, which will give rise to a new species.

(2) *Changes in the environment* cause shifts in the gene pool. The genes that ensure the survival of organisms increase in the pool. Thus, the environment changes the characteristics of the original population, and a new species is formed.

(3) *Genetic recombination* occurs during meiosis and fertilization. The shuffling of genes results in the appearance of new characteristics in a population.

99. **3** The *recombination of genes* is one factor that tends to cause species to change. Mutations are changes in genes. When like mutations combine in the fertilized egg, the new characteristic will be expressed in the offspring. If this mutation adds to the survival value of the organism, the gene change will be passed on to progeny because individuals having this mutation will live to reproduce.

WRONG CHOICES EXPLAINED:

(1) A *stable environment* will probably not cause species to change. Beneficial mutations become effective in changing environments. For example, the mutations that produced white fur in the polar bear were beneficial. At one time, the polar regions were tropical. A change to a glacial environment was accompanied by a change or changes in the animal species that inhabited the region. Bears with a dark coat color became immediate targets for natural enemies.

(2) *Migration* aids the recombination of genes because organisms have greater opportunities for interbreeding.

(4) A *decrease in mutations* does not aid speciation but slows it.

100. **3** *Biochemistry* is the study of the chemistry of living organisms. Proteins are compounds found only in living organisms.

WRONG CHOICES EXPLAINED:

(1) *Anatomy* is the study of the structure of organisms.
(2) *Embryology* is the study of the development of embryos.
(4) *Ecology* is the study of the relationship of living organisms to each other and to their environment.

101. **3** Geographic isolation involves the separation of organisms by natural barriers. Each group of isolated individuals develops its own gene pool because each group lives under different environmental conditions. In the case of members of the same species who have been geographically isolated from each other, the *selection for individuals with special survival traits* is different in each environment.

WRONG CHOICES EXPLAINED:

(1) *Mutations* are the raw materials for evolution. Although mutations might have been the same in each group, they do not have the same adaptive value in each group.
(2) *Random recombination* occurs in the same manner in each group. However, the recombination process operates on two distinctly different gene pools.
(4) Usually, *different species cannot mate*. If their mating happens to be successful, their offspring will not be fertile.

102. **3** Morphology is the study of the structure and form of living things. When the arm of a human and the wing of a bird are studied, they are seen to have similar bone structure. This indicates that both organisms descended from a *common ancestor*. They have both undergone many changes since then and are now very different from each other. However, they still retain some of the same genes and therefore show a similarity in many parts of their bodies, including the arrangement of the bones in their forelimbs.

103. **1** A close evolutionary relationship between animals can be shown by a study of their *blood proteins*. The precipitin test is used to show such a relationship. A rabbit can be sensitized to human blood by being injected with human serum. When the sensitized rabbit serum is mixed with human serum, a white precipitate forms. If the sensitized rabbit serum is mixed with serum from a chicken, there is no reaction. However, a precipitate does form when the sensitized rabbit serum is mixed with the serum of a chimpanzee. In a like manner, the serum of a dog and a wolf show precipitation with serum sensitized to dog serum.

WRONG CHOICES EXPLAINED:

(2) The *forelimbs* of many unrelated animals may be used for the same purpose.
(3) Many unrelated animals *consume similar food*.
(4) Many unrelated animals *live in the same habitat*.

104. **2** There are many variations among the organisms of a species. Some variations allow an organism to survive best in a particular environment. Such factors as climate, food supply, and type of predators determine which organisms are *best adapted to that environment*.

WRONG CHOICES EXPLAINED:

(1) *Mutations* occur naturally and randomly. Mutations increase the variations among organisms.

(3) *Fossils* are the remains of organisms that lived in the past. They present evidence that evolution has occurred.

(4) There are many factors in an environment that influence the survival of different species. A *stable environment* preserves the species that have already adapted to that environment.

105. **3** Stable gene pools are a hallmark of nonevolving populations. The rate of evolution in such populations neither increases nor decreases, but *remain the same.*

WRONG CHOICES EXPLAINED:

(1), (2), and (4) all refer to changing rates of evolution that do not occur.

106. **3** Bacteria resistant to penicillin developed as a result of mutation. Organisms that did not receive the mutated gene were killed by the antibiotic. Those in which gene mutation occurred survived and *passed the mutation on to succeeding generations.*

WRONG CHOICES EXPLAINED:

(1) *The mutation rate did not increase*. The survivors had the mutated gene that allowed the bacteria to resist the effects of penicillin. These resistant strains reproduced, creating populations that replaced the nonresistant strains.

(2) *Need does not determine mutation*. Mutations are chance occurrences.

(4) Survival of a species depends on the ability of its members to obtain food, carry out respiration, and *reproduce successfully*. *Isolation* does not increase species survival.

107. **2** As long as environmental conditions remain stable, the alleles controlling traits that promote individual survival in that environment tend to be maintained at a high level in the population's gene pool. When *the environment changes,* however, the factors that promoted survival may no longer be present. Selection pressure may then operate to reduce the frequency of the once-prevailing alleles in favor of alleles controlling other, contrasting traits that increase individuals' chances for survival in the new environment.

WRONG CHOICES EXPLAINED:

(1) As long as *conditions remain stable,* selection pressures on individuals displaying favorable traits remain low, promoting the maintenance of a high frequency of alleles controlling these traits.

(3) If *all organisms with these traits survive,* their genes will be passed on to future generations at a high rate. This will help to maintain a high frequency of alleles controlling these traits in the population.

(4) If *mating remains random,* the probability that alleles will pair in unrestricted combinations will remain high, helping to ensure that the laws of probability will operate freely and that gene frequencies for existing traits will remain stable.

108. **4** Darwin proposed his theory of evolution in 1856. His theory did not explain how variation arose in organisms. In 1901, Hugo De Vries discovered the existence of mutations. *Mutations accounted for the rise of variations in organisms.*

WRONG CHOICES EXPLAINED:

(1) One of the principles of Darwin's theory of evolution stated that a species produced *more offspring than could possibly survive*.

(2) Another principle stated that the *individuals that survived were those best suited to the environment*.

(3) *The variations favored by the environment are retained within a species.* It is the environment that determines which variations are favorable.

109. **4** Darwin's theory of evolution was completed and published before Mendel completed his study of inheritance in the garden pea. Darwin could not explain *how variations occurred* or how they were passed on from parent to offspring.

WRONG CHOICES EXPLAINED:

(1) Darwin's theory of *natural selection* was divided into five distinct principles that formulated his concept of evolution. These ideas were set forth in his book *The Origin of the Species by Natural Selection.*

(2) *Overproduction* was one of the principles of Darwin's theory. He explained that for a species to continue in existence, it must overproduce in order to maintain the species number. For example, one female codfish lays about 9 million eggs. Not all of these eggs are fertilized and not all codfish lay about 9 million eggs. Not all of these eggs are fertilized and not all codfish fry reach adulthood. If the 9 million eggs per female were fertilized and if these zygotes developed into adult fish, the seas would be overrun with codfish. However, if the number of gametes produced by codfish were greatly reduced, the species would die out. This overproduction of gametes is necessary to maintain codfish survival.

(3) Another of Darwin's principles was *survival of the fittest*. No two organisms are alike; each has variations. These variations may either help or hinder the organism in its struggle for existence. An organism with variations that help it reach food faster is more fit and has a better potential for survival than a slower, less fit member of the species.

Key Idea 4—Reproductive Continuity: The continuity of life is sustained through reproduction and development.

Performance Indicator	Description
4.1	The student should be able to explain how organisms, including humans, reproduce their own kind.

110. Which occurs in a plant cell but *not* in an animal cell during mitotic cell division?
(1) formation of spindle fibers (3) formation of a cell plate
(2) chromosome duplication (4) cytoplasmic division 110____

111. A plant cell with 12 chromosomes undergoes normal mitosis. What is the total number of chromosomes in each of the resulting daughter cells?
(1) 24 (2) 12 (3) 6 (4) 4 111____

112. Asexual reproduction *differs* from sexual reproduction in that, in asexual reproduction,
(1) new organisms are usually genetically identical to the parent
(2) the reproductive cycle involves the production of gametes
(3) nuclei of sex cells fuse to form a zygote
(4) offspring show much genetic variation 112____

113. In most multicellular animals, meiotic cell division occurs in specialized organs known as
(1) gonads (3) kidneys
(2) gametes (4) cytoplasmic organelles 113____

114. Which is an important adaptation for reproduction among land animals?
(1) fertilization of gametes outside the body of the female
(2) fertilization of gametes within the body of the female
(3) production of sperm cells with thick cell walls
(4) production of sperm cells with thin cell walls 114____

115. In humans, a single primary sex cell may produce four gametes. These gametes are known as
(1) diploid egg cells (3) polar bodies
(2) monoploid egg cells (4) sperm cells 115____

116. In sexual reproduction, the $2n$ chromosome number is restored as a direct result of
(1) fertilization
(2) gamete formation
(3) cleavage
(4) meiosis
116____

117. In human females, the main function of the follicle-stimulating hormone (FSH) secreted by the pituitary gland is to
(1) stimulate the adrenal glands to produce cortisone
(2) stimulate activity in the ovaries
(3) control the metabolism of calcium
(4) regulate the rate of oxidation in the body
117____

118. If the first stage of an uninterrupted human menstrual cycle is the follicle stage, the last stage includes the
(1) formation of sperm cells in the testis
(2) release of a mature egg
(3) buildup of the uterine lining
(4) shedding of the uterine lining
118____

119. Which statement best describes internal fertilization?
(1) It does not require motile gametes.
(2) It helps to make terrestrial life possible.
(3) It requires the presence of many eggs.
(4) It normally occurs in the male.
119____

120. What are the normal chromosome numbers of a sperm, egg, and zygote, respectively?
(1) monoploid, monoploid, and monoploid
(2) monoploid, diploid, and diploid
(3) diploid, diploid, and diploid
(4) monoploid, monoploid, and diploid
120____

121. When compared with the number of gametes produced from a single primary sex cell during oogenesis, the number of gametes produced from a single human primary sex cell during spermatogenesis is usually
(1) four times as great
(2) twice as great
(3) half as great
(4) the same
121____

122. In human males, sperm cells are suspended in a fluid medium. The main advantage gained from this adaptation is that the fluid

(1) removes polar bodies from the surface of the sperm
(2) activates the egg nucleus so that it begins to divide
(3) acts as a transport medium for sperm
(4) provides currents that propel the egg down the oviduct 122_____

Base your answers to questions 123 and 124 on the diagrams and the information below.

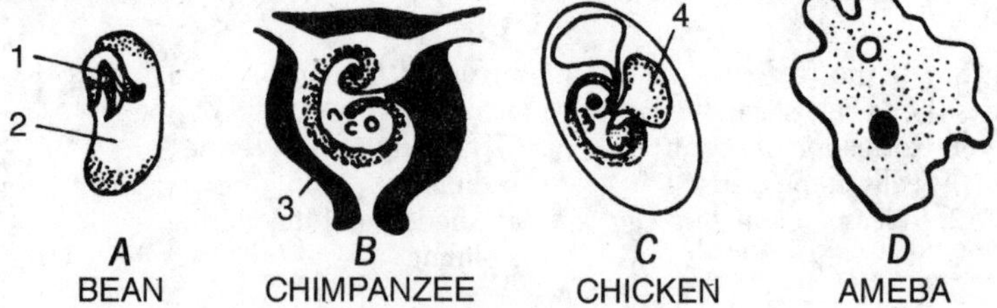

123. Which organisms were produced as a result of fertilization?

(1) *A, B,* and *C,* only
(2) *B* and *C,* only
(3) *C* and *D,* only
(4) *B, C,* and *D,* only 123_____

124. Structures that function in the storage of food to be used by growing embryonic cells are indicated by

(1) 1 and 3
(2) 2 and 3
(3) 2 and 4
(4) 3 and 4 124_____

Base your answers to questions 125 through 127 on your knowledge of biology and the information below.

> A biologist cut a flap of ectoderm from the top of a developing embryo. He did not remove the piece of ectoderm but just folded it back. Then he cut out the mesoderm underneath and completely removed it. He folded the flap of ectoderm back in place. The ectoderm healed; however, a complete nervous system did not develop.

125. This experiment was most likely performed immediately after

(1) cleavage
(2) gestation
(3) fertilization
(4) gastrulation 125_____

126. This experiment interfered with the process of
(1) differentiation (3) cleavage
(2) zygote formation (4) ovulation 126____

127. This experiment demonstrates that the
(1) ectoderm is solely responsible for development of the nervous system
(2) nervous system is destroyed during surgical operations
(3) mesoderm influences the development of the nervous system
(4) digestive enzymes have a major role in the development of embryonic layers 127____

128. In a developing embryo, the mesoderm layer normally gives rise to
(1) epidermal tissue (3) digestive tract lining
(2) skeletal tissue (4) respiratory tract lining 128____

129. What is the function of the placenta in a mammal?
(1) It surrounds the embryo and protects it from shock.
(2) It allows mixing of the maternal and fetal blood.
(3) It permits the passage of nutrients and oxygen from the mother to the fetus.
(4) It replaces the heart of the fetus until the fetus is born. 129____

ANSWERS EXPLAINED Questions 110–129

Standard 4, Key Idea 4, Performance Indicator 1

110. **3** Mitosis is the process by which two identical nuclei are formed. Mitotic cell division is usually followed by cytoplasmic division. A plant cell has a rigid cell wall. Division of the cytoplasm begins with the appearance of a *cell plate* between the two nuclei. The cell plate is composed of membrane fragments from the endoplasmic reticulum.

WRONG CHOICES EXPLAINED:

(1) *Spindle fibers* are elastic-like protein fibers. Chromosome movement is controlled by spindle fibers.

(2) In order for two nuclei to be identical, they must have the same number and kind of chromosomes. The *chromosomes duplicate* before the nucleus divides. The mitotic process is the same in both plant and animal cells.

(4) *Cytoplasmic division*, or cytokinesis, usually follows nuclear division in both plant and animal cells.

111. **2** Chromosomes are structures in the nucleus. During mitosis, two cells with identical chromosomes are formed. Because the cell had 12 chromosomes, the daughter cells must also have *12* chromosomes.

WRONG CHOICES EXPLAINED:

(1) A cell with *24* chromosomes has twice the diploid number. The condition in which there are extra sets of chromosomes is known as polyploidy.

(3) A cell with *6* chromosomes has one-half the diploid number. Monoploid cells arise through meiosis.

(4) A cell with *4* chromosomes can only arise through a complete breakdown of the mitotic or meiotic process.

112. **1** In asexual reproduction, new organisms are produced by a single parent. Asexual reproduction involves the mitotic process. The *genetic material of the offspring is identical to that of the parent*.

WRONG CHOICES EXPLAINED:

(2) *Gametes* are produced by sexually reproducing organisms.

(3) The fusion of sex cells (gametes) to *form a zygote* is characteristic of sexually reproducing organisms.

(4) *Genetic variation* among offspring is characteristic of sexually reproducing organisms. The process of meiosis through synapsis and segregation ensures new combinations of genetic material.

113. **1** *Gonads* are sex glands. In these glands, gametes, or sex cells, are produced from primary sex cells that undergo meiosis, also known as reduction division. Male gonads are called testes, and female gonads, ovaries.

WRONG CHOICES EXPLAINED:

(2) *Gametes* are sex cells and not organs. Sex cells are specialized for fertilization.

(3) *Kidneys* are organs of excretion and are specialized for filtering metabolic wastes out of the blood. The nephron is the unit of structure and function in the kidney. Meiosis does not take place in kidney cells.

(4) *Cytoplasmic organelles* such as mitochondria, ribosomes, lysosomes, and endoplasmic reticula are not organs.

114. **2** A gamete is a reproductive cell that must fuse with another gamete to produce a new individual. Sperm cells and egg cells are gametes. Fertilization is the fusion of an egg cell and a sperm cell. In land animals, *fertilization occurs within the body of the female* and is known as internal fertilization.

WRONG CHOICES EXPLAINED:

(1) External fertilization, the *union of gametes outside the female's body*, occurs in animals that live in a watery environment. Fish and amphibians reproduce by external fertilization.

(3), (4) The question refers to reproduction in animals. Animal cells, including gametes, do not have *cell walls*.

115. **4** Primary sex cells give rise to gametes. Gametes are formed by the process of meiosis. In meiosis, a diploid cell divides twice to form four monoploid cells. In humans the four gametes, which are identical in size, are known as *sperm cells.*

WRONG CHOICES EXPLAINED:

(1) Chromosomes occur in pairs. The diploid number of chromosomes is the full number of chromosomes of all the pairs. Meiosis is cell division in which the nucleus receives one member of each pair of chromosomes. The nucleus of an *egg cell* thus contains half the diploid chromosome number, or the monoploid number.

(2) In formation of the egg cell, the cytoplasm does not divide equally. One large monoploid cell, the egg cell, and three very small cells (polar bodies) are produced from one primary sex cell.

(3) The three small monoploid cells accompanying the egg cell are known as *polar bodies.* Polar bodies degenerate and do *not* function in fertilization.

116. **1** The diploid chromosome number is represented as $2n$, and the monoploid number as n. When two gametes in the n condition combine, a $2n$ cell is produced. *Fertilization* is the union of two gametes.

WRONG CHOICES EXPLAINED:

(2) *Gamete formation* reduces the chromosome number from $2n$ to n.

(3) *Cleavage* is mitotic cell division without growth. It is the process by which a fertilized egg cell becomes a multicellular embryo.

(4) *Meiosis,* or reduction division, reduces the chromosome number of diploid cells.

117. **2** Follicles contain immature egg cells. The follicles are found in the ovary. FSH *stimulates ovarian* follicle development.

WRONG CHOICES EXPLAINED:

(1) ACTH stimulates the *adrenal glands* to produce cortisone. ACTH is secreted by the pituitary gland.

(3) *Calcium metabolism* is controlled by parathormone. The parathyroid gland secretes parathormone.

(4) Thyroxin secreted by the thyroid gland is the major regulator of the rate of *oxidation*. The hormones from the adrenal glands and the pancreas also play a role in the oxidation of glucose.

118. **4** The menstrual cycle is a series of changes that occur within the female reproductive system. The events of the cycle prepare the uterus to receive an embryo. The lining of the uterus is built up. If the cycle is not interrupted, the egg is not fertilized and no embryo is formed. In the last stage of the cycle, the *lining of the uterus disintegrates and is shed.*

WRONG CHOICES EXPLAINED:

(1) *Sperm cells* are produced by the male.

(2) Ovulation (*release of a mature egg*) occurs midway through the menstrual cycle.

(3) Once a month the *uterus* is prepared to receive an embryo. What happens to the lining of the uterus depends upon presence or absence of an embryo. If an embryo is present, the uterus continues to develop and the menstrual cycle is interrupted.

119. **2** *It helps to make terrestrial life possible* is the correct response. Internal fertilization, as its name implies, occurs within the body of the parent (usually female). The conditions in the female reproductive tract provide an ideal environment for the survival and pairing of gametes, helping to ensure that fertilization occurs successfully. This method of reproduction is especially helpful in the survival of terrestrial animal species, who live where harsh conditions (such as drying, heat, and cold) can easily damage or kill gametes released into the environment for external fertilization.

WRONG CHOICES EXPLAINED:

(1) *Motile gametes* (such as human sperm cells) are common in species employing both external and internal fertilization. Motility (ability to move) enables the sperm cells to swim toward the egg cell in either environment.

(3) Because of the dangers posed to fragile gametes in any environment, the *presence of many eggs* is characteristic of species employing external fertilization. Species using internal fertilization produce relatively few eggs in the reproductive process.

(4) In most species internal *fertilization occurs within the body of the female*, not that of the *male.*

120. **4** *Monoploid, monoploid, and diploid* is the correct combination. Sperm cells and egg cells are monoploid (n) gametes formed during the process of meiotic cell division. A zygote results from the fusion of two monoploid nuclei in fertilization and so must be diploid ($2n$) in chromosome number.

WRONG CHOICES EXPLAINED:

(1), (2), (3) Each of these distracters contains an incorrect combination of choices (see above).

121. **1** In the process of oogenesis, a single primary sex cell gives rise to a single monoploid egg cell and three nonfunctional monoploid polar bodies. The process of spermatogenesis yields four functional monoploid sperm cells for each primary sex cell. Therefore, a comparison of these two processes leads to the conclusion that, per primary sex cell, spermatogenesis yields *four times* as many gametes as oogenesis does.

WRONG CHOICES EXPLAINED:

(2), (3), (4) Each of these distracters contains a mathematical comparison that is not consistent with the explanation above.

122. **3** The fluid surrounding human sperm cells *acts as a transport medium for sperm.* This fluid is known as semen. Its primary function is to provide a protective watery medium for sperm cells as they enter the female reproductive tract.

WRONG CHOICES EXPLAINED:

(1) *Removes polar bodies from the surface of the sperm* is a "nonsense" distracter. Polar bodies are not associated with sperm production.

(2) *Activates the egg nucleus so that it begins to divide* is not a function of semen. The egg is stimulated to divide by the act of fertilization. Semen is not directly involved in this process.

(4) *Provides currents that propel the egg down the oviduct* is not a function of semen. Cilia that line the oviduct are responsible for establishing fluid currents that both carry the egg downward toward the uterus and carry sperm upward toward the ovary. Semen is not directly involved in this process.

123. **1** Fertilization is one of the processes in sexual reproduction. Sexual reproduction is the method of reproduction in the bean plant, chimpanzee, and chicken (*A, B, and C, only*). The ameba reproduces asexually by binary fission.

WRONG CHOICES EXPLAINED:

(2) The bean plant (*B*) was omitted in this choice.

(3) The ameba (*D*) is an incorrect answer. Both the bean (*A*) and the chimpanzee (*B*) were omitted in this choice.

(4) The bean plant (*A*) was omitted and the ameba (*D*) is an incorrect answer.

124. **3** The structures that function in the storage of food for the embryonic cells are labeled 2 and *4*. Structure 2 is the cotyledon of the seed. The yolk *sac* is structure *4*.

WRONG CHOICES EXPLAINED:

(1) Structure *3* refers to the wall of the uterus. The embryo of a chimpanzee is nourished through the placenta, not the uterine wall. Structure *1* is the leaf of the embryo bean plant.

(2) Although structure 2 is a correct answer, structure *3* is incorrect.

(4) Although structure *4* is a correct answer, structure *3* is incorrect.

125. **4** The experiment was performed after *gastrulation.* Gastrulation is a stage in embryonic development that gives rise to three germ layers of cells. The three germ layers are the ectoderm, mesoderm, and endoderm.

WRONG CHOICES EXPLAINED:

(1) *Cleavage* is a stage of embryonic development in which the zygote undergoes rapid mitotic divisions. The final result is a ball of cells.

(2) *Gestation* is a prebirth period. It is the time a developing embryo spends in the uterus.

(3) The union of a sperm cell nucleus with an egg cell nucleus is called *fertilization.* The result of the process is a zygote.

126. **1** The experiment interfered with the development of a nervous system. The development of special tissues and organisms is known as *differentiation.*

WRONG CHOICES EXPLAINED:

(2) *Zygote formation* must occur before an embryo can develop.

(3) The process of *cleavage* provides the embryo with hundreds of undifferentiated cells.

(4) *Ovulation* is the release of an egg from the ovary.

127. **3** The experiment demonstrates that the development of the *nervous system is influenced by the presence of the mesoderm.* The nervous system does not develop when the mesoderm is removed.

WRONG CHOICES EXPLAINED:

(1) If the *ectoderm was solely responsible for the development of the nervous system,* the nervous system would have developed after the mesoderm was removed.

(2) There was no *nervous system* present when the surgery was performed.

(4) The experiment was not concerned with the reasons for the *development of the embryonic layers.*

128. **2** Each of the three germ layers of the embryo is responsible for the development of the systems of the body. The *skeletal system* develops from the mesoderm. The muscle system, circulatory system, and excretory system also evolve from the mesoderm.

WRONG CHOICES EXPLAINED:

(1) *Epidermal cells* form the outer covering or skin of the body. The skin and nervous system develop from the ectoderm.

(3), (4) The linings of the *digestive* and *respiratory tracts* develop from the endoderm.

129. **3** The placenta is an area of spongy tissue in the uterus. It is very rich in blood vessels. The placenta functions as a respiratory and excretory organ of the fetus. *Vital materials are exchanged between the capillaries of the fetus and the capillaries of the mother.*

WRONG CHOICES EXPLAINED:

(1) The amnion is a fluid-filled sac surrounding the embryo. The fluid bathes the cells of the fetus and *protects it against shock.*

(2) The circulatory systems of the mother and the fetus are separate from each other. *Blood does not flow from one system into the other.*

(4) The *embryo develops its own heart.* The placenta provides an area for the diffusion of materials into and out of the fetus.

Key Idea 5—Dynamic Equilibrium and Homeostasis: Organisms maintain a dynamic equilibrium that sustains life.

Performance Indicator	Description
5.1	The student should be able to explain the basic biochemical processes in living organisms and their importance in maintaining dynamic equilibrium.
5.2	The student should be able to explain disease as a failure of homeostasis.
5.3	The student should be able to relate processes at the system level to the cellular level in order to explain dynamic equilibrium.

130. Which energy conversion occurs in the process of photosynthesis?

(1) Light energy is converted to nuclear energy.
(2) Chemical bond energy is converted to nuclear energy.
(3) Light energy is converted to chemical bond energy.
(4) Mechanical energy is converted to light energy 130_____

131. An environmental change that would most likely increase the rate of photosynthesis in a bean plant would be an increase in the
(1) intensity of green light
(2) concentration of nitrogen in the air
(3) concentration of oxygen in the air
(4) concentration of carbon dioxide in the air 131____

132. During photosynthesis, molecules of oxygen are released as a result of the "splitting" of water molecules. This is a direct result of the
(1) dark reaction
(2) light reaction
(3) formation of PGAL
(4) formation of CO_2 132____

133. An organism that makes its own food without the direct need for any light energy is known as a
(1) chemosynthetic heterotroph
(2) chemosynthetic autotroph
(3) photosynthetic heterotroph
(4) photosynthetic autotroph 133____

134. While looking through a microscope at a section of a leaf from a freshwater plant, a student observed some cells in which chloroplasts were moving around with the cytoplasm. This type of movement is known as
(1) pinocytosis
(2) synapsis
(3) osmosis
(4) cyclosis 134____

135. By what process does carbon dioxide pass through the stomates into the leaf?
(1) diffusion
(2) osmosis
(3) respiration
(4) pinocytosis 135____

136. Two end products of aerobic respiration are
(1) oxygen and alcohol
(2) oxygen and water
(3) carbon dioxide and water
(4) carbon dioxide and oxygen 136____

137. Homeostatic regulation of the body is made possible through the coordination of all body systems. This coordination is achieved mainly by
(1) respiratory and reproductive systems
(2) skeletal and excretory systems
(3) nervous and endocrine systems
(4) circulatory and digestive systems 137____

138. Phenylketonuria (PKU) is an inherited condition characterized by mental retardation. The symptoms of the disorder result from an inability to synthesize a single type of

(1) enzyme (3) blood cell
(2) nutrient (4) brain cell 138____

139. All the children of a hemophiliac male and a normal female are normal with respect to blood clotting. However, some of their grandsons are hemophiliacs. This is an example of the pattern of hereditary known as

(1) sex determination (3) incomplete dominance
(2) sex linkage (4) multiple alleles 139____

140. What is the total number of chromosomes in a typical body cell of a person with Down's syndrome?

(1) 22 (2) 23 (3) 44 (4) 47 140____

ANSWERS EXPLAINED Questions 130–140

Standard 4, Key Idea 5, Performance Indicators 1–3

130. **3** In the process of photosynthesis, carbon dioxide and water molecules are converted to glucose. Light is the energy source for the reaction. *Light energy is transformed into the chemical bond energy* of the glucose molecules.

WRONG CHOICES EXPLAINED:

(1), (2), (4) The Law of Conservation of Energy states that energy cannot be created or destroyed, but it can be changed from one form to another. All the choices refer to this law, but only choice (3) occurs in living organisms.

131. **4** Experiments have shown that the rate of photosynthesis depends on the availability of carbon dioxide. The greater the *concentration of carbon dioxide*, the greater the rate of photosynthesis.

WRONG CHOICES EXPLAINED:

(1) Chlorophyll is a light-absorbing pigment found in chloroplasts. However, it does not absorb much of the green wavelength of light. An increase in the *intensity of green light* has no effect on the photosynthetic rate.

(2) Carbohydrates are the products of photosynthesis. They do not contain atoms of nitrogen. The nitrogen needed by plants comes from nitrates in the soil, not from *the concentration of nitrogen in the air*.

(3) Oxygen is released during photosynthesis. Therefore, *the concentration of oxygen in the air* has no direct effect on photosynthesis.

132. **2** Sunlight provides the energy needed to split water molecules into hydrogen and oxygen. Because light is required, this part of photosynthesis is known as the *light reaction.*

WRONG CHOICES EXPLAINED:

(1) The *dark reaction* does not use light energy. In this reaction, the hydrogen released from the light reaction is combined with carbon dioxide.

(3) *PGAL*, phosphoglyceric aldehyde, is the first stable compound formed during the dark reaction. This compound is later converted to glucose.

(4) *Carbon dioxide is not formed* but is used during photosynthesis.

133. **2** An autotroph manufactures its own food. A *chemosynthetic autotroph* produces its own food without the use of light energy. It obtains its energy from certain chemical reactions that take place in the cell.

WRONG CHOICES EXPLAINED:

(1) A *chemosynthetic organism* cannot be a heterotroph. Heterotrophs do not have the ability to manufacture their own food. All animals are heterotrophs.

(3) *Photosynthetic organisms* manufacture their own food. They cannot be heterotrophs.

(4) A *photosynthetic autotroph* utilizes light energy. All green plants are photosynthetic autotrophs.

134. **4** The movement of chloroplasts in the plant cell was due to the movement of cytoplasm in the cell. *Cyclosis* is the streaming of cytoplasm in a cell.

WRONG CHOICES EXPLAINED:

(1) *Pinocytosis* is the formation of a pocket by an infolding of the cell membrane. Large molecules are brought into the cell by this process.

(2) *Synapsis* is the pairing of homologous chromosomes during meiosis.

(3) *Osmosis* is the movement of water across a selectively permeable membrane. Osmosis is the diffusion of water.

135. **1** Carbon dioxide passes through the stomates of a leaf by *diffusion.* Diffusion is passive transport. Molecules move along a concentration gradient from an area of high density to an area of lower density.

WRONG CHOICES EXPLAINED:

(2) *Osmosis* is the diffusion of water.

(3) *Respiration* is an energy-releasing process.

(4) *Pinocytosis* is active transport. Cells use energy to draw in large molecules by the infolding of their cell membranes.

136. **3** Respiration is a process by which cells release energy from glucose molecules. Aerobic respiration requires the presence of oxygen. In the process of aerobic respiration, glucose is oxidized to *carbon dioxide and water*. Both compounds are end products, the results of a chemical reaction.

WRONG CHOICES EXPLAINED:

(1), (2), (4) All three choices are incorrect because oxygen is consumed in aerobic respiration. Oxygen is not the end product of the reaction.

137. **3** Homeostasis refers to the steady state of control of the cell and, in turn, the entire body. The biochemical processes that take place in body cells occur in even and regular sequences. The cells, tissues, and organs in all body systems must function cooperatively so that the organism can carry out its life functions effectively. The coordination of all these biochemical activities is made possible by the work of the *nervous and endocrine systems*. The nervous system carries impulses from sense organs to the brain or spinal cord and then to effector organs such as muscles or glands. The endocrine system secretes hormones that control the functions of certain glands, tissues, and organs. Together the nervous and endocrine systems maintain the homeostasis of the body.

WRONG CHOICES EXPLAINED:

(1) The *respiratory system* is specialized for the intake and distribution of oxygen. It also expels waste gases. The *reproductive system* is specialized for the developing of embryos. Both systems are controlled by the nervous and endocrine systems. In mammals, hormones control gestation and birth.

(2) The *skeletal system* gives support to the body. Hormones control the growth of long bones. Nerve cell fibers help the muscles to function. The *excretory system* coordinates waste removal from the body. Hormones control water loss and reabsorption by the kidney tubules.

(4) Blood and lymph circulate by way of the *circulatory system*. Hydrolysis of food takes place in the *digestive system*. Both systems depend on the nervous and endocrine systems for the coordination of mechanical and biochemical activities.

138. **1** According to the one gene–one enzyme theory, a single gene is responsible for the production of a single enzyme. Because PKU is an inherited defect, the gene for an *enzyme* is absent in the victim.

WRONG CHOICES EXPLAINED:

(2) *Nutrients* are *not* synthesized by animals. Nutrients must be taken in from organic sources.

(3) *Blood cells* are *not* affected nor involved in the PKU disorder.

(4) The development of *brain cells* is affected by the PKU condition. Because of the lack of an enzyme, phenylalanine is converted to phenylpyruvic acid, which accumulates in brain tissue. The result is mental retardation.

139. **2** A trait that appears more often in one sex than in the other sex is said to be *sex-linked*. The gene for the trait is located on a sex chromosome. Hemophilia and color blindness are examples of sex-linked traits.

WRONG CHOICES EXPLAINED:

(1) *Sex determination* is controlled by a pair of sex chromosomes. There are two kinds of sex chromosomes, an X and a Y chromosome. A female has two X chromosomes (XX). A male has one X chromosome and one Y chromosome (XY).

(3) *Incomplete dominance* is a type of inheritance in which neither allele in a hybrid is dominant. The hybrid shows a trait completely different from either parent. The inheritance of color in a Japanese four-o'clock flower is an example of incomplete dominance, or blending.

(4) Alleles are different forms of the same gene. The term *multiple alleles* implies that a gene has more than two forms. Inheritance of human blood type involves multiple alleles.

140. **4** The normal number of chromosomes in human body cells is 46. Down's syndrome results from the presence of an extra chromosome. The extra or *47*th chromosome is due to meiotic nondisjunction.

WRONG CHOICES EXPLAINED:

(1), (3) The numbers *22* and *44* do not apply to any known normal human chromosome number.

(2) The number *23* is the monoploid number of chromosomes in human gametes.

Key Idea 6—Interdependence of Living Things: Plants and animals depend on each other and their physical environment.

Performance Indicator	Description
6.1	The student should be able to explain factors that limit growth of individuals and populations.
6.2	The student should be able to explain the importance of preserving diversity of species and habitats.
6.3	The student should be able to explain how the living and nonliving environments change over time and respond to disturbances.

141. Animals *cannot* synthesize nutrients from inorganic raw materials. Therefore, animals obtain their nutrients by
(1) combining carbon dioxide with water
(2) consuming preformed organic compounds
(3) hydrolyzing large quantities of simple sugars
(4) oxidizing inorganic molecules for energy 141____

142. Which organisms carry out heterotrophic nutrition?
(1) ferns (3) fungi
(2) grasses (4) mosses 142____

143. Which activity is an example of intracellular digestion?
(1) a grasshopper chewing blades of grass
(2) a maple tree converting starch to sugar in its roots
(3) an earthworm digesting proteins in its intestine
(4) a fungus digesting dead leaves 143____

144. A hydra ingests a daphnia, digests it, and later egests some materials. All these events are most closely associated with the life process known as
(1) transport (3) growth
(2) synthesis (4) nutrition 144____

145. Some bacteria are classified as saprophytes because they are organisms that
(1) feed on other living things
(2) feed on dead organic matter
(3) manufacture food by photosynthesis
(4) contain vascular bundles 145____

146. Aerobic organisms are dependent on autotrophs. One reason for this dependency is that most autotrophs provide the aerobic organisms with
(1) oxygen (3) nitrogen gas
(2) carbon dioxide (4) hydrogen 146____

147. Of the following, the greatest amount of the Earth's food production is thought to occur in
(1) coastal ocean waters (3) taiga forests
(2) desert biomes (4) tundra biomes 147____

148. Most of the minerals within an ecosystem are recycled and returned to the environment by the direct activities of organisms known as

(1) producers (3) decomposers
(2) secondary consumers (4) primary consumers 148_____

149. Which type of organism is *not* shown in the following representation of a food chain?

grass → mouse → snake → hawk

(1) herbivore (3) producer
(2) decomposer (4) carnivore 149_____

150. In the food chain shown below, which organism represents a primary consumer?

(1) grasshopper (3) frog
(2) grass (4) snake 150_____

151. A lake contains minnows, mosquito larvae, sunfish, algae, and pike. Which of these organisms would probably be present in the largest number?

(1) minnows (3) sunfish
(2) larvae (4) algae 151_____

152. An abiotic factor that affects the ability of pioneer organisms such as lichens to survive is the

(1) type of climax vegetation (3) type of substratum
(2) species of algae (4) species of bacteria 152_____

153. In order to avoid predators, the clown fish hides unharmed in the stinging tentacles of the sea anemone. The clown fish attracts food to the sea anemone. This is an example of a type of relationship known as

(1) mutualism (3) predator–prey
(2) commensalism (4) parasitism 153_____

154. Which world biome has the greatest number of organisms?

(1) tundra
(2) tropical forest
(3) temperate deciduous forest
(4) marine 154____

155. In a particular area, living organisms and the nonliving environment function together as

(1) a population
(2) a community
(3) an ecosystem
(4) a species 155____

ANSWERS EXPLAINED Questions 141–155

Standard 4, Key Idea 6, Performance Indicators 1–3

141. **2** The portions of food that are usable to an animal are known as nutrients. Carbohydrates, lipids, and proteins are the organic nutrients needed by all organisms. Animals cannot synthesize their own nutrients. They must eat other organisms that contain the *preformed organic nutrients*.

WRONG CHOICES EXPLAINED:

(1) *Carbon dioxide and water* are inorganic compounds. These compounds are converted to nutrients by plants only. The process is called photosynthesis.

(3) Hydrolysis is the breakdown of compounds to simpler molecules through the action of enzymes in the presence of water. Glucose is a *simple sugar*. The hydrolyzing of glucose results in the release of energy in a cell.

(4) The use of *inorganic molecules* for the production of energy occurs only in certain species of bacteria. Chemosynthesis does not occur in members of the animal kingdom.

142. **3** *Fungi* are nongreen plants. They are heterotrophs, which means that they cannot manufacture their own food. Organisms that carry out heterotrophic nutrition must take in preformed organic molecules.

WRONG CHOICES EXPLAINED:

(1), (2), (4) *Ferns*, *grasses*, and *mosses* are green plants or bryophytes. Green plants are *autotrophs,* which means that they manufacture their own food.

143. **2** Digestion that occurs within a cell is known as intracellular digestion. Plants do not have special digestive systems. Digestion, or the *conversion of starch to sugar*, occurs within the individual cells of a plant, including those of the root.

WRONG CHOICES EXPLAINED:

(1) *Grasshoppers* have a digestive system. Digestion is extracellular and takes place in a digestive tube outside the body cells. Chewing a blade of grass is an example of mechanical digestion taking place in the mouth.

(3) *Earthworms* also have a digestive system. Proteins are digested outside the body cells in a portion of the digestive system known as the intestine.

(4) *Fungi* demonstrate a special form of extracellular digestion. Digestive enzymes are secreted into the external environment. The nutrients from the digested food diffuse into the cells.

144. **4** *Nutrition* is the life process most closely associated with a hydra ingesting, digesting, and egesting a daphnia. Ingestion is the process by which food materials are taken into the body of an organism such as a hydra. Digestion is the process by which the complex food molecules within the daphnia are hydrolyzed to soluble end products. Egestion is the process by which the undigestible materials of the daphnia's body are expelled from the body of the hydra.

WRONG CHOICES EXPLAINED:

(1) *Transport* is the life process by which soluble foods and other materials are circulated through the body of an organism such that they reach all parts of the organism's body.

(2) *Synthesis* is a process by which complex materials are constructed from simpler chemical components. The processes described in the question represent hydrolysis, the opposite of synthesis.

(3) *Growth* is a process that involves an increase in cell number and cell size, leading to an increase in the size of the organism. The end products of digestion can be used to supply raw materials for such growth.

145. **2** *Saprophytes* are organisms that *feed on dead organic matter*. Fungi and the bacteria of decay are examples of saprophytes.

WRONG CHOICES EXPLAINED:

(1) Heterotrophs *live on or off other living organisms*. A dog flea is an example of a heterotroph also known as a parasite.

(3) Autotrophs *manufacture their own food by photosynthesis*. Algae, mosses, and grasses are examples of autotrophs.

(4) Higher plants *contain vascular bundles*. These plants, called *tracheophytes*, include ferns, conifers, and flowering plants.

146. **1** Aerobic organisms need *oxygen* for cellular respiration. Some autotrophs are photosynthetic organisms. Oxygen is released by photosynthesis. Aerobic organisms depend on the autotrophs to release oxygen into the environment.

WRONG CHOICES EXPLAINED:

(2) *Carbon dioxide* is a waste product from the cellular respiration of aerobic organisms.

(3) *Nitrogen gas* makes up 78% of the atmosphere. However, the nitrogen cannot be used in the gaseous form by aerobic organisms and most autotrophs.

(4) *Hydrogen* does *not* exist as a gas on our planet. It is combined with other elements. The hydrogen needed by organisms comes mostly from water and organic compounds.

147. **1** The area of greatest food production is in the region where the greatest rate of photosynthesis occurs. The area must be rich in minerals, water, gases, and light. The *coastal ocean waters* meet these requirements.

WRONG CHOICES EXPLAINED:

(2) There is very little precipitation in the *desert*. Water is the factor that limits plant growth.

(3) The water in the *taiga* is frozen part of the year.

(4) The *tundra* is a frozen plain. Water is frozen almost all year long on the tundra.

148. **3** Bacteria of decay are *decomposers* that release minerals from decaying plant and animal bodies and return them to the environment.

WRONG CHOICES EXPLAINED:

(1) *Producers* are autotrophs, that is, green plants that synthesize food by photosynthesis from carbon dioxide and water in the presence of sunlight.

(2) *Secondary consumers* are animals that eat other animals. For example, a frog feeds on flies. Thus, a frog is a secondary consumer.

(4) *Primary consumers* are organisms that feed on plants only. Herbivores are primary consumers.

149. **2** *Decomposers* are not shown in the food chain. They are organisms that live on dead things. Fungi and bacteria are decomposers.

WRONG CHOICES EXPLAINED:

(1) An *herbivore* is a primary consumer; that is, it eats vegetation. The mouse is the herbivore in the food chain.

(3) A *producer* is a green plant. It depends on sunlight to synthesize its own food. The grass is the producer in the food chain.

(4) A *carnivore* is an animal that eats the flesh of other animals. Both the snake and the hawk are carnivores.

150. **1** The *grasshopper* is a primary consumer because it feeds on vegetation.

WRONG CHOICES EXPLAINED:

(2) *Grass* is an autotroph, that is, a producer or self-feeder. Grass is a green plant that can make its own food.

(3) A *frog* is a secondary consumer; it eats insects that are plant eaters.

(4) A *snake* is a secondary consumer; it eats animals that are primary consumers.

151. **4** The organisms in the question make up a food chain. The number of organisms at each level of the food chain decreases as one moves down the chain. The pyramid of energy shown below represents this fact. The producers, which form the base of the pyramid, are the most numerous. *Algae* are the producers in this food chain.

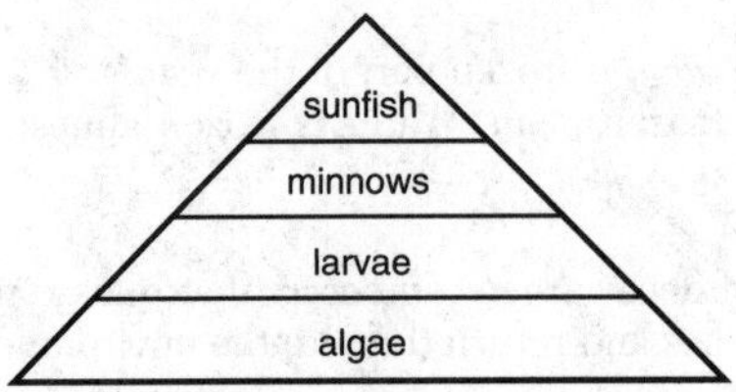

WRONG CHOICES EXPLAINED:

(1) The *minnows* are carnivores. They are the secondary consumers.

(2) The *larvae* are herbivores. They are primary consumers.

(3) The *sunfish* are tertiary consumers and occupy the top level of the pyramid. The organisms at the top of the pyramid are the least numerous.

152. **3** A *substratum* is the surface on which organisms grow. Lichens grow on rocks. Rocks are nonliving. The nonliving parts of the environment make up the abiotic factors.

WRONG CHOICES EXPLAINED:

(1) *Climax vegetation* is the type of vegetation that occupies an area in its final stage of succession. Plants make up the biotic, or living, environment.

(2) *Algae* are living organisms. They are part of the biotic (living) environment.

(4) *Bacteria* also make up the biotic environment.

153. **1** A relationship between two organisms in which both benefit from the association is known as *mutualism.* The clown fish is protected by the sea anemone. The sea anemone is sessile. The clown fish draws food to the sea anemone.

WRONG CHOICES EXPLAINED:

(2) *Commensalism* is a relationship between two organisms in which one organism is benefited by the association. The second organism is neither harmed nor benefited by the association. Barnacles attached to a whale are an example of a commensal relationship.

(3) A predator is a carnivore that hunts, kills, and eats its prey. The prey is the hunted organism. A *predator-prey* relationship is important in controlling the population of both.

(4) *Parasitism* is a relationship between two organisms in which one organism is benefited by the association whereas the second organism is harmed by the association. The parasitized organism is called the host.

154. **4** A biome is a large area dominated by one major type of vegetation and one type of climate. The *marine* biome has the greatest number of organisms.

WRONG CHOICES EXPLAINED:

(1) The *tundra* is a region where the ground is frozen all year long. Mosses and lichens are the dominant vegetation.

(2) The *tropical rain forest* is dominated by broadleaf plants. The region is always warm, and the rainfall is abundant and continuous.

(3) The *temperate deciduous forest* is made up of trees that shed their leaves once a year. The winters are cold, and the summers are warm. The rainfall is distributed throughout the year.

155. **3** *An ecosystem* is an area in which communities of living organisms interact with the nonliving environment.

WRONG CHOICES EXPLAINED:

(1) *A population* is all the organisms of a particular species living in a given area.

(2) *A community* is made up of populations of different species that interact with each other.

(4) *A species* is a group of organisms whose members are able to interbreed with each other. The offspring resulting from the matings are fertile and can reproduce.

Key Idea 7—Human Impact on the Environment: Human decisions and activities have a profound impact on the physical and living environment.

Performance Indicator	Description
7.1	The student should be able to describe the range of interrelationships of humans with the living and nonliving environment.
7.2	The student should be able to explain the impact of technological development and growth in the human population on the living and nonliving environment.
7.3	The student should be able to explain how individual choices and societal actions can contribute to improving the environment.

156. Human impact on the environment is most often more dramatic than the impact of most other living things because humans have a greater

(1) need for water
(2) need for food
(3) ability to adapt to change
(4) ability to alter the environment

156____

157. Which human activity would have the most direct impact on the oxygen–carbon dioxide cycle?

(1) reducing the rate of ecological succession
(2) decreasing the use of water
(3) destroying large forest areas
(4) enforcing laws that prevent the use of leaded gasoline

157____

158. Fertilizers used to improve lawns and gardens may interfere with the equilibrium of an ecosystem because they

(1) cause mutations in all plants
(2) cannot be absorbed by roots
(3) can be carried into local water supplies
(4) cause atmospheric pollution

158____

159. The tall wetland plant purple loosestrife was brought from Europe to the United States in the early 1800s as a garden plant. The plant's growth is now so widespread across the United States that it is crowding out a number of native plants. This situation is an example of

(1) the results of the use of pesticides
(2) the recycling of nutrients
(3) the flow of energy present in all ecosystems
(4) an unintended effect of adding a species to an ecosystem 159____

160. Choose *one* ecological problem from the list below.

Ecological Problems
Global warming
Destruction of the ozone shield
Loss of biodiversity

Discuss the ecological problem you chose. In your answer be sure to state:

- the problem you selected and *one* human action that may have caused the problem [1]
- *one* way in which the problem can negatively affect humans [1]
- *one* positive action that could be taken to reduce the problem [1]

____

____

____

Base your answers to questions 161 through 163 on the information below and on your knowledge of biology.

The planning board of a community held a public hearing in response to complaints by residents concerning a waste-recycling plant. The residents claimed that the waste-hauling trucks were polluting air, land, and water and that the garbage has brought an increase in rats, mice, and pathogenic bacteria to the area. The residents insisted that the waste-recycling plant be closed permanently.

Other residents recognized the health risks but felt that the benefits of waste recycling outweighed the health issues.

161. Identify two specific health problems that could result from living near the waste-recycling plant [2]

____

162. Identify one specific contaminant that might be released into the environment from operation of the waste-recycling plant. [1]

____

163. State one ecological benefit of recycling wastes. [1]

____

164. Which organism is a near-extinct species?
(1) Japanese beetle (3) blue whale
(2) dodo bird (4) passenger pigeon 164____

165. Which human activity has probably contributed most to the acidification of lakes in the Adirondack region?
(1) passing environmental protection laws
(2) establishing reforestation projects in lumbered areas
(3) burning fossil fuels that produce air pollutants containing sulfur and nitrogen
(4) using pesticides for the control of insects that feed on trees 165____

166. Compared to a natural forest, the wheat field of a farmer lacks
(1) heterotrophs (3) autotrophs
(2) significant biodiversity (4) stored energy 166____

167. Which factor is not considered by ecologists when they evaluate the impact of human activities on an ecosystem?
(1) amount of energy released from the Sun
(2) quality of the atmosphere
(3) degree of biodiversity
(4) location of power plants 167____

168. A new type of fuel gives off excessive amounts of smoke. Before this type of fuel is widely used, an ecologist would most likely want to know
(1) what effect the smoke will have on the environment
(2) how much it will cost to produce the fuel
(3) how long it will take to produce the fuel
(4) if the fuel will be widely accepted by consumers 168____

169. Which of the following is the most ecologically promising method of insect control?

(1) interference with insect reproductive processes
(2) stronger insecticides designed to kill higher percentages of insects
(3) physical barriers to insect pests
(4) draining marshes and other insect habitats 169____

170. Which is an example of biological control of a pest species?

(1) DDT was used to destroy the red mite.
(2) Most of the predators of a deer population were destroyed by humans.
(3) Gypsy moth larvae (tree defoliators) are destroyed by beetle predators that were cultured and released.
(4) Drugs were used in the control of certain pathogenic bacteria. 170____

171. To ensure environmental quality for the future, each individual should

(1) acquire and apply knowledge of ecological principles
(2) continue to take part in deforestation
(3) use Earth's finite resources
(4) add and take away organisms from ecosystems 171____

172. Ladybugs were introduced as predators into an agricultural area of the United States to reduce the number of aphids (pests that feed on grain crops). Describe the positive and negative effects of this method of pest control. Your response must include at least:

- two advantages of this method of pest control [2]
- two possible dangers of using this method of pest control [2]

____

____

____

____

____

173. Some people claim that certain carnivores should be destroyed because they kill beneficial animals. Explain why these carnivores should be protected. Your answer must include information concerning:

- prey population growth [1]
- extinction [1]
- the importance of carnivores in an ecosystem [1]

____

____

____

____

ANSWERS EXPLAINED Questions 156–173

Standard 4, Key Idea 7, Performance Indicators 1–3

156. **4** The fact that humans have a greater *ability to alter the environment* means that human impact on the environment is often more dramatic than that of most other living things. In addition to our ability to make physical changes in the environment, humans have the unique ability to alter the environment chemically, introducing many materials that are not found in nature and that cannot be converted to useful products by nature.

WRONG CHOICES EXPLAINED:

(1), (2) On an individual basis, humans' *need for water* and *need for food* are not significantly greater than those of other living things. However, the fact is that our large population places incredible demands on the environment to supply these basic resources. As a result, our tendency to destroy natural habitats to create additional water and agricultural resources is a significant factor affecting the natural world.

(3) On an individual basis, humans' *ability to adapt to change* is not significantly greater than that of other living things. However, as a species, we have created artificial environments to protect ourselves from harsh environmental conditions. To the extent that these artificial environments are dependent on energy and other natural resources, their construction and maintenance have resulted in significant alterations of the natural world.

157. **3** *Destroying large forest areas* is the human activity that would have the most direct impact on the oxygen-carbon cycle. Reducing the number of trees over a large area would decrease the forest's ability to absorb carbon dioxide and water and convert them to atmospheric oxygen and glucose. The millions of leaves in a forest are capable of releasing many tons of oxygen gas to the atmosphere. The massive bodies of forest trees can likewise store tons of carbon in the form of complex carbohydrates such as cellulose.

WRONG CHOICES EXPLAINED:

(1) *Reducing the rate of ecological succession* is not the human activity that would have the most direct impact on the oxygen-carbon cycle. Ecological succession is a process by which one plant-animal community is replaced over time by other plant-animal communities until a stable climax community is established. Reducing its rate would only have the effect of prolonging each successive community longer than might otherwise be expected but would not directly alter the cycling of carbon and oxygen.

(2) *Decreasing the use of water* is not the human activity that would have the most direct impact on the oxygen-carbon cycle. Water is a precious resource in many parts of the world. Reducing water use so as to conserve it would represent a positive impact of human activity on the environment but would not directly alter the cycling of carbon and oxygen.

(4) *Enforcing laws that prevent the use of leaded gasoline* is not the human activity that would have the most direct impact on the oxygen-carbon cycle. Lead is a dangerous heavy metal pollutant released when leaded gasoline is burned. Enforcing laws that limit its use would represent a positive impact of human activity on the environment, but would not directly alter the cycling of carbon and oxygen.

158. **3** Fertilizers used to improve lawns and gardens may interfere with the equilibrium of an ecosystem because they *can be carried into local water supplies*. Once dissolved fertilizers enter streams, ponds, wetlands, or lakes, they provide an abundant nutrient source for the growth of algae. As masses of algae die off in the water environment, their decomposition can rob the water of oxygen needed for the survival of fish and other water-dwelling populations, causing their elimination from the habitat. When these species disappear, other species that depend on them for food must migrate or starve. Because the changes caused by the entry of fertilizers into water environments are so significant, it can be said that ecosystem equilibrium is destroyed.

WRONG CHOICES EXPLAINED:

(1) It is not true that fertilizers *cause mutations in all plants*. Some compounds with chemical structures similar to that of fertilizers are known to stimulate rapid gene mutation in plant cells that may lead to the death of the plant. However, the class of chemical compounds known as fertilizers do not have this effect on all plants.

(2) It is not true that fertilizers *cannot be absorbed by all plants*. When dissolved in water, fertilizers can easily enter plants by being absorbed via simple diffusion into root hairs.

(4) It is not normally true that fertilizers *cause atmospheric pollution*. Most fertilizers are relatively stable chemical compounds that are solids at normal temperatures. For this reason fertilizers are not normally responsible for atmospheric pollution unless they are applied in a gaseous form (such as ammonia) or become airborne (when attached to dry soil particles).

159. **4** The situation described in the question is an example of *an unintended effect of adding a species to an ecosystem*. Although purple loosestrife has adapted well to North American habitats, its rapid growth in wetland environments has stressed or eliminated populations of cattail, pickerelweed, and other native plant species. The introduction of nonnative purple loosestrife to the North American continent has had an unintended negative effect on these native species and on the balance of nature established over many centuries.

WRONG CHOICES EXPLAINED:

(1), (2), (3) The situation described in the question is not an example of *the results of the use of pesticides*, *the recycling of nutrients*, or *the flow of energy present in all ecosystems*. The introduction of a nonnative plant (a living thing) is not the same as the introduction of a chemical pesticide, the recycling of nutrients, or the flow of energy (nonliving things).

160. A three-part response is required that must include the following points:

- One human activity that may have caused the ecological problem selected from the list [1]
- One way the problem may negatively affect humans [1]
- One positive action that could be taken to reduce the problem [1]

Note: No credit is awarded for discussing an ecological problem not on the list.

Acceptable responses include: [3]

- *Global warming is a worldwide ecological problem that may be caused by the release of carbon dioxide and other gases in automobile exhaust. [1] This problem may negatively affect humans if the warming conditions disrupt weather patterns and lead to droughts, floods, or other natural disasters. [1] One positive action that could be taken to help the problem would be to find an energy source for automobiles that would not release carbon dioxide into the atmosphere. [1]*
- *An ecological problem affecting humans is destruction of the ozone layer, which is caused by the use of chemicals known as CFCs as propellants in aerosol sprays. [1] This is a problem for*

humans because the ozone layer protects us from ultraviolet radiation from the sun; without this protection we would have an increased chance of getting skin cancer. [1] A way to help solve this problem would be to ban the use of CFCs in aerosols. [1]

- *Loss of biodiversity is an ecological problem that negatively impacts humans. This problem is caused whenever humans destroy a natural habitat and convert it to other uses. [1] The overall health of our environment depends on the diversity of species that fill different roles in nature. When species diversity and environmental health are reduced, our health is threatened as well. [1] This problem can be reversed only if we use education to learn that protecting natural species is just as important as protecting our own. [1]*

161. Two responses are required. Acceptable responses include:

- *Asthma*
- *Respiratory infections*
- *Allergic reactions*
- *Cancer*
- *Bacterial infections*
- *Viral infections*
- *Disease linked to a pathogen*
- *Poisoning linked to toxic contamination of groundwater*

162. One response is required. Acceptable responses include:

- *Particles in the air*
- *Presence of viruses or bacteria on trucks*
- *Chemicals in air or water*
- *Carcinogens*
- *Mold and fungus spores*

163. One response is required. Acceptable responses include:

- *Conservation of natural resources*
- *Protection of finite resources*
- *Energy conservation*
- *Reduction in pollution*
- *Landfills last longer*
- *Preservation of open space resources*

164. **3** The blue whale is near extinction because of uncontrolled hunting by humans.

WRONG CHOICES EXPLAINED:

(1) The Japanese beetle is a plant pest that was accidentally introduced into the United States. Its population is kept in check by the praying mantis, its predator.

(2) The dodo bird became extinct because of hunting by humans.

(4) The passenger pigeon became extinct in the 1900s due to hunting by humans.

165. **3** *Burning fossil fuels that produce air pollutants containing sulfur and nitrogen* is the human activity that has probably contributed the most to the acidification of lakes in the Adirondack region. These pollutants combine with water in the atmosphere to form sulfuric and nitric acids. These acids then enter lakes in rainfall and runoff, adding to the acidic quality of the lake water and killing many susceptible species.

WRONG CHOICES EXPLAINED:

(1) *Passing environmental protection laws* is not an activity that results in the acidification of lakes. In fact, it is a positive human activity that can help to limit the production and release of such gases into the atmosphere.

(2) *Establishing reforestation projects in lumbered areas* is not an activity that results in the acidification of lakes. In fact, it is a positive human activity that can help to replace trees lost because of the acidification of soils by acid rain

(4) *Using pesticides for the control of insects that feed on trees* is not an activity that results in the acidification of lakes. It is a negative human activity carried out to protect commercial crops from destruction and does not normally result in the production of sulfur and nitrogen gases.

166. **2** *Significant biodiversity* is the factor lacking in a wheat field as compared to a natural forest. *Biodiversity* is a term relating to the variety of life forms in an environment. Natural environments, including forests, are typically made up of thousands of species that interact to provide a balanced, ecologically responsive community. By contrast, farm fields are often limited to a small number of different species, and predominantly a single species. Communities lacking in biodiversity are unstable and prone to collapse when environmental conditions change.

WRONG CHOICES EXPLAINED:

(1) *Heterotrophs* are not lacking in a farm field compared to a forest. Heterotrophs are found within a wheat field, although their number and variety are normally limited to those that use wheat or its by-products as food.

(3) *Autotrophs* are not lacking in a farm field compared to a forest. Wheat is a type of autotroph, as are the various weed species that may be interspersed among the wheat plants in the field.

(4) *Stored energy* is not lacking in a farm field compared to a forest. As the wheat grows in the field, it absorbs the Sun's energy and stores it as the chemical bond energy of carbohydrates and other organic compounds.

167. **1** The *amount of energy released from the Sun* is not normally considered by an ecologist when evaluating the impact of human activities on an ecosystem. The amount of solar energy emitted by the Sun is generally constant and out of our direct control. Because it is not a variable that can be directly affected by human activities, it is usually not a consideration in decisions of this kind.

WRONG CHOICES EXPLAINED:

(2) The *quality of the atmosphere* is often a factor considered by ecologists in evaluating the impact of human activities on an ecosystem. Many human activities introduce chemical contaminants into the atmosphere. These chemicals may have a negative impact on the health and survival of humans and other species.

(3) The *degree of biodiversity* is often a factor considered by ecologists in evaluating the impact of human activities on an ecosystem. Human activities often put pressure on natural species, eliminating those unable to migrate or adapt. As biodiversity in an area declines, so does environmental stability. This situation threatens the health and survival of humans and other species.

(4) The *location of power plants* is often a factor considered by ecologists in evaluating the impact of human activities on an ecosystem. Fossil fuel plants can pollute the atmosphere and consume valuable petroleum products. Nuclear plants can release radiation and heat into the environment. Hydroelectric, solar, wind, and geothermal plants can destroy natural habitats because of space considerations. Each of these consequences can affect the health and survival of humans and other species.

168. **1** An ecologist would want to know *what effect the smoke will have on the environment* before a new type of fuel is widely used. By understanding this effect, the ecologist can make more informed judgments about whether the smoke will harm the environment and human health.

WRONG CHOICES EXPLAINED:

(2), (3), (4) An ecologist is less likely to want to know *how much it will cost to produce the fuel*, *how long it will take to produce the fuel*, and *if the fuel will be widely accepted by consumers*. Although these are important questions for the manufacturer, they do not provide critical information for the ecologist, whose main concern is the protection of environmental quality for humans and other organisms.

169. **1** Interference with insect reproductive processes is known as biological control. It is the most promising method of controlling insects because it is the least ecologically damaging.

WRONG CHOICES EXPLAINED:

(2) The use of insecticides is a chemical control of insects. Insecticides kill both harmful and helpful insects. The chemicals accumulate in the bodies of birds, fish, and mammals and interfere with their normal life activities.

(3) It is impossible to set up physical barriers for insects because they are motile and are also carried from place to place by animals and humans.

(4) Draining marshes and other insect habitats has helped to control many insects such as mosquitoes. However, this method interferes with the life cycles of useful organisms living in the area.

170. **3** Insecticides are chemical pest controls. Biological controls are other insect species that feed on or in some way prey on an insect pest species. The example given here is control of the gypsy moth larvae by a certain species of beetle.

WRONG CHOICES EXPLAINED:

(1) DDT is an insecticide and represents chemical control. DDT is no longer used because it destroyed the insect food of birds and other wildlife.

(2) Humans upset the balance of nature (the balance of natural communities) by killing off deer predators. The deer population then increased so dramatically that deer starved to death because there was not enough food to support them.

(4) The use of drugs to cure disease is an example of chemical control of pathogens.

171. **1** Each individual should *acquire and apply knowledge of ecological principles* in order to ensure environmental quality for the future. By understanding how environmental principles operate, we can make more informed judgments about activities that may harm the environment and human health.

WRONG CHOICES EXPLAINED:

(2) If each individual were to *continue to take part in deforestation*, environmental quality would be degraded. Because forests are a natural part of the environment, eliminating them disturbs the balance of nature and can have significant negative consequences for environmental quality.

(3) If each individual were to *use Earth's finite resources*, environmental quality would be threatened. As these resources are used up, fewer remain for future generations. In addition, processing these resources consumes energy, produces pollutants, and adds to the solid waste problem.

(4) If each individual were to *add and take away organisms from ecosystems*, environmental quality would be diminished. Each natural community has established itself based on the particular niches filled by each type of organism. Adding to or taking away from this community upsets the balance of nature and would likely cause negative consequences.

172. Write one or more paragraphs describing positive and negative effects of this method of pest control. Include the following points:

- Two advantages of this method of pest control [2]
  - ➤ Chemicals are not added to the environment.
  - ➤ Biological controls are more specific than chemical controls.
  - ➤ Ladybugs are less likely to kill beneficial organisms.
  - ➤ Desirable garden plants are protected from aphid attacks.
  - ➤ Birds and other unintended victims of pesticide use are spared.
  - ➤ Human health is protected against the toxic effects of pesticides.
- Two possible dangers of using this method of pest control [2]
  - ➤ The control insects may eat the food of other organisms.
  - ➤ The population of natural predators of the aphids may be eliminated or greatly reduced.
  - ➤ The control organism may become overpopulated.
  - ➤ The control organisms may themselves become pests.

Sample paragraph: *The method of pest control described is known as "biological control." This method of insect control has some distinct advantages over chemical controls: First, biological controls don't release toxic chemicals into the air and water, a fact that helps to protect wildlife and humans from being unintended victims of chemical pesticides. Second, biological controls are usually specific, which means that beneficial insects such as ladybugs and preying mantises aren't harmed. [2] There are also some things we should be careful of in the use of biological controls: First, we should know a lot about the control organism to be sure that it doesn't crowd out our native beneficial organisms. Second, we should remember that the control organism could become a pest, too, if it gets too numerous in the environment. [2]*

173. Write one or more paragraphs explaining why carnivores should be protected. Include the following points:

- Information concerning prey population growth [1]
  - ➤ If predators are destroyed, the prey population will increase.
  - ➤ If unchecked by predation or disease, a natural population will tend to increase in number geometrically.
- Information concerning extinction [1]
  - ➤ If too many carnivores of a particular species are killed, the species may become extinct.
  - ➤ Extinction is a definite possibility when any species has too few members alive to carry out effective breeding.
  - ➤ Complete elimination of any species from its natural range can destabilize the ecosystem.
- Information concerning the importance of carnivores in an environment [1]
  - ➤ By feeding on herbivores, carnivores help keep certain species of plants from being eliminated because of overgrazing in a particular area.
  - ➤ Without predators to limit its number, a prey population could exceed the capacity of its range, resulting in widespread starvation and death of the prey population.
  - ➤ Carnivorous animals are part of the natural scheme that promotes ecological equilibrium.

Sample paragraph: *Carnivores are important in an ecosystem because by reducing the number of prey organisms, the food organisms of the prey are kept from being eliminated from the environment. [1] If the predators were destroyed, the prey population would increase [1], perhaps to the point of consuming so many of the plants that the prey feed on that these plants would become extinct. [1]*

STANDARD/KEY IDEA	QUESTION NUMBERS	NUMBER OF CORRECT RESPONSES	NUMBER OF INCORRECT RESPONSES
1.1 Purpose of Scientific Inquiry	1–8		
1.2 Methods of Scientific Inquiry	9–28		
1.3 Analysis in Scientific Inquiry	29–47		
4.1 Application of Scientific Principles	48–75		
4.2 Genetic Continuity	76–90		
4.3 Organic Evolution	91–109		
4.4 Reproductive Continuity	110–129		
4.5 Dynamic Equilibrium and Homeostasis	130–140		
4.6 Interdependence of Living Things	141–155		
4.7 Human Impact on the Environment	156–173		

Glossary

PROMINENT SCIENTISTS

Crick, Francis A 20th-century British scientist who, with James Watson, developed the first workable model of DNA structure and function.

Darwin, Charles A 19th-century British naturalist whose theory of organic evolution by natural selection forms the basis for the modern scientific theory of evolution.

Fox, Sidney A 20th-century American scientist whose experiments showed that Stanley Miller's simple chemical precursors could be joined to form more complex biochemicals.

Hardy, G. H. A 20th-century British mathematician who, with W. Weinberg, developed the Hardy-Weinberg principle of gene frequencies.

Lamarck, Jean An 18th-century French scientist who devised an early theory of organic evolution based on the concept of "use and disuse."

Linnaeus, Carl An 18th-century Dutch scientist who developed the first scientific system of classification, based on similarity of structure.

Mendel, Gregor A 19th-century Austrian monk and teacher who was the first to describe many of the fundamental concepts of genetic inheritance through his work with garden peas.

Miller, Stanley A 20th-century American scientist whose experiments showed that the simple chemical precursors of life could be produced in the laboratory.

Morgan, Thomas Hunt A 20th-century American geneticist whose pioneering work with Drosophila led to the discovery of several genetic principles, including sex linkage.

Watson, James A 20th-century American scientist who, with Francis Crick, developed the first workable model of DNA structure and function.

Weinberg, W. A 20th-century German physician who, with G. H. Hardy, developed the Hardy-Weinberg principle of gene frequencies.

Weismann, August A 19th-century German biologist who tested Lamarck's theory of use and disuse and found it to be unsupportable by scientific methods.

BIOLOGICAL TERMS

abiotic factor Any of several nonliving, physical conditions that affect the survival of an organism in its environment.

absorption The process by which water and dissolved solids, liquids, and gases are taken in by the cell through the cell membrane.

accessory organ In human beings, any organ that has a digestive function but is not part of the food tube. (See **liver; gallbladder; pancreas.**)

acid A chemical that releases hydrogen ion (H+) in solution with water.

acid precipitation A phenomenon in which there is thought to be an interaction between atmospheric moisture and the oxides of sulfur and nitrogen that results in rainfall with low pH values.

active immunity The immunity that develops when the body's immune system is stimulated by a disease organism or a vaccination.

active site The specific area of an enzyme molecule that links to the substrate molecule and catalyzes its metabolism.

active transport A process by which materials are absorbed or released by cells against the concentration gradient (from low to high concentration) with the expenditure of cell energy.

adaptation Any structural, biochemical, or behavioral characteristic of an organism that helps it to survive potentially harsh environmental conditions.

addition A type of chromosome mutation in which a section of a chromosome is transferred to a homologous chromosome.

adenine A nitrogenous base found in DNA and RNA molecules.

adenosine triphosphate (ATP) An organic compound that stores respiratory energy in the form of chemical-bond energy for transport from one part of the cell to another.

adrenal cortex A portion of the adrenal gland that secretes steroid hormones which regulate various aspects of blood composition.

adrenal gland An endocrine gland that produces several hormones, including **adrenaline**. (See **adrenal cortex; adrenal medulla**.)

adrenal medulla A portion of the adrenal gland that secretes the hormone adrenaline, which regulates various aspects of the body's metabolic rate.

adrenaline A hormone of the adrenal medulla that regulates general metabolic rate, the rates of heartbeat and breathing, and the conversion of glycogen to glucose.

aerobic phase of respiration The reactions of aerobic respiration in which two pyruvic acid molecules are converted to six molecules of water and six molecules of carbon dioxide.

aerobic respiration A type of respiration in which energy is released from organic molecules with the aid of oxygen.

aging A stage of postnatal development that involves differentiation, maturation, and eventual deterioration of the body's tissues.

air pollution The addition, due to technological oversight, of some unwanted factor (e.g., chemical oxides, hydrocarbons, particulates) to our air resources.

albinism A condition, controlled by a single mutant gene, in which the skin lacks the ability to produce skin pigments.

alcoholic fermentation A type of anaerobic respiration in which glucose is converted to ethyl alcohol and carbon dioxide.

allantois A membrane that serves as a reservoir for wastes and as a respiratory surface for the embryos of many animal species.

allele One of a pair of genes that exist at the same location on a pair of homologous chromosomes and exert parallel control over the same genetic trait.

allergy A reaction of the body's immune system to the chemical composition of various substances.

alveolus One of many "air sacs" within the lung that function to absorb atmospheric gases and pass them on to the bloodstream.

amino acid An organic compound that is the component unit of proteins.

amino group A chemical group having the formula $—NH_2$ that is found as a part of all amino acid molecules.

ammonia A type of nitrogenous waste with high solubility and high toxicity.

amniocentesis A technique for the detection of genetic disorders in human beings in which a small amount of amniotic fluid is removed and the chromosome content of its cells analyzed. (See **karyotyping**.)

amnion A membrane that surrounds the embryo in many animal species and contains a fluid to protect the developing embryo from mechanical shock.

amniotic fluid The fluid within the amnion membrane that bathes the developing embryo.

amylase An enzyme specific for the hydrolysis of starch.

anaerobic phase of respiration The reactions of aerobic respiration in which glucose is converted to two pyruvic acid molecules.

anaerobic respiration A type of respiration in which energy is released from organic molecules without the aid of oxygen.

anal pore The egestive organ of the paramecium.

anemia A disorder of the human transport system in which the ability of the blood to carry oxygen is impaired, usually because of reduced numbers of red blood cells.

angina pectoris A disorder of the human transport system in which chest pain signals potential damage to the heart muscle due to narrowing of the opening of the coronary artery.

Animal One of the five biological kingdoms; it includes multicellular organisms whose cells are not bounded by cell walls and which are incapable of photosynthesis (e.g., human being).

Annelida A phylum of the Animal Kingdom whose members (annelids) include the segmented worms (e.g., earthworm).

antenna A receptor organ found in many arthropods (e.g., grasshopper), which is specialized for detecting chemical stimuli.

anther The portion of the stamen that produces pollen.

antibody A chemical substance, produced in response to the presence of a specific antigen, which neutralizes that antigen in the immune response.

antigen A chemical substance, usually a protein, that is recognized by the immune system as a foreign "invader" and is neutralized by a specific antibody.

anus The organ of egestion of the digestive tract.

aorta The principal artery carrying blood from the heart to the body tissues.

aortic arches A specialized part of the earthworm's transport system that serves as a pumping mechanism for the blood fluid.

apical meristem A plant growth region located at the tip of the root or tip of the stem.

appendicitis A disorder of the human digestive tract in which the appendix becomes inflamed as a result of bacterial infection.

aquatic biome An ecological biome composed of many different water environments.

artery A thick-walled blood vessel that carries blood away from the heart under pressure.

arthritis A disorder of the human locomotor system in which skeletal joints become inflamed, swollen, and painful.

Arthropoda A phylum of the Animal Kingdom whose members (arthropods) have bodies with chitinous exoskeletons and jointed appendages (e.g., grasshopper).

artificial selection A technique of plant/animal breeding in which individual organisms displaying desirable characteristics are chosen for breeding purposes.

asexual reproduction A type of reproduction in which new organisms are formed from a single parent organism.

asthma A disorder of the human respiratory system in which the respiratory tube becomes constricted by swelling brought on by some irritant.

atrium In human beings, one of the two thin-walled upper chambers of the heart that receive blood.

autonomic nervous system A subdivision of the peripheral nervous system consisting of nerves associated with automatic functions (e.g., heartbeat, breathing).

autosome One of several chromosomes present in the cell that carry genes controlling "body" traits not associated with primary and secondary sex characteristics.

autotroph An organism capable of carrying on autotrophic nutrition. Self feeder.

autotrophic nutrition A type of nutrition in which organisms manufacture their own organic foods from inorganic raw materials.

auxin A biochemical substance, plant hormone, produced by plants that regulates growth patterns.

axon An elongated portion of a neuron that conducts nerve impulses, usually away from the cell body of the neuron.

base A chemical that releases hydroxyl ion (OH^-) in solution with water.

bicarbonate ion The chemical formed in the blood plasma when carbon dioxide is absorbed from body tissues.

bile In human beings, a secretion of the liver that is stored in the gallbladder and that emulsifies fats.

binary fission A type of cell division in which mitosis is followed by equal cytoplasmic division.

binomial nomenclature A system of naming, used in biological classification, that consists of the genus and species names (e.g., *Homo sapiens*).

biocide use The use of pesticides that eliminate one undesirable organism but that have, due to technological oversight, unanticipated effects on beneficial species as well.

biological controls The use of natural enemies of various agricultural pests for pest control, thereby eliminating the need for biocide use—a positive aspect of human involvement with the environment.

biomass The total mass of living material present at the various trophic levels in a food chain.

biome A major geographical grouping of similar ecosystems, usually named for the climax flora in the region (e.g., Northeast Deciduous Forest).

biosphere The portion of the earth in which living things exist, including all land and water environments.

biotic factor Any of several conditions associated with life and living things that affect the survival of living things in the environment.

birth In placental mammals, a stage of embryonic development in which the baby passes through the vaginal canal to the outside of the mother's body.
blastula In certain animals, a stage of embryonic development in which the embryo resembles a hollow ball of undifferentiated cells.
blood The complex fluid tissue that functions to transport nutrients and respiratory gases to all parts of the body.
blood typing An application of the study of immunity in which the blood of a person is characterized by its antigen composition.
bone A tissue that provides mechanical support and protection for bodily organs, and levers for the body's locomotive activities.
Bowman's capsule A cup-shaped portion of the nephron responsible for the filtration of soluble blood components.
brain An organ of the central nervous system that is responsible for regulating conscious and much unconscious activity in the body.
breathing A mechanical process by which air is forced into the lungs by means of muscular contraction of the diaphragm and rib muscles.
bronchiole One of several subdivisions of the bronchi that penetrate the lung interior and terminate in alveoli.
bronchitis A disorder of the human respiratory system in which the bronchi become inflamed.
bronchus One of the two major subdivisions of the breathing tube; the bronchi are ringed with cartilage and conduct air from the trachea to the lung interior.
Bryophyta A phylum of the Plant Kingdom that consists of organisms lacking vascular tissues (e.g., moss).
budding A type of asexual reproduction in which mitosis is followed by unequal cytoplasmic division.
bulb A type of vegetative propagation in which a plant bulb produces new bulbs that may be established as independent organisms with identical characteristics.

cambium The lateral meristem tissue in woody plants responsible for annual growth in stem diameter.
cancer Any of a number of conditions characterized by rapid, abnormal, and uncontrolled division of affected cells.
capillary A very small, thin-walled blood vessel that connects an artery to a vein and through which all absorption into the blood fluid occurs.
carbohydrate An organic compound composed of carbon, hydrogen, and oxygen in a 1:2:1 ratio (e.g., $C_6H_{12}O_6$).
carbon-14 A radioactive isotope of carbon used to trace the movement of carbon in various biochemical reactions, and also used in the "carbon dating" of fossils.

carbon-fixation reactions A set of biochemical reactions in photosynthesis in which hydrogen atoms are combined with carbon and oxygen atoms to form PGAL and glucose.

carbon-hydrogen-oxygen cycle A process by which these three elements are made available for use by other organisms through the chemical reactions of respiration and photosynthesis.

carboxyl group A chemical group having the formula—COOH and found as part of all amino acid and fatty acid molecules.

cardiac muscle A type of muscle tissue in the heart and arteries that is associated with the rhythmic nature of the pulse and heartbeat.

cardiovascular disease In human beings, any disease of the circulatory organs.

carnivore A heterotrophic organism that consumes animal tissue as its primary source of nutrition. (See **secondary consumer**.)

carrier An individual who, though not expressing a particular recessive trait, carries this gene as part of his/her heterozygous genotype.

carrier protein A specialized molecule embedded in the cell membrane that aids the movement of materials across the membrane.

cartilage A flexible connective tissue found in many flexible parts of the body (e.g., knee); common in the embryonic stages of development.

catalyst Any substance that speeds up or slows down the rate of a chemical reaction. (See **enzyme.**)

cell plate A structure that forms during cytoplasmic division in plant cells and serves to separate the cytoplasm into two roughly equal parts.

cell theory A scientific theory that states, "All cells arise from previously existing cells" and "Cells are the unit of structure and function of living things."

cell wall A cell organelle that surrounds and gives structural support to plant cells; cell walls are composed of cellulose.

central nervous system The portion of the vertebrate nervous system that consists of the brain and the spinal cord.

centriole A cell organelle found in animal cells that functions in the process of cell division.

centromere The area of attachment of two chromatids in a double-stranded chromosome.

cerebellum The portion of the human brain responsible for the coordination of muscular activity.

cerebral hemorrhage A disorder of the human regulatory system in which a broken blood vessel in the brain may result in severe dysfunction or death.

cerebral palsy A disorder of the human regulatory system in which the motor and speech centers of the brain are impaired.

cerebrum The portion of the human brain responsible for thought, reasoning, sense interpretation, learning, and other conscious activities.

cervix A structure that bounds the lower end of the uterus and through which sperm must pass in order to fertilize the egg.

chemical digestion The process by which nutrient molecules are converted by chemical means into a form usable by the cells.

chemosynthesis A type of autotrophic nutrition in which certain bacteria use the energy of chemical oxidation to convert inorganic raw materials to organic food molecules.

chitin A polysaccharide substance that forms the exoskeleton of the grasshopper and other arthropods.

chlorophyll A green pigment in plant cells that absorbs sunlight and makes possible certain aspects of the photosynthetic process.

chloroplast A cell organelle found in plant cells that contains chlorophyll and functions in photosynthesis.

Chordata A phylum of the Animal Kingdom whose members (chordates) have internal skeletons made of cartilage and/or bone (e.g., human being).

chorion A membrane that surrounds all other embryonic membranes in many animal species, protecting them from mechanical damage.

chromatid One strand of a double-stranded chromosome.

chromosome mutation An alteration in the structure of a chromosome involving many genes. (See **nondisjunction; translocation; addition; deletion**.)

cilia Small, hairlike structures in paramecia and other unicellular organisms that aid in nutrition and locomotion.

classification A technique by which scientists sort, group, and name organisms for easier study.

cleavage A series of rapid mitotic divisions that increase cell number in a developing embryo without corresponding increase in cell size.

climax community A stable, self-perpetuating community that results from an ecological succession.

cloning A technique of genetic investigation in which undifferentiated cells of an organism are used to produce new organisms with the same set of traits as the original cells.

closed transport system A type of circulatory system in which the transport fluid is always enclosed within blood vessels (e.g., earthworm, human).

clot A structure that forms as a result of enzyme-controlled reactions following the rupturing of a blood vessel and serves as a plug to prevent blood loss.

codominance A type of intermediate inheritance that results from the simultaneous expression of two dominant alleles with contrasting effects.

codon See **triplet codon**.

Coelenterata A phylum of the Animal Kingdom whose members (coelenterates) have bodies that resemble a sack (e.g., hydra, jellyfish).

coenzyme A chemical substance or chemical subunit that functions to aid the action of a particular enzyme. (See **vitamin.**)

cohesion A force binding water molecules together that aids in the upward conduction of materials in the xylem.

commensalism A type of symbiosis in which one organism in the relationship benefits and the other is neither helped nor harmed.

common ancestry A concept central to the science of evolution which postulates that all organisms share a common ancestry whose closeness varies with the degree of shared similarity.

community A level of biological organization that includes all of the species populations inhabiting a particular geographic area.

comparative anatomy The study of similarities in the anatomical structures of organisms, and their use as an indicator of common ancestry and as evidence of organic evolution.

comparative biochemistry The study of similarities in the biochemical makeups of organisms, and their use as an indicator of common ancestry and as evidence of organic evolution.

comparative cytology The study of similarities in the cell structures of organisms, and their use as an indicator of common ancestry and as evidence of organic evolution.

comparative embryology The study of similarities in the patterns of embryological development of organisms, and their use as an indicator of common ancestry and as evidence of organic evolution.

competition A condition that arises when different species in the same habitat attempt to use the same limited resources.

complete protein A protein that contains all eight essential amino acids.

compound A substance composed of two or more different kinds of atom (e.g., water: H_2O).

compound light microscope A tool of biological study capable of producing a magnified image of a biological specimen by using a focused beam of light.

conditioned behavior A type of response that is learned, but that becomes automatic with repetition.

conservation of resources The development and application of practices to protect valuable and irreplaceable soil and mineral resources—a positive aspect of human involvement with the environment.

constipation A disorder of the human digestive tract in which fecal matter solidifies and becomes difficult to egest.

consumer Any heterotrophic animal organism (e.g., human being).

coronary artery An artery that branches off the aorta to feed the heart muscle.

coronary thrombosis A disorder of the human transport system in which the heart muscle becomes damaged as a result of blockage of the coronary artery.

corpus luteum A structure resulting from the hormone-controlled transformation of the ovarian follicle that produces the hormone progesterone.

corpus luteum stage A stage of the menstrual cycle in which the cells of the follicle are transformed into the corpus luteum under the influence of the hormone LH.

cotyledon A portion of the plant embryo that serves as a source of nutrition for the young plant before photosynthesis begins.

cover-cropping A proper agricultural practice in which a temporary planting (cover crop) is used to limit soil erosion between seasonal plantings of main crops.

crop A portion of the digestive tract of certain animals that stores food temporarily before digestion.

cross-pollination A type of pollination in which pollen from one flower pollinates flowers of a different plant of the same species.

crossing-over A pattern of inheritance in which linked genes may be separated during synapsis in the first meiotic division, when sections of homologous chromosomes may be exchanged.

cuticle A waxy coating that covers the upper epidermis of most leaves and acts to help the leaf retain water.

cutting A technique of plant propagation in which vegetative parts of the parent plant are cut and rooted to establish new plant organisms with identical characteristics.

cyclosis The circulation of the cell fluid (cytoplasm) within the cell interior.

cyton The "cell body" of the neuron, which generates the nerve impulse.

cytoplasm The watery fluid that provides a medium for the suspension of organelles within the cell.

cytoplasmic division The separation of daughter nuclei into two new daughter cells.

cytosine A nitrogenous base found in both DNA and RNA molecules.

daughter cell A cell that results from mitotic cell division.

daughter nucleus One of two nuclei that form as a result of mitosis.

deamination A process by which amino acids are broken down into their component parts for conversion into urea.

death The irreversible cessation of bodily functions and cellular activities.

deciduous A term relating to broadleaf trees which shed their leaves in the fall.

decomposer Any saprophytic organism that derives its energy from the decay of plant and animal tissues (e.g., bacteria of decay, fungus); the final stage of a food chain.

decomposition bacteria In the nitrogen cycle, bacteria that break down plant and animal protein and produce ammonia as a by-product.
dehydration synthesis A chemical process in which two organic molecules may be joined after removing the atoms needed to form a molecule of water as a by-product.
deletion A type of chromosome mutation in which a section of a chromosome is separated and lost.
dendrite A cytoplasmic extension of a neuron that serves to detect an environmental stimulus and carry an impulse to the cell body of the neuron.
denitrifying bacteria In the nitrogen cycle, bacteria that convert excess nitrate salts into gaseous nitrogen.
deoxygenated blood Blood that has released its transported oxygen to the body tissues.
deoxyribonucleic acid (DNA) A nucleic acid molecule known to be the chemically active agent of the gene; the fundamental hereditary material of living organisms.
deoxyribose A five-carbon sugar that is a component part of the nucleotide unit in DNA only.
desert A terrestrial biome characterized by sparse rainfall, extreme temperature variation, and a climax flora that includes cactus.
diabetes A disorder of the human regulatory system in which insufficient insulin production leads to elevated blood sugar concentrations.
diarrhea A disorder of the human digestive tract in which the large intestine fails to absorb water from the waste matter, resulting in watery feces.
diastole The lower pressure registered during blood pressure testing. (See **systole.**)
differentiation The process by which embryonic cells become specialized to perform the various tasks of particular tissues throughout the body.
diffusion A form of passive transport by which soluble substances are absorbed or released by cells.
digestion The process by which complex foods are broken down by mechanical or chemical means for use by the body.
dipeptide A chemical unit composed of two amino acid units linked by a peptide bond.
diploid chromosome number The number of chromosomes found characteristically in the cells (except gametes) of sexually reproducing species.
disaccharidase Any disaccharide-hydrolyzing enzyme.
disaccharide A type of carbohydrate known also as a "double sugar"; all disaccharides have the molecular formula $C_{12}H_{22}O_{11}$.
disjunction The separation of homologous chromosome pairs at the end of the first meiotic division.
disposal problems Problems, due to technological oversight, that result when commercial and technological activities produce solid and/or chemical wastes that must be disposed of.

dissecting microscope A tool of biological study that magnifies the image of a biological specimen up to 20 times normal size for purposes of gross dissection.
dominance A pattern of genetic inheritance in which the effects of a dominant allele mask those of a recessive allele.
dominant allele (gene) An allele (gene) whose effect masks that of its recessive allele.
double-stranded chromosome The two-stranded structure that results from chromosomal replication.
Down's syndrome In human beings, a condition, characterized by mental and physical retardation, that may be caused by the nondisjunction of chromosome number 21.
Drosophila The common fruit fly, an organism that has served as an object of genetic research in the development of the gene-chromosome theory.
ductless gland See **endocrine gland**.

ecology The science that studies the interactions of living things with each other and with the nonliving environment.
ecosystem The basic unit of study in ecology, including the plant and animal community in interaction with the nonliving environment.
ectoderm An embryonic tissue that differentiates into skin and nerve tissue in the adult animal.
effector An organ specialized to produce a response to an environmental stimulus: effectors may be muscles or glands.
egestion The process by which undigested food materials are eliminated from the body.
electron microscope A tool of biological study that uses a focused beam of electrons to produce an image of a biological specimen magnified up to 25,000 times normal size.
element The simplest form of matter; an element is a substance (e.g., nitrogen) made up of a single type of atom.
embryo An organism in the early stages of development following fertilization.
embryonic development A series of complex processes by which animal and plant embryos develop into adult organisms.
emphysema A disorder of the human respiratory system in which lung tissue deteriorates, leaving the lung with diminished capacity and efficiency.
emulsification A process by which fat globules are surrounded by bile to form fat droplets.
endocrine ("ductless") gland A gland (e.g., thyroid, pituitary) specialized for the production of hormones and their secretion directly into the bloodstream; such glands lack ducts.
endoderm An embryonic tissue that differentiates into the digestive and respiratory tract lining in the adult animal.

endoplasmic reticulum (ER) A cell organelle known to function in the transport of cell products from place to place within the cell.

environmental laws Federal, state, and local legislation enacted in an attempt to protect environmental resources—a positive aspect of human involvement with the environment.

enzymatic hydrolysis An enzyme-controlled reaction by which complex food molecules are broken down chemically into simpler subunits.

enzyme An organic catalyst that controls the rate of metabolism of a single type of substrate; enzymes are protein in nature.

enzyme-substrate complex A physical association between an enzyme molecule and its substrate within which the substrate is metabolized.

epicotyl A portion of the plant embryo that specializes to become the upper stem, leaves, and flowers of the adult plant.

epidermis The outermost cell layer in a plant or animal.

epiglottis In a human being, a flap of tissue that covers the upper end of the trachea during swallowing and prevents inhalation of food.

esophagus A structure in the upper portion of the digestive tract that conducts the food from the pharynx to the midgut.

essential amino acid An amino acid that cannot be synthesized by the human body, but must be obtained by means of the diet.

estrogen A hormone, secreted by the ovary, that regulates the production of female secondary sex characteristics.

evolution Any process of gradual change through time.

excretion The life function by which living things eliminate metabolic wastes from their cells.

exoskeleton A chitinous material that covers the outside of the bodies of most arthropods and provides protection for internal organs and anchorage for muscles.

exploitation of organisms Systematic removal of animals and plants with commercial value from their environments, for sale—a negative aspect of human involvement with the environment.

extensor A skeletal muscle that extends (opens) a joint.

external development Embryonic development that occurs outside the body of the female parent (e.g., birds).

external fertilization Fertilization that occurs outside the body of the female parent (e.g., fish).

extracellular digestion Digestion that occurs outside the cell.

fallopian tube See **oviduct.**

fatty acid An organic molecule that is a component of certain lipids.

fauna The animal species comprising an ecological community.

feces The semisolid material that results from the solidification of undigested foods in the large intestine.

fertilization The fusion of gametic nuclei in the process of sexual reproduction.
filament The portion of the stamen that supports the anther.
flagella Microscopic, whiplike structures found on certain cells that aid in locomotion and circulation.
flexor A skeletal muscle that flexes (closes) a joint.
flora The plant species comprising an ecological community.
flower The portion of a flowering plant that is specialized for sexual reproduction.
fluid-mosaic model A model of the structure of the cell membrane in which large protein molecules are thought to be embedded in a bilipid layer.
follicle One of many areas within the ovary that serve as sites for the periodic maturation of ova.
follicle stage The stage of the menstrual cycle in which an ovum reaches its final maturity under the influence of the hormone FSH.
follicle-stimulating hormone (FSH) A pituitary hormone that regulates the maturation of, and the secretion of estrogen by, the ovarian follicle.
food chain A series of nutritional relationships in which food energy is passed from producer to herbivore to carnivore to decomposer; a segment of a food web.
food web A construct showing a series of interrelated food chains and illustrating the complex nutritional interrelationships that exist in an ecosystem.
fossil The preserved direct or indirect remains of an organism that lived in the past, as found in the geologic record.
fraternal twins In human beings, twin offspring that result from the simultaneous fertilization of two ova by two sperm; such twins are not genetically identical.
freshwater biome An aquatic biome made up of many separate freshwater systems that vary in size and stability and may be closely associated with terrestrial biomes.
fruit Any plant structure that contains seeds; a mechanism of seed dispersal.
Fungi One of the five biological kingdoms; it includes organisms unable to manufacture their own organic foods (e.g., mushroom).

gallbladder An accessory organ that stores bile.
gallstones A disorder of the human digestive tract in which deposits of hardened cholesterol lodge in the gallbladder.
gamete A specialized reproductive cell produced by organisms of sexually reproducing species. (See **sperm; ovum; pollen; ovule**.)

gametogenesis The process of cell division by which gametes are produced. (See **meiosis; spermatogenesis; oogenesis**.)
ganglion An area of bunched nerve cells that acts as a switching point for nerve impulses traveling from receptors and to effectors.
garden pea The research organism used by Mendel in his early scientific work in genetic inheritance.
gastric cecum A gland in the grasshopper that secretes digestive enzymes.
gastrula A stage of embryonic development in animals in which the embryo assumes a tube-within-a-tube structure and distinct embryonic tissues (ectoderm, mesoderm, endoderm) begin to differentiate.
gastrulation The process by which a blastula becomes progressively more indented, forming a gastrula.
gene A unit of heredity; a discrete portion of a chromosome thought to be responsible for the production of a single type of polypeptide; the "factor" responsible for the inheritance of a genetic trait.
gene frequency The proportion (percentage) of each allele for a particular trait that is present in the gene pool of a population.
gene linkage A pattern of inheritance in which genes located along the same chromosome are prevented from assorting independently, but are linked together in their inheritance.
gene mutation An alteration of the chemical nature of a gene that changes its ability to control the production of a polypeptide chain.
gene pool The sum total of all the inheritable genes for the traits in a given sexually reproducing population.
gene-chromosome theory A theory of genetic inheritance that is based on current understanding of the relationships between the biochemical control of traits and the process of cell division.
genetic counseling Clinical discussions concerning inheritance patterns that are designed to inform prospective parents of the potential for expression of a genetic disorder in their offspring.
genetic engineering The use of various techniques to move genes from one organism to another.
genetic screening A technique for the detection of human genetic disorders in which bodily fluids are analyzed for the presence of certain marker chemicals.
genotype The particular combination of genes in an allele pair.
genus A level of biological classification that represents a subdivision of the phylum level; having fewer organisms with great similarity (e.g., *Drosophila*, paramecium).
geographic isolation The separation of species populations by geographical barriers, facilitating the evolutionary process.
geologic record A supporting item of evidence of organic evolution, supplied within the earth's rock and other geological deposits.

germination The growth of the pollen tube from a pollen grain; the growth of the embryonic root and stem from a seed.

gestation The period of prenatal development of a placental mammal; human gestation requires approximately 9 months.

gizzard A portion of the digestive tract of certain organisms, including the earthworm and the grasshopper, in which food is ground into smaller fragments.

glomerulus A capillary network lying within Bowman's capsule of the nephron.

glucagon A hormone, secreted by the islets of Langerhans, that regulates the release of blood sugar from stored glycogen.

glucose A monosaccharide produced commonly in photosynthesis and used by both plants and animals as a "fuel" in the process of respiration.

glycerol An organic compound that is a component of certain lipids.

glycogen A polysaccharide synthesized in animals as a means of storing glucose; glycogen is stored in the liver and in the muscles.

goiter A disorder of the human regulatory system in which the thyroid gland enlarges because of a deficiency of dietary iodine.

Golgi complex Cell organelles that package cell products and move them to the plasma membrane for secretion.

gonad An endocrine gland that produces the hormones responsible for the production of various secondary sex characteristics. (See **ovary; testis**.)

gout A disorder of the human excretory system in which uric acid accumulates in the joints, causing severe pain.

gradualism A theory of the time frame required for organic evolution which assumes that evolutionary change is slow, gradual, and continuous.

grafting A technique of plant propagation in which the stems of desirable plants are attached (grafted) to rootstocks of related varieties to produce new plants for commercial purposes.

grana The portion of the chloroplast within which chlorophyll molecules are concentrated.

grassland A terrestrial biome characterized by wide variation in temperature and a climax flora that includes grasses.

growth A process by which cells increase in number and size, resulting in an increase in size of the organism.

growth-stimulating hormone (GSH) A pituitary hormone regulating the elongation of the long bones of the body.

guanine A nitrogenous base found in both DNA and RNA molecules.

guard cell One of a pair of cells that surround the leaf stomate and regulate its size.

habitat The environment or set of ecological conditions within which an organism lives.

Hardy-Weinberg principle A hypothesis, advanced by G. H. Hardy and W. Weinberg, which states that the gene pool of a population should remain stable as long as a set of "ideal" conditions is met.

heart In human beings, a four-chambered muscular pump that facilitates the movement of blood throughout the body.

helix Literally a spiral; a term used to describe the "twisted ladder" shape of the DNA molecule.

hemoglobin A type of protein specialized for the transport of respiratory oxygen in certain organisms, including earthworms and human beings.

herbivore A heterotrophic organism that consumes plant matter as its primary source of nutrition. (See **primary consumer**.)

hermaphrodite An animal organism that produces both male and female gametes.

heterotroph An organism that typically carries on heterotrophic nutrition.

heterotroph hypothesis A scientific hypothesis devised to explain the probable origin and early evolution of life on earth.

heterotrophic nutrition A type of nutrition in which organisms must obtain their foods from outside sources of organic nutrients.

heterozygous A term used to refer to an allele pair in which the alleles have different contrasting effects (e.g., *Aa, RW*).

high blood pressure A disorder of the human transport system in which systolic and diastolic pressures register higher than normal because of narrowing of the artery opening.

histamine A chemical product of the body that causes irritation and swelling of the mucous membranes.

homeostasis The condition of balance and dynamic stability that characterizes living systems under normal conditions.

homologous chromosomes A pair of chromosomes that carry corresponding genes for the same traits.

homologous structures Structures present within different species that can be shown to have had a common origin, but that may or may not share a common function.

homozygous A term used to refer to an allele pair in which the alleles are identical in terms of effect (e.g., *AA, aa*).

hormone A chemical product of an endocrine gland which has a regulatory effect on the cell's metabolism.

host The organism that is harmed in a parasitic relationship.

hybrid A term used to describe a heterozygous genotype. (See **heterozygous**.)

hybridization A technique of plant/animal breeding in which two varieties of the same species are crossbred in the hope of producing offspring with the favorable traits of both varieties.

hydrogen bond A weak electrostatic bond that holds together the twisted strands of DNA and RNA molecules.

hydrolysis The chemical process by which a complex food molecule is split into simpler components through the addition of a molecule of water to the bonds holding it together.

hypocotyl A portion of the plant embryo that specializes to become the root and lower stem of the adult plant.

hypothalamus An endocrine gland whose secretions affect the pituitary gland.

identical twins In human beings, twin offspring resulting from the separation of the embryonic cell mass of a single fertilization into two separate masses; such twins are genetically identical.

importation of organisms The introduction of nonactive plants and animals into new areas where they compete strongly with native species—a negative aspect of human involvement with the environment.

in vitro fertilization A laboratory technique in which fertilization is accomplished outside the mother's body using mature ova and sperm extracted from the parents' bodies.

inbreeding A technique of plant/animal breeding in which a "purebred" variety is bred only with its own members, so as to maintain a set of desired characteristics.

independent assortment A pattern of inheritance in which genes on different, nonhomologous chromosomes are free to be inherited randomly and regardless of the inheritance of the others.

ingestion The mechanism by which an organism takes in food from its environment.

inorganic compound A chemical compound that lacks the element carbon or hydrogen (e.g., table salt: NaCl).

insulin A hormone, secreted by the islets of Langerhans, that regulates the storage of blood sugar as glycogen.

intercellular fluid (ICF) The fluid that bathes cells and fills intercellular spaces.

interferon A substance, important in the fight against human cancer, that may now be produced in large quantities through techniques of genetic engineering.

intermediate inheritance Any pattern of inheritance in which the offspring expresses a phenotype different from the phenotypes of its parents and usually representing a form intermediate between them.

internal development Embryonic development that occurs within the body of the female parent.

internal fertilization Fertilization that occurs inside the body of the female parent.

interneuron A type of neuron, located in the central nervous system, that is responsible for the interpretation of impulses received from sensory neurons.

intestine A portion of the digestive tract in which chemical digestion and absorption of digestive end-products occur.

intracellular digestion A type of chemical digestion carried out within the cell.

iodine A chemical stain used in cell study; an indicator used to detect the presence of starch. (See **staining**.)

islets of Langerhans An endocrine gland, located within the pancreas, that produces the hormones insulin and glucagon.

karyotype An enlarged photograph of the paired homologous chromosomes of an individual cell that is used in the detection of certain genetic disorders involving chromosome mutation.

karyotyping A technique for the detection of human genetic disorders in which a karyotype is analyzed for abnormalities in chromosome structure or number.

kidney The excretory organ responsible for maintaining the chemical composition of the blood. (See **nephron**.)

kidney failure A disorder of the human excretory system in which there is a general breakdown of the kidney's ability to filter blood components.

kingdom A level of biological classification that includes a broad grouping of organisms displaying general structural similarity; five kingdoms have been named by scientists.

lacteal A small extension of the lymphatic system, found inside the villus, that absorbs fatty acids and glycerol resulting from lipid hydrolysis.

lactic acid fermentation A type of anaerobic respiration in which glucose is converted to two lactic acid molecules.

large intestine A portion of the digestive tract in which undigested foods are solidified by means of water absorption to form feces.

lateral meristem A plant growth region located under the epidermis or bark of a stem. (See **cambium**.)

Latin The language used in biological classification for naming organisms by means of binomial nomenclature.

lenticel A small pore in the stem surface that permits the absorption and release of respiratory gases within stem tissues.

leukemia A disorder of the human transport system in which the bone marrow produces large numbers of abnormal white blood cells. (See **cancer.**)

lichen A symbiosis of alga and fungus that frequently acts as a pioneer species on bare rock.

limiting factor Any abiotic or biotic condition that places limits on the survival of organisms and on the growth of species populations in the environment.

lipase Any lipid-hydrolyzing enzyme.
lipid An organic compound composed of carbon, hydrogen, and oxygen in which hydrogen and oxygen are *not* in a 2:1 ratio (e.g., a wax, plant oil); many lipids are constructed of a glycerol and three fatty acids.
liver An accessory organ that stores glycogen, produces bile, destroys old red blood cells, deaminates amino acids, and produces urea.
lock-and-key model A theoretical model of enzyme action that attempts to explain the concept of enzyme specificity.
lung The major organ of respiratory gas exchange.
luteinizing hormone (LH) A pituitary hormone that regulates the conversion of the ovarian follicle into the corpus luteum.
lymph Intercellular fluid (ICF) that has passed into the lymph vessels.
lymph node One of a series of structures in the body that act as reservoirs of lymph and also contain white blood cells as part of the body's immune system.
lymph vessel One of a branching series of tubes that collect ICF from the tissues and redistribute it as lymph.
lymphatic circulation The movement of lymph throughout the body.
lymphocyte A type of white blood cell that produces antibodies.
lysosome A cell organelle that houses hydrolytic enzymes used by the cell in the process of chemical digestion.

Malpighian tubules In arthropods (e.g., grasshopper), an organ specialized for the removal of metabolic wastes.
maltase A specific enzyme that catalyzes the hydrolysis (and dehydration synthesis) of maltose.
maltose A type of disaccharide; a maltose molecule is composed of two units of glucose joined together by dehydration synthesis.
marine biome An aquatic biome characterized by relatively stable conditions of moisture, salinity, and temperature.
marsupial mammal See **nonplacental mammal**.
mechanical digestion Any of the processes by which foods are broken apart physically into smaller particles.
medulla The portion of the human brain responsible for regulating the automatic processes of the body.
meiosis The process by which four monoploid nuclei are formed from a single diploid nucleus.
meningitis A disorder of the human regulatory system in which the membranes of the brain or spinal cord become inflamed.
menstrual cycle A hormone-controlled process responsible for the monthly release of mature ova.
menstruation The stage of the menstrual cycle in which the lining of the uterus breaks down and is expelled from the body via the vaginal canal.

meristem A plant tissue specialized for embryonic development. (See **apical meristem; lateral meristem; cambium**.)

mesoderm An embryonic tissue that differentiates into muscle, bone, the excretory system, and most of the reproductive system in the adult animal.

messenger RNA (m-RNA) A type of RNA that carries the genetic code from the nuclear DNA to the ribosome for transcription.

metabolism All of the chemical processes of life considered together; the sum total of all the cell's chemical activity.

methylene blue A chemical stain used in cell study. (See **staining**.)

microdissection instruments Tools of biological study that are used to remove certain cell organelles from within cells for examination.

micrometer (μm) A unit of linear measurement equal in length to 0.001 millimeter (0.000001 meter), used for expressing the dimensions of cells and cell organelles.

mitochondrion A cell organelle that contains the enzymes necessary for aerobic respiration.

mitosis A precise duplication of the contents of a parent cell nucleus, followed by an orderly separation of these contents into two new, identical daughter nuclei.

mitotic cell division A type of cell division that results in the production of two daughter cells identical to each other and to the parent cell.

Monera One of the five biological kingdoms; it includes simple unicellular forms lacking nuclear membranes (e.g., bacteria).

monohybrid cross A genetic cross between two organisms both heterozygous for a trait controlled by a single allele pair. The phenotypic ratio resulting is 3:1; the genotypic ratio is 1:2:1.

monoploid chromosome number The number of chromosomes commonly found in the gametes of sexually reproducing species.

monosaccharide A type of carbohydrate known also as a "simple sugar"; all monosaccharides have the molecular formula $C_6H_{22}O_6$.

motor neuron A type of neuron that carries "command" impulses from the central nervous system to an effector organ.

mucus A protein-rich mixture that bathes and moistens the respiratory surfaces.

multicellular Having a body that consists of large groupings of specialized cells (e.g., human being).

multiple alleles A pattern of inheritance in which the existence of more than two alleles is hypothesized, only two of which are present in the genotype of any one individual.

muscle A type of tissue specialized to produce movement of body parts.

mutagenic agent Any environmental condition that initiates or accelerates genetic mutation.

mutation Any alteration of the genetic material, either a chromosome or a gene, in an organism.
mutualism A type of symbiosis beneficial to both organisms in the relationship.

nasal cavity A series of channels through which outside air is admitted to the body interior and is warmed and moistened before entering the lung.
natural selection A concept, central to Darwin's theory of evolution, to the effect that the individuals best adapted to their environment tend to survive and to pass their favorable traits on to the next generation.
negative feedback A type of endocrine regulation in which the effects of one gland may inhibit its own secretory activity, while stimulating the secretory activity of another gland.
nephridium An organ found in certain organisms, including the earthworm, specialized for the removal of metabolic wastes.
nephron The functional unit of the kidney. (See **glomerulus; Bowman's capsule**.)
nerve A structure formed from the bundling of neurons carrying sensory or motor impulses.
nerve impulse An electrochemical change in the surface of the nerve cell.
nerve net A network of "nerve" cells in coelenterates such as the hydra.
neuron A cell specialized for the transmission of nerve impulses.
neurotransmitter A chemical substance secreted by a neuron that aids in the transmission of the nerve impulse to an adjacent neuron.
niche The role that an organism plays in its environment.
nitrifying bacteria In the nitrogen cycle, bacteria that absorb ammonia and convert it into nitrate salts.
nitrogen cycle The process by which nitrogen is recycled and made available for use by other organisms.
nitrogen-fixing bacteria A type of bacteria responsible for absorbing atmospheric nitrogen and converting it to nitrate salts in the soil.
nitrogenous base A chemical unit composed of carbon, hydrogen, and nitrogen that is a component part of the nucleotide unit.
nitrogenous waste Any of a number of nitrogen-rich compounds that result from the metabolism of proteins and amino acids in the cell. (See **ammonia; urea; uric acid**.)
nondisjunction A type of chromosome mutation in which the members of one or more pairs of homologous chromosomes fail to separate during the disjunction phase of the first meiotic division.
nonplacental mammal A species of mammal in which internal development is accomplished without the aid of a placental connection (marsupial mammals).

nucleic acid An organic compound composed of repeating units of nucleotide.

nucleolus A cell organelle located within the nucleus that is known to function in protein synthesis.

nucleotide The repeating unit making up the nucleic acid polymer (e.g., DNA, RNA).

nucleus A cell organelle that contains the cell's genetic information in the form of chromosomes.

nutrition The life function by which living things obtain food and process it for their use.

omnivore A heterotrophic organism that consumes both plant and animal matter as sources of nutrition.

one gene-one polypeptide A scientific hypothesis concerning the role of the individual gene in protein synthesis.

oogenesis A type of meiotic cell division in which one ovum and three polar bodies are produced from each primary sex cell.

open transport system A type of circulatory system in which the transport fluid is *not* always enclosed within blood vessels (e.g., grasshopper).

oral cavity In human beings, the organ used for the ingestion of foods.

oral groove The ingestive organ of the paramecium.

organ transplant An application of the study of immunity in which an organ or tissue of a donor is transplanted into a compatible recipient.

organelle A small, functional part of a cell specialized to perform a specific life function (e.g., nucleus, mitochondrion).

organic compound A chemical compound that contains the elements carbon and hydrogen (e.g., carbohydrate, protein).

organic evolution The mechanism thought to govern the changes in living species over geologic time.

osmosis A form of passive transport by which water is absorbed or released by cells.

ovary A female gonad that secretes the hormone estrogen, which regulates female secondary sex characteristics; the ovary also produces ova, which are used in reproduction.

overcropping A negative aspect of human involvement with the environment in which soil is overused for the production of crops, leading to exhaustion of soil nutrients.

overgrazing The exposure of soil to erosion due to the loss of stabilizing grasses when it is overused by domestic animals—a negative aspect of human involvement with the environment.

overhunting A negative aspect of human involvement with the environment in which certain species have been greatly reduced or made extinct by uncontrolled hunting practices.

oviduct A tube that serves as a channel for conducting mature ova from the ovary to the uterus; the site of fertilization and the earliest stages of embryonic development.

ovulation The stage of the menstrual cycle in which the mature ovum is released from the follicle into the oviduct.

ovule A structure located within the flower ovary that contains a monoploid egg nucleus and serves as the site of fertilization.

ovum A type of gamete produced as a result of oogenesis in female animals; the egg, the female sex cell.

oxygen-18 A radioactive isotope of oxygen that is used to trace the movement of this element in biochemical reaction sequences.

oxygenated blood Blood that contains a high percentage of oxyhemoglobin.

oxyhemoglobin Hemoglobin that is loosely bound to oxygen for purposes of oxygen transport.

palisade layer A cell layer found in most leaves that contains high concentrations of chloroplasts.

pancreas An accessory organ which produces enzymes that complete the hydrolysis of foods to soluble end-products; also the site of insulin and glucagon production.

parasitism A type of symbiosis from which one organism in the relationship benefits, while the other (the "host") is harmed, but not ordinarily killed.

parathormone A hormone of the parathyroid gland that regulates the metabolism of calcium in the body.

parathyroid gland An endocrine gland whose secretion, parathormone, regulates the metabolism of calcium in the body.

passive immunity A temporary immunity produced as a result of the injection of preformed antibodies.

passive transport Any process by which materials are absorbed into the cell interior from an area of high concentration to an area of low concentration, without the expenditure of cell energy (e.g., osmosis, diffusion).

penis A structure that permits internal fertilization through direct implantation of sperm into the female reproductive tract.

peptide bond A type of chemical bond that links the nitrogen atom of one amino acid with the terminal carbon atom of a second amino acid in the formation of a dipeptide.

peripheral nerves Nerves in the earthworm and grasshopper that branch from the ventral nerve cord to other parts of the body.

peripheral nervous system A major subdivision of the nervous system that consists of all the nerves of all types branching through the body. (See **autonomic nervous system; somatic nervous system**.)

peristalsis A wave of contraction of the smooth muscle lining; the digestive tract that causes ingested food to pass along the food tube.

petal An accessory part of the flower that is thought to attract pollinating insects.

pH A chemical unit used to express the concentration of hydrogen ion (H^+), or the acidity, of a solution.

phagocyte A type of white blood cell that engulfs and destroys bacteria.

phagocytosis The process by which the ameba surrounds and ingests large food particles for intracellular digestion.

pharynx The upper part of the digestive tube that temporarily stores food before digestion.

phenotype The observable trait that results from the action of an allele pair.

phenylketonuria (PKU) A genetically related human disorder in which the homozygous combination of a particular mutant gene prevents the normal metabolism of the amino acid phenylalanine.

phloem A type of vascular tissue through which water and dissolved sugars are transported in plants from the leaf downward to the roots for storage.

phosphate group A chemical group made up of phosphorus and oxygen that is a component part of the nucleotide unit.

phosphoglyceraldehyde (PGAL) An intermediate product formed during photosynthesis that acts as the precursor of glucose formation.

photochemical reactions A set of biochemical reactions in photosynthesis in which light is absorbed and water molecules are split. (See **photolysis**.)

photolysis The portion of the photochemical reactions in which water molecules are split into hydrogen atoms and made available to the carbon fixation reactions.

photosynthesis A type of autotrophic nutrition in which green plants use the energy of sunlight to convert carbon dioxide and water into glucose.

phylum A level of biological classification that is a major subdivision of the kingdom level, containing fewer organisms with greater similarity (e.g., Chordata).

pinocytosis A special type of absorption by which liquids and particles too large to diffuse through the cell membrane may be taken in by vacuoles formed at the cell surface.

pioneer autotrophs The organisms supposed by the heterotroph hypothesis to have been the first to evolve the ability to carry on autotrophic nutrition.

pioneer species In an ecological succession, the first organisms to inhabit a barren environment.

pistil The female sex organ of the flower. (See **stigma; style; ovary**.)

pituitary gland An endocrine gland that produces hormones regulating the secretions of other endocrine glands; the "master gland."

placenta In placental mammals, a structure composed of both embryonic and maternal tissues that permits the diffusion of soluble substances to and from the fetus for nourishment and the elimination of fetal waste.

placental mammal A mammal species in which embryonic development occurs internally with the aid of a placental connection to the female parent's body.

Plant One of the five biological kingdoms; it includes multicellular organisms whose cells are bounded by cell walls and which are capable of photosynthesis (e.g., maple tree).

plasma The liquid fraction of blood, containing water and dissolved proteins.

plasma membrane A cell organelle that encloses the cytoplasm and other cell organelles and regulates the passage of materials into and out of the cell.

platelet A cell-like component of the blood that is important in clot formation.

polar body One of three nonfunctional cells produced during oogenesis that contain monoploid nuclei and disintegrate soon after completion of the process.

polio A disorder of the human regulatory system in which viral infection of the central nervous system may result in severe paralysis.

pollen The male gamete of the flowering plant.

pollen tube A structure produced by the germinating pollen grain that grows through the style to the ovary and carries the sperm nucleus to the ovule for fertilization.

pollination The transfer of pollen grains from anther to stigma.

pollution control The development of new procedures to reduce the incidence of air, water, and soil pollution—a positive aspect of human involvement with the environment.

polyploidy A type of chromosome mutation in which an entire set of homologous chromosomes fail to separate during the disjunction phase of the first meiotic division.

polysaccharide A type of carbohydrate composed of repeating units of monosaccharide that form a polymeric chain.

polyunsaturated fat A type of fat in which many bonding sites are unavailable for the addition of hydrogen atoms.

population All the members of a particular species in a given geographical location at a given time.

population control The use of various practices to slow the rapid growth in the human population—a positive aspect of human interaction with the environment.

population genetics A science that studies the genetic characteristics of a sexually reproducing species and the factors that affect its gene frequencies.

postnatal development The growth and maturation of an individual from birth, through aging, to death.

prenatal development The embryonic development that occurs within the uterus before birth. (See **gestation.**)

primary consumer Any herbivorous organism that receives food energy from the producer level (e.g., mouse); the second stage of a food chain.

primary sex cell The diploid cell that undergoes meiotic cell division to produce monoploid gametes.

producer Any autotrophic organism capable of trapping light energy and converting it to the chemical bond energy of food (e.g., green plants); the organisms forming the basis of the food chain.

progesterone A hormone produced by the corpus luteum and/or placenta that has the effect of maintaining the uterine lining and suppressing ovulation during gestation.

protease Any protein-hydrolyzing enzyme.

protein A complex organic compound composed of repeating units of amino acid.

Protista One of the five biological kingdoms; it includes simple unicellular forms whose nuclei are surrounded by nuclear membranes (e.g., ameba, paramecium).

pseudopod A temporary, flowing extension of the cytoplasm of an ameba that is used in nutrition and locomotion.

pulmonary artery One of two arteries that carry blood from the heart to the lungs for reoxygenation.

pulmonary circulation Circulation of blood from the heart through the lungs and back to the heart.

pulmonary vein One of four veins that carry oxygenated blood from the lungs to the heart.

pulse Rhythmic contractions of the artery walls that help to push the blood fluid through the capillary networks of the body.

punctuated equilibrium A theory of the time frame required for evolution which assumes that evolutionary change occurs in "bursts" with long periods of relative stability intervening.

pyramid of biomass A construct used to illustrate the fact that the total biomass available in each stage of a food chain diminishes from producer level to consumer level.

pyramid of energy A construct used to illustrate the fact that energy is lost at each trophic level in a food chain, being most abundant at the producer level.

pyruvic acid An intermediate product in the aerobic or anaerobic respiration of glucose.

receptor An organ specialized to receive a particular type of environmental stimulus.

recessive allele (gene) An allele (gene) whose effect is masked by that of its dominant allele.

recombinant DNA DNA molecules that have been moved from one cell to another in order to give the recipient cell a genetic characteristic of the donor cell.

recombination The process by which the members of segregated allele pairs are randomly recombined in the zygote as a result of fertilization.

rectum The portion of the digestive tract in which digestive wastes are stored until they can be released to the environment.

red blood cell Small, nonnucleated cells in the blood that contain hemoglobin and carry oxygen to bodily tissues.

reduction division See **meiosis**.

reflex A simple, inborn, involuntary response to an environmental stimulus.

reflex arc The complete path, involving a series of three neurons (sensory, interneuron, and motor), working together, in a reflex action.

regeneration A type of asexual reproduction in which new organisms are produced from the severed parts of a single parent organism; the replacement of lost or damaged tissues.

regulation The life process by which living things respond to changes within and around them, and by which all life processes are coordinated.

replication An exact self-duplication of the chromosome during the early stages of cell division; the exact self-duplication of a molecule of DNA.

reproduction The life process by which new cells arise from preexisting cells by cell division.

reproductive isolation The inability of species varieties to interbreed and produce fertile offspring, because of variations in behavior or chromosome structure.

respiration The life function by which living things convert the energy of organic foods into a form more easily used by the cell.

response The reaction of an organism to an environmental stimulus.

rhizoid A rootlike fiber produced by fungi that secrete hydrolytic enzymes and absorb digested nutrients.

ribonucleic acid (RNA) A type of nucleic acid that operates in various ways to facilitate protein synthesis.

ribose A five-carbon sugar found as a component part of the nucleotides of RNA molecules only.

ribosomal RNA (r-RNA) The type of RNA that makes up the ribosome.

ribosome A cell organelle that serves as the site of protein synthesis in the cell.

root A plant organ specialized to absorb water and dissolved substances from the soil, as well as to anchor the plant to the soil.

root hair A small projection of the growing root that serves to increase the surface area of the root for absorption.

roughage A variety of undigestible carbohydrates that add bulk to the diet and facilitate the movement of foods through the intestine.
runner A type of vegetative propagation in which an above-ground stem (runner) produces roots and leaves and establishes new organisms with identical characteristics.

saliva A fluid secreted by salivary glands that contains hydrolytic enzymes specific to the digestion of starches.
salivary gland The gland that secretes saliva, important in the chemical digestion of certain foods.
salt A chemical composed of a metal and a nonmetal joined by means of an ionic bond (e.g., sodium chloride).
saprophyte A heterotrophic organism that obtains its nutrition from the decomposing remains of dead plant and animal tissues (e.g., fungus, bacteria).
saturated fat A type of fat molecule in which all available bonding sites on the hydrocarbon chains are taken up with hydrogen atoms.
scrotum A pouch extending from the wall of the lower abdomen that houses the testes at a temperature optimum for sperm production.
secondary consumer Any carnivorous animal that derives its food energy from the primary consumer level (e.g., a snake); the third level of a food chain.
secondary sex characteristics The physical features, different in males and females, that appear with the onset of sexual maturity.
seed A structure that develops from the fertilized ovule of the flower and germinates to produce a new plant.
seed dispersal Any mechanism by which seeds are distributed in the environment so as to widen the range of a plant species. (See **fruit**.)
segregation The random separation of the members of allele pairs that occurs during meiotic cell division.
self-pollination A type of pollination in which the pollen of a flower pollinates another flower located on the same plant organism.
sensory neuron A type of neuron specialized for receiving environmental stimuli, which are detected by receptor organs.
sepal An accessory part of the flower that functions to protect the bud during development.
sessile A term that relates to the "unmoving" state of certain organisms, including the hydra.
seta One of several small, chitinous structures (setae) that aid the earthworm in its locomotor function.
sex chromosomes A pair of homologous chromosomes carrying genes that determine the sex of an individual; these chromosomes are designated as X and Y.

sex determination A pattern of inheritance in which the conditions of maleness and femaleness are determined by the inheritance of a pair of sex chromosomes (XX = female; XY = male).
sex linkage A pattern of inheritance in which certain nonsex genes are located on the X sex chromosome, but have no corresponding alleles on the Y sex chromosome.
sex-linked trait A genetic trait whose inheritance is controlled by the genetic pattern of sex linkage (e.g., color blindness).
sexual reproduction A type of reproduction in which new organisms are formed as a result of the fusion of gametes from two parent organisms.
shell An adaptation for embryonic development in many terrestrial, externally developing species that protects the developing embryo from drying and physical damage (e.g., birds).
sickle-cell anemia A genetically related human disorder in which the homozygous combination of a mutant gene leads to the production of abnormal hemoglobin and crescent-shaped red blood cells.
skeletal muscle A type of muscle tissue associated with the voluntary movements of skeletal levers in locomotion.
small intestine In human beings, the longest portion of the food tube, in which final digestion and absorption of soluble end-products occur.
smooth muscle See **visceral muscle**.
somatic nervous system A subdivision of the peripheral nervous system that is made up of nerves associated with voluntary actions.
speciation The process by which new species are thought to arise from previously existing species.
species A biological grouping of organisms so closely related that they are capable of interbreeding and producing fertile offspring (e.g., human being).
species presentation The establishment of game lands and wildlife refuges that have permitted the recovery of certain endangered species—a positive aspect of human involvement with the environment.
sperm A type of gamete produced as a result of spermatogenesis in male animals; the male reproductive cell.
spermatogenesis A type of meiotic cell division in which four sperm cells are produced for each primary sex cell.
spinal cord The part of the central nervous system responsible for reflex action, as well as impulse conduction between the peripheral nervous system and the brain.
spindle apparatus A network of fibers that form during cell division and to which centromeres attach during the separation of chromosomes.
spiracle One of several small pores in arthropods, including the grasshopper, that serve as points of entry of respiratory gases from the atmosphere to the tracheal tubes.

spongy layer A cell layer found in most leaves that is loosely packed and contains many air spaces to aid in gas exchange.

spore A specialized asexual reproductive cell produced by certain plants.

sporulation A type of asexual reproduction in which spores released from special spore cases on the parent plant germinate and grow into new adult organisms of the species.

staining A technique of cell study in which chemical stains are used to make cell parts more visible for microscopic study.

stamen The male reproductive structure in a flower. (See **anther; filament**.)

starch A type of polysaccharide produced and stored by plants.

stem A plant organ specialized to support the leaves and flowers of a plant, as well as to conduct materials between the roots and the leaves.

stigma The sticky upper portion of the pistil, which serves to receive pollen.

stimulus Any change in the environment to which an organism responds.

stomach A muscular organ that acts to liquefy food and that produces gastric protease for the hydrolysis of protein.

stomate A small opening that penetrates the lower epidermis of a leaf and through which respiratory and photosynthetic gases diffuse.

strata The layers of sedimentary rock that contain fossils, whose ages may be determined by studying the patterns of sedimentation.

stroke A disorder of the human regulatory system in which brain function is impaired because of oxygen starvation of brain centers.

stroma An area of the chloroplast within which the carbon-fixation reactions occur; stroma lie between pairs of grana.

style The portion of the pistil that connects the stigma to the ovary.

substrate A chemical that is metabolized by the action of a specific enzyme.

succession A situation in which an established ecological community is gradually replaced by another until a climax community is established.

survival of the fittest The concept, frequently associated with Darwin's theory of evolution, that in the intraspecies competition among naturally occurring species the organisms best adapted to the particular environment will survive.

sweat glands In human beings, the glands responsible for the production of perspiration.

symbiosis A term which refers to a variety of biotic relationships in which organisms of different species live together in close physical association.

synapse The gap that separates the terminal branches of one neuron from the dendrites of an adjacent neuron.

synapsis The intimate, highly specific pairing of homologous chromosomes that occurs in the first meiotic division, forming tetrads.

synthesis The life function by which living things manufacture the complex compounds required to sustain life.
systemic circulation The circulation of blood from the heart through the body tissues (except the lungs) and back to the heart.
systole The higher pressure registered during blood pressure testing. (See **diastole**.)

taiga A terrestrial biome characterized by long, severe winters and climax flora that includes coniferous trees.
Tay-Sachs A genetically related human disorder in which fatty deposits in the cells, particularly of the brain, inhibit proper functioning of the nervous system.
technological oversight A term relating to human activities that adversely affect environmental quality due to failure to adequately assess the environmental impact of a technological development.
teeth Structures located in the mouth that are specialized to aid in the mechanical digestion of foods.
temperate deciduous forest A terrestrial biome characterized by moderate climatic conditions and climax flora that includes deciduous trees.
template A pattern or design provided by the DNA molecule for the synthesis of protein molecules.
tendon A type of connective tissue that attaches a skeletal muscle to a bone.
tendonitis A disorder of the human locomotor system in which the junction between a tendon and a bone becomes irritated and inflamed.
tentacle A grasping structure in certain organisms, including the hydra, that contains stinging cells and is used for capturing prey.
terminal branch A cytoplasmic extension of the neuron that transmits a nerve impulse to adjacent neurons via the secretion of neurotransmitters.
terrestrial biome A biome that comprises primarily land ecosystems, the characteristics of which are determined by the major climate zone of the earth.
test cross A genetic cross accomplished for the purpose of determining the genotype of an organism expressing a dominant phenotype; the unknown is crossed with a homozygous recessive.
testis A gonad in human males that secretes the hormone testosterone, which regulates male secondary sex characteristics; the testis also produces sperm cells for reproduction.
testosterone A hormone secreted by the testis that regulates the production of male secondary sex characteristics.
tetrad A grouping of four chromatids that results from synapsis.
thymine A nitrogenous base found only in DNA.

thyroid gland An endocrine gland that regulates the body's general rate of metabolism through secretion of the hormone thyroxin.

thyroid-stimulating hormone (TSH) A pituitary hormone that regulates the secretions of the thyroid gland.

thyroxin A thyroid hormone that regulates the body's general metabolic rate.

tongue A structure that aids in the mechanical digestion of foods.

trachea A cartilage-ringed tube that conducts air from the mouth to the bronchi.

tracheal tube An adaptation in arthropods (e.g., grasshopper) which functions to conduct respiratory gases from the environment to the moist internal tissues.

Tracheophyta A phylum of the Plant Kingdom whose members (tracheophytes) contain vascular tissues and true roots, stems, and leaves (e.g., geranium, fern, bean, maple tree, corn).

transfer RNA (t-RNA) A type of RNA that functions to transport specific amino acids from the cytoplasm to the ribosome for protein synthesis.

translocation A type of chromosome mutation in which a section of a chromosome is transferred to a nonhomologous chromosome.

transpiration The evaporation of water from leaf stomates.

transpiration pull A force that aids the upward conduction of materials in the xylem by means of the evaporation of water (transpiration) from leaf surfaces.

transport The life function by which substances are absorbed, circulated, and released by living things.

triplet codon A group of three nitrogenous bases that provide information for the placement of amino acids in the synthesis of proteins.

tropical forest A terrestrial biome characterized by a warm, moist climate and a climax flora that includes many species of broadleaf trees.

tropism A plant growth response to an environmental stimulus.

tuber A type of vegetative propagation in which an underground stem (tuber) produces new tubers, each of which is capable of producing new organisms with identical characteristics.

tundra A terrestrial biome characterized by permanently frozen soil and climax flora that includes lichens and mosses.

tympanum A receptor organ in arthropods (e.g., grasshopper) which is specialized to detect vibrational stimuli.

ulcer A disorder of the human digestive tract in which a portion of its lining erodes and becomes irritated.

ultracentrifuge A tool of biological study that uses very high speeds of centrifugation to separate cell parts for examination.

umbilical cord In placental mammals, a structure containing blood vessels that connects the placenta to the embryo.

unicellular Having a body that consists of a single cell (e.g., paramecium).
uracil A nitrogenous base that is a component part of the nucleotides of RNA molecules only.
urea A type of nitrogenous waste with moderate solubility and moderate toxicity.
ureter In human beings, a tube that conducts urine from the kidney to the urinary bladder.
urethra In human beings, a tube that conducts urine from the urinary bladder to the exterior of the body.
uric acid A type of nitrogenous waste with low solubility and low toxicity.
urinary bladder An organ responsible for the temporary storage of urine.
urine A mixture of water, salts, and urea excreted from the kidney.
use and disuse A term associated with the evolutionary theory of Lamarck, since proved incorrect.
uterus In female placental mammals, the organ within which embryonic development occurs.

vaccination An inoculation of dead or weakened disease organisms that stimulates the body's immune system to produce active immunity.
vacuole A cell organelle that contains storage materials (e.g., starch, water) housed inside the cell.
vagina In female placental mammals, the portion of the reproductive tract into which sperm are implanted during sexual intercourse and through which the baby passes during birth.
variation A concept, central to Darwin's theory of evolution, that refers to the range of adaptation which can be observed in all species.
vascular tissues Tubelike plant tissues specialized for the conduction of water and dissolved materials within the plant. (See **xylem; phloem**.)
vegetative propagation A type of asexual reproduction in which new plant organisms are produced from the vegetative (nonfloral) parts of the parent plant.
vein (human) A relatively thin-walled blood vessel that carries blood from capillary networks back toward the heart.
vein (plant) An area of vascular tissues located in the leaf that aid the upward transport of water and minerals through the leaf and the transport of dissolved sugars to the stem and roots.
vena cava One of two major arteries that return blood to the heart from the body tissues.
ventral nerve cord The main pathway for nerve impulses between the brain and peripheral nerves of the grasshopper and earthworm.
ventricle One of two thick-walled, muscular chambers of the heart that pump blood out to the lungs and body.
villi Microscopic projections of the lining of the small intestine that absorb the soluble end-products of digestion. (See **lacteal**.)

visceral muscle A type of muscle tissue associated with the involuntary movements of internal organs (e.g., peristalsis in the small intestine).

vitamin a type of nutrient that acts as a coenzyme in various enzyme-controlled reactions.

water cycle The mechanism by which water is made available to living things in the environment through the processes of precipitation, evaporation, runoff, and percolation.

water pollution A type of technological oversight that involves the addition of some unwanted factor (e.g., sewage, heavy metals, heat, toxic chemicals) to our water resources.

Watson-Crick model A model of DNA structure devised by J. Watson and F. Crick that hypothesizes a "twisted ladder" arrangement for the DNA molecule.

white blood cell A type of blood cell that functions in disease control. (See **phagocyte; lymphocyte**.)

xylem A type of vascular tissue through which water and dissolved minerals are transported upward through a plant from the root to the stems and leaves.

yolk A food substance, rich in protein and lipid, found in the eggs of many animal species.

yolk sac The membrane that surrounds the yolk food supply of the embryos of many animal species.

zygote The single diploid cell that results from the fusion of gametes in sexual reproduction; a fertilized egg.

Regents Examinations, Answers, and Student Self-Appraisal Guides

Examination August 2016
Living Environment

PART A

Answer all questions in this part. [30]

Directions **(1–30): For *each* statement or question, record in the space provided the *number* of the word or expression that, of those given, best completes the statement or answers the question.**

1 Which sequence represents structures organized from most complex to least complex?

1 chloroplast → guard cell → leaf → oak tree
2 guard cell → chloroplast → leaf → oak tree
3 oak tree → guard cell → leaf → chloroplast
4 oak tree → leaf → guard cell → chloroplast 1 _____

2 Autotrophs differ from heterotrophs in that only autotrophs

1 require carbon dioxide for cellular respiration
2 release oxygen as a product of cellular respiration
3 synthesize nutrients using carbon dioxide and water
4 break down sugars to assemble other molecules 2 _____

3 Burmese pythons are large snakes that have been introduced into the Florida Everglades ecosystem. Burmese pythons and alligators hunt the same prey. One likely effect of the introduction of the pythons is that

1 alligators will have more prey available
2 pythons will become native to the Everglades
3 alligator populations will decline
4 pythons will become an endangered species 3____

4 Which activity enables humans to produce new genetic combinations in other organisms?

1 selecting and breeding the organisms for specific traits
2 increasing the number of enzymes available to the organisms
3 growing organisms that reproduce asexually
4 decreasing the amount of DNA in the diet of the organisms 4____

5 Before a cell divides, an exact copy of each chromosome is made by the process of

1 genetic engineering
2 replication
3 mutation
4 recombination 5____

6 Some time ago, there were thousands of California condors in North America. Large numbers were poisoned from lead in bullets that were used to kill the animals the condors fed on. An effort was made to help save this large scavenger. There are now more than 350 California condors in North America. The condors most likely increased in number because humans decided to

1 produce lead-resistant condors through asexual reproduction
2 pass laws against using lead bullets to kill animals used by condors for food
3 introduce plants that didn't absorb the lead from discharged bullets
4 produce lead-resistant prey for the condors through genetic engineering 6_____

7 In humans, the maintenance of a stable internal temperature is a direct result of

1 detection of and reaction to stimuli in the environment
2 digestion of starches and absorption of protein from the internal environment
3 diffusion of water and excretion of glucose to the external environment
4 transport of ATP and locomotion through the environment 7_____

8 Which molecules are needed to cut and copy segments of DNA?

1 reproductive hormones
2 carbohydrates
3 antibodies
4 biological catalysts

8____

9 Evolution can occur at different rates; however, for evolution to occur, there must be

1 variations within a species
2 extinction of the species
3 asexual reproduction
4 no change in the genes of an organism

9____

10 The two reactions illustrated in the diagrams below often occur when a foreign substance enters the body.

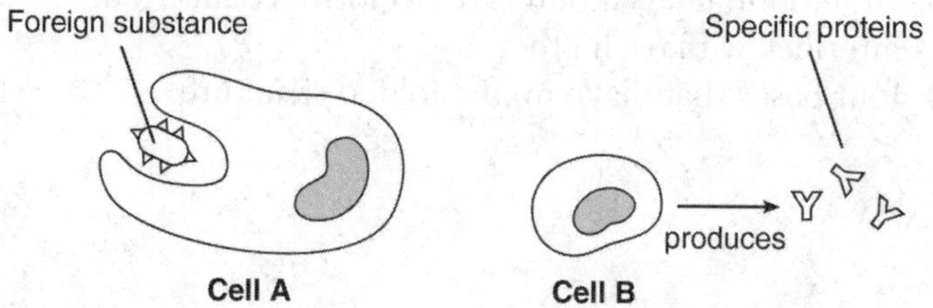

The cells labeled *A* and *B* are examples of cells known as

1 guard cells
2 reproductive cells
3 white blood cells
4 specialized skin cells

10____

11 The diagram below represents a protein molecule present in some living things.

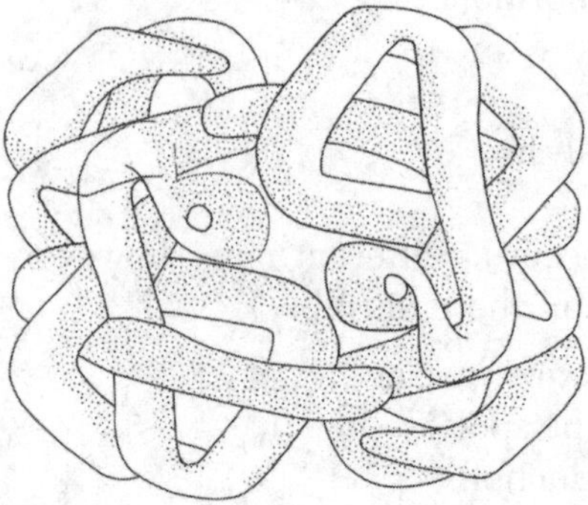

This type of molecule is composed of a sequence of

1 amino acids arranged in a specific order
2 simple sugars alternating with starches arranged in a folded pattern
3 large inorganic subunits that form chains that interlock with each other
4 four bases that make up the folded structure 11 ____

12 Three human actions that have been made possible in recent times are:

- Doctors are able to diagnose and treat some fetal problems prior to the birth of a child.
- Cloning can produce large numbers of plants that are resistant to drought.
- Male insects can be sterilized with radiation to prevent them from mating successfully.

Which statement summarizes these three actions?

1 Reproductive technology has medical, agricultural, and ecological applications.
2 Development is a highly regulated process involving mitosis and differentiation.
3 Reproduction and development are subject to environmental effects.
4 Human development, birth, and aging should be viewed as a predictable pattern of events. 12_____

13 Natural selection is best described as

1 a change in an organism in response to a need of that organism
2 a process of nearly constant improvement that leads to an organism that is nearly perfect
3 differences in survival rates as a result of different inherited characteristics
4 inheritance of characteristics acquired during the life of an organism 13_____

14 Which statement best describes a situation where competition occurs in an ecosystem?

1 A deer outruns an attacking wolf.
2 A deer, during the winter, consumes tree bark.
3 A deer and a rabbit consume grass in a field.
4 A deer and a rabbit are both startled by a hawk flying overhead. 14____

15 A woman changes her hair color to red; however, her children will not inherit this red hair color because the woman does not have

1 genes for red hair in her skin
2 genes for red hair in her sex cells
3 proteins for red hair in the placenta
4 proteins for red hair in her egg cells 15____

16 Fossils provide evidence that

1 life on Earth millions of years ago was more complex than life is today
2 the changes that will occur in species in the future are easy to predict
3 many species of organisms that lived long ago are now extinct
4 most species of organisms that lived long ago are exactly the same today 16____

17 A male frigatebird displays to the female by inflating its large red throat sac, throwing its head back, vibrating its wings, and producing a "drumming" sound with its throat sac. For the frigatebird, this behavior has most likely resulted in

1 hiding from predators
2 greater reproductive success
3 locating new sources of food
4 reduced population growth 17____

18 Plant species *X* lives in a hot, dry environment. Slowly, over hundreds of years, the climate becomes wetter. Fungi attack species *X* and cause the population of species *X* to decrease. However, plant species *X* could survive if the plants

1 try to mutate quickly and synthesize new proteins
2 are watered often and fertilized with extra nutrients
3 can adapt to the new conditions by mating with the fungus
4 have a few members of the population that are fungus-resistant 18____

19 The diagram below represents reproduction in a yeast cell. The genes in the bud are identical to the genes in the parent.

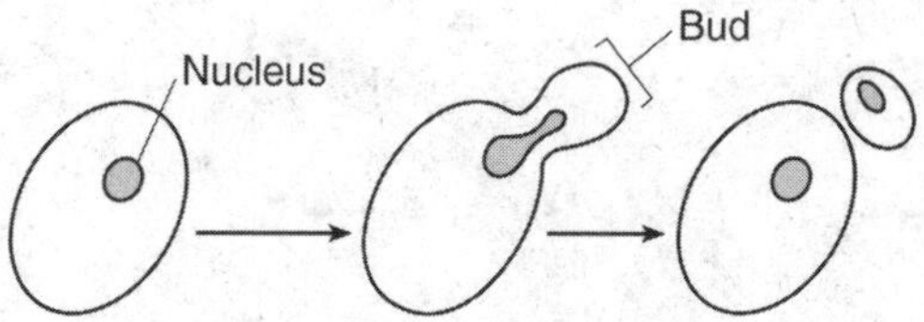

This type of production of offspring is a form of

1 sexual reproduction
2 asexual reproduction
3 gene manipulation
4 genetic engineering 19____

20 A human cell that contains all of the information necessary for the growth and development of a complete organism is

1 a sperm cell
2 a gamete
3 a zygote
4 an egg cell

20 _____

21 When would exposure to a potentially harmful substance be most likely to damage many organs in a developing embryo?

1 during the last three months of pregnancy
2 during the early stages of pregnancy
3 during the formation of the zygote
4 during meiosis in both males and females

21 _____

22 The human female reproductive system is represented below.

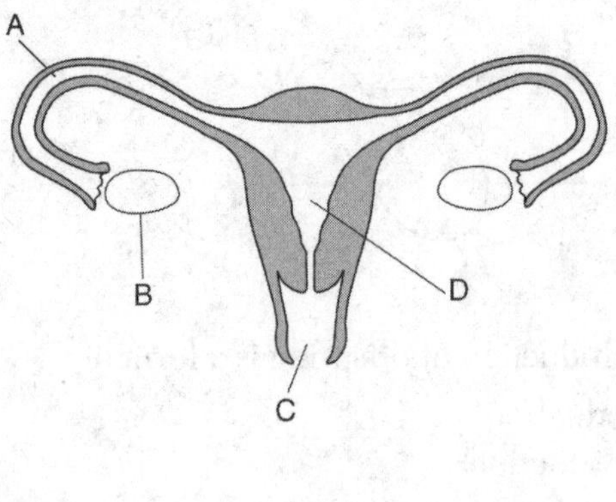

Within which structure does the placenta normally develop?

1 *A*
2 *B*
3 *C*
4 *D*

22 _____

23 An energy-rich organic compound needed by organisms is

1 water
2 salt
3 oxygen
4 glucose

23____

24 SCIDS (Severe Combined Immunodeficiency Syndrome) is a disorder where a genetic mutation inhibits the production and functioning of T-cells. T-cells are special types of white blood cells that play a role in the body's immune response. A possible symptom of SCIDS would be an increase in the

1 number of antigens produced
2 red blood cell count
3 number of infections by pathogens
4 ability to maintain homeostasis

24____

25 An organ, such as a kidney, used for transplant needs to be tested for compatibility with the person who is to receive the organ. If this is not done, the

1 donated organ might attack the body
2 donated organ might attack the immune system
3 immune system might attack its own body cells
4 immune system might attack the donated organ

25____

26 A researcher concludes from a 10-year study that the biodiversity of an ecosystem had increased. Which set of observations represents evidence for this claim?

1 There were more niches and greater stability in the ecosystem.
2 There were more niches and less energy lost as heat in the ecosystem.
3 There were fewer niches for decomposers and less stability.
4 There were fewer niches for consumers and greater cycling of materials.

26____

27 The final consumers in many food webs are

1 autotrophs
2 hosts
3 herbivores
4 carnivores 27____

28 Abandoned farmland that once grew corn is now covered with bushes and small trees. These observed changes resulted directly from

1 evolutionary change
2 ecological succession
3 loss of biodiversity
4 selective breeding 28____

29 The overuse of chemical fertilizers has resulted in the growth of some lawns in which decomposers cannot live. This would interfere most directly with the ability of the lawn ecosystem to

1 recycle energy
2 recycle nutrients
3 maintain atmospheric pH
4 reduce biodiversity 29____

30 On Long Island, several businesses use geothermal technology. This involves taking heat from within Earth and using it to heat buildings. One benefit of this technology is that it

1 contributes to global warming
2 reduces the ozone shield
3 reduces dependence on fossil fuels
4 decreases resources for many species 30____

PART B–1

Answer all questions in this part. [13]

Directions **(31–43): For *each* statement or question, record in the space provided the *number* of the word or expression that, of those given, best completes the statement or answers the question.**

31 Recently, researchers from Stanford University have changed mouse skin cells into mouse nerve cells. This was accomplished by inserting genes that control the synthesis of certain proteins into the skin cells. This type of research is often successful in advancing knowledge regarding the functioning of human cells because

1 cells present in humans often function in similar ways to cells present in other organisms
2 cells from different types of organisms function differently when transplanted into humans
3 the cells in all complex organisms contain the same genes and function in similar ways
4 cellular research using mice can always be applied to human cells since all complex organisms produce the same proteins

31 ____

32 In the experimental setup below, which substance would be used to prove that the gas produced by the yeast in the vacuum bottle could change the pH of the liquid in the flask?

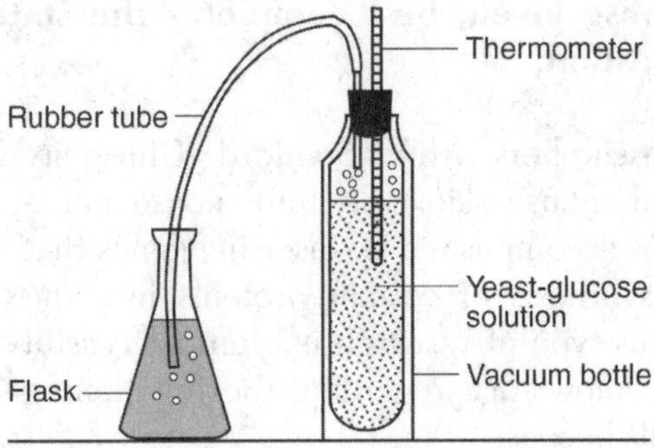

1 an indicator
2 a chemical messenger
3 an enzyme
4 a salt solution 32____

33 Which statement best expresses a basic scientific assumption?

1 Interpretation of experimental results has provided explanations for all natural phenomena.
2 If a conclusion is valid, similar investigations by other scientists should result in the same conclusion.
3 For any conclusion to be valid, the design of the experiment requires that only two groups be compared.
4 After a scientist formulates a conclusion based on an experiment, no further investigation is necessary. 33____

34 The diagram below represents a cycle that occurs in nature.

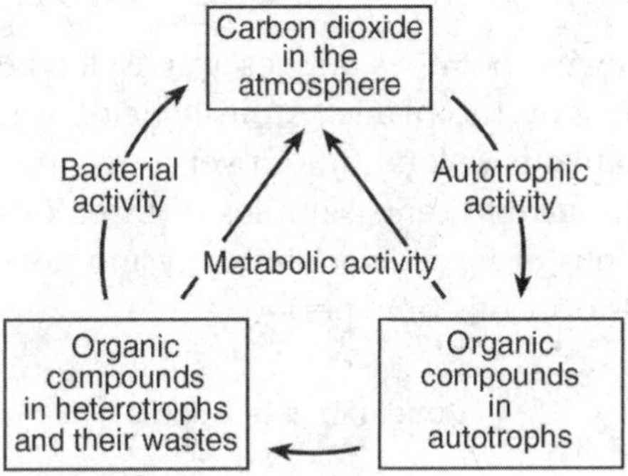

Which phrase describes a human activity that could have a *negative* effect on this cycle?

1 a decrease in the amount of sulfates given off by motor vehicles
2 an increase in recycling programs for plastics and metals
3 the continued deforestation and removal of forest resources
4 development of programs to conserve wildlife 34____

Base your answers to questions 35 through 37 on the information and graph below and on your knowledge of biology.

A farmer growing potatoes notices aphids, a type of insect, feeding on the plants. An insecticide was sprayed on the plants several times over a two-year period. The graph represents samples of three different generations of insecticide-resistant and nonresistant aphids over this time period.

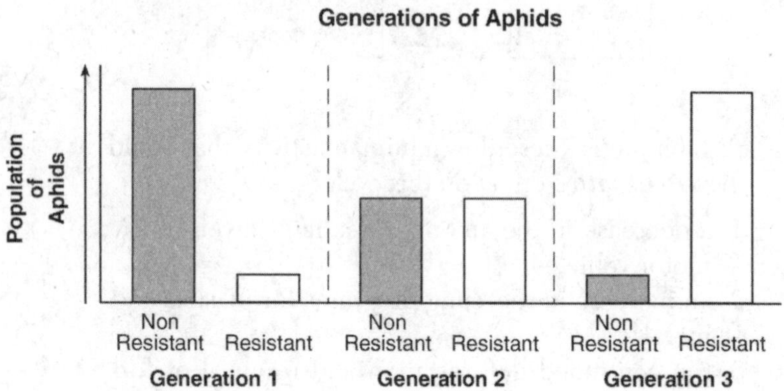

35 The resistance gene was present in the aphid population as a result of

1 the need of the potatoes to become resistant to the insecticide
2 changes in the aphids' local habitat by the insecticide
3 a recombination of the proteins in the potato cells
4 a random change in the aphids' DNA sequence 35_____

36 In year three, the farmer discontinued the use of the insecticide. Which statement would best predict the population in generation 4?

1 The nonresistant aphid would become extinct.
2 The nonresistant aphid population would likely increase.
3 The resistant aphid would mutate to a nonresistant aphid.
4 The plants would be free of insect populations. 36_____

37 One *negative* consequence of using an insecticide is that it

1 selects for insecticide-resistant organisms
2 keeps a balance of organic compounds
3 encourages biodiversity in plants
4 gives the nonresistant aphids a survival advantage 37_____

38 The activity of a single-celled organism is represented in the diagram below.

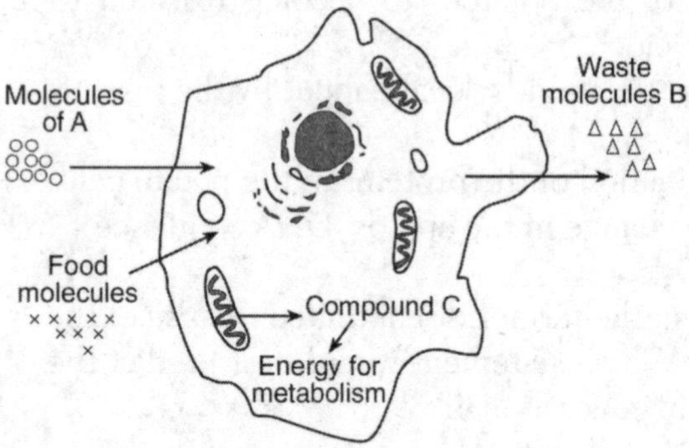

Which concept is best illustrated by this diagram?

1 The life functions performed by single-celled organisms are different from the life functions performed by complex organisms.
2 Single-celled organisms carry out life functions that are essential for survival.
3 Since single-celled organisms lack organs, they can survive only in moist environments.
4 Single-celled organisms contain one organelle that performs all the life functions. 38 ____

39 The chart below provides information about two scientific discoveries in the field of biology.

Early Discovery	Later Discovery
People living near swamps are more likely to get malaria than people who do not live near swamps. Burning swamps early in the summer reduces the amount of malaria.	Mosquitoes breed and lay their eggs in swamps and other pools of still water. Mosquitoes are the carriers of the organisms that cause malaria.
Dark-staining bodies called chromosomes can be seen only in dividing cells. The number of chromosomes doubles during cell division.	Chromosomes contain DNA, which is able to copy itself. DNA carries the genetic code, which is passed from a parent cell to two or more daughter cells.

Which statement is the best interpretation of the material presented in the chart?

1 Scientific explanations are built by combining evidence that can be observed with what people already know.
2 Inquiry involves making judgments about the reliability of the source and relevance of the information.
3 Science provides information, but values are also essential to making ethical decisions.
4 Hypotheses are valuable even if they turn out not to be true, because they may lead to further investigation. 39 _____

Base your answers to questions 40 through 43 on the diagram below and on your knowledge of biology. This diagram represents the roles of different parts of the human body in keeping blood sugar at a balanced, normal level over time.

Homeostasis of Blood Sugar Level

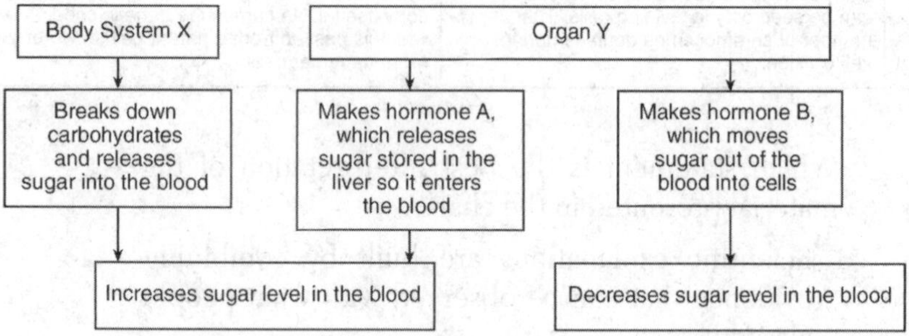

40 The diagram shows human body structures that are coordinated to maintain homeostasis. Which row correctly identifies the functions of these structures?

Row	Body System X	Organ Y
(1)	Digestion	Regulation
(2)	Circulation	Synthesis
(3)	Excretion	Transport
(4)	Locomotion	Nutrition

40 _____

41 When body system *X* releases too much sugar into the blood, the body can maintain homeostasis by making

1 more hormone *A*, only
2 more hormone *B*, only
3 more hormone *A* and more hormone *B*
4 no hormone *A* and no hormone *B* 41_____

42 If organ *Y* becomes unable to produce enough hormone *B*, then homeostasis would be disrupted. To restore homeostasis and compensate for the lack of hormone *B*, one useful action would be to

1 increase the production of hormone *A*
2 remove organ *Y* from the body surgically
3 reduce the carbohydrates in body system *X*
4 reduce the synthesis of enzymes in organ *Y* 42_____

43 If body system *X* temporarily stops releasing sugar into the blood, a likely response of the body would be to

1 stop using enzymes in body system *X*
2 stop organ Y from producing hormone *A*
3 start to increase synthesis of hormone *B*
4 start to increase synthesis of hormone *A* 43_____

PART B–2

Answer all questions in this part. [12]

Directions **(44–55): For those questions that are multiple choice, record in the space provided the *number* of the choice that, of those given, best completes each statement or answers each question. For all other questions in this part, follow the directions given and record your answers in the spaces provided.**

Base your answers to questions 44 through 47 on the information, diagram, and table below and on your knowledge of biology.

A concentrated starch solution was placed in a thistle tube with a semi-permeable membrane covering the wide opening. It was then placed in a beaker of water. The height of the solution in the tube was measured every 5 minutes for 25 minutes. The setup and the data collected are shown below.

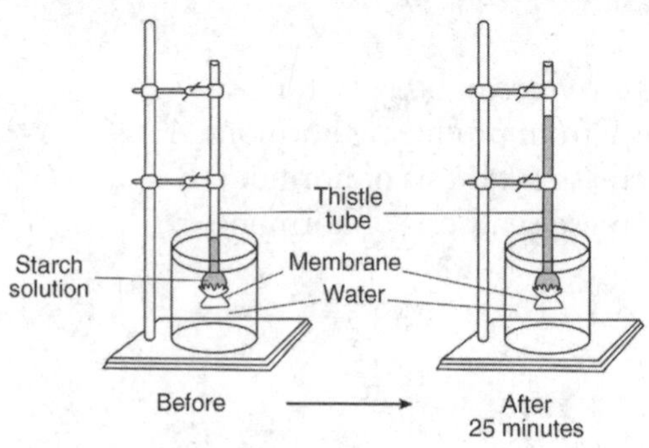

Height of Liquid in Thistle Tube

Time (min)	Height (cm)
0	2
5	3
10	6
15	8
20	10
25	11

***Directions* (44–45): Using the information in the data table, construct a line graph on the grid below, following the directions below.**

44 Mark an appropriate scale, without any breaks in the data, on each labeled axis. [1]

45 Plot the data for height on the grid. Connect the points and surround each point with a small circle. [1]

Example:

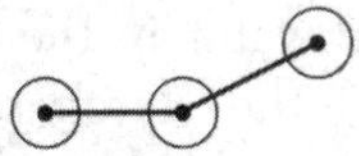

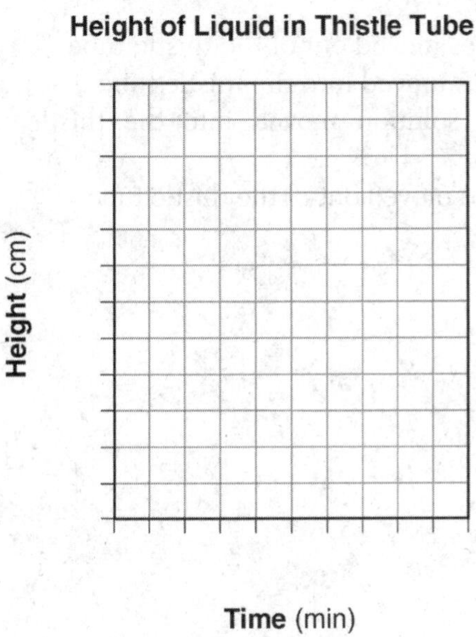

46 Explain why the height of the solution in the thistle tube increased during the 25-minute period. [1]

____

____

Note: The answer to question 47 should be recorded in the space provided.

47 The experiment was repeated, and an amber-colored solution was added to the water in the beaker. After 10 minutes, the water in the beaker remained amber-colored and the starch solution had turned blue-black. The most likely reason for this observation is that

1 starch molecules moved out of the thistle tube
2 water molecules moved into the thistle tube
3 amber-colored solution moved into the thistle tube
4 water molecules moved out of the thistle tube 47 _____

Base your answer to question 48 on the information and diagrams below and on your knowledge of biology.

When fish of certain species are injured, a chemical substance stored in skin cells of the fish is released into the water. This chemical causes an alarm response among other fish of the same species in the area. Nearby fish of this species become more alert and group together near the bottom.

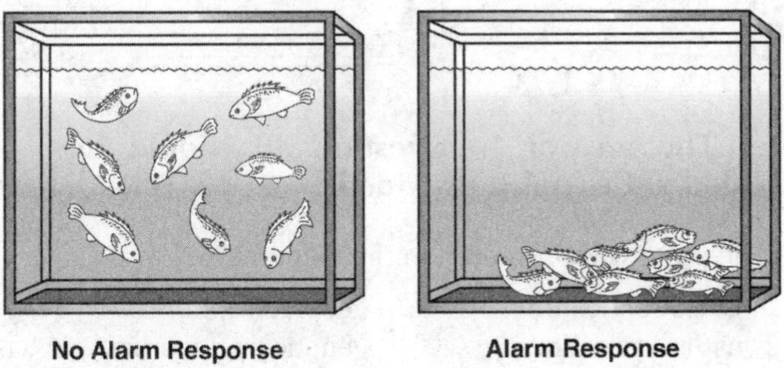

48 Explain why the chemical released from the injured fish may *not* cause an alarm response in other fish species. [1]

____

____

Base your answers to questions 49 through 51 on the diagram below and on your knowledge of biology.

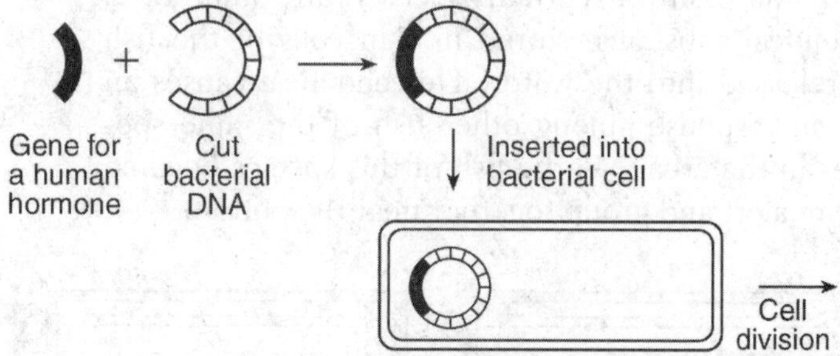

Note: The answer to question 49 should be recorded in the space provided.

49 The process represented in the diagram is

1 DNA replication
2 natural selection
3 gel electrophoresis
4 genetic engineering 49____

Note: The answer to question 50 should be recorded in the space provided.

50 The original gene for the production of a human hormone was most likely removed from a

1 chromosome
2 ribosome
3 mitochondrion
4 cell membrane 50____

51 State *one* possible reason why a gene for the production of a human hormone would be placed in bacterial DNA. [1]

____

____

Base your answers to questions 52 through 55 on the passage below and on your knowledge of biology.

The lake sturgeon is a fish that often grows over six feet long and can weigh close to two hundred pounds. It is currently an endangered species in the Great Lakes area, although the species has lived in those lakes and rivers for millions of years. Now, there is a program to increase the sturgeon population by reintroducing lake sturgeon to areas where they have disappeared.

Like the lake sturgeon, bloater fish are also found in the Great Lakes. Both find their food on or near the bottoms of lakes. They eat a variety of small organisms, including insect larvae, worms, and clams. These small organisms feed on algae.

52 Identify *one* population that will *decrease* in size after the lake sturgeon are added to the new ecosystems. Support your answer. [1]

____

____

53 Part of the food web of a lake ecosystem is represented in the diagram below. Indicate the position in the food web where each organism listed below would be placed, by writing the name of each in the appropriate box. [1]

algae
bloater fish
clams

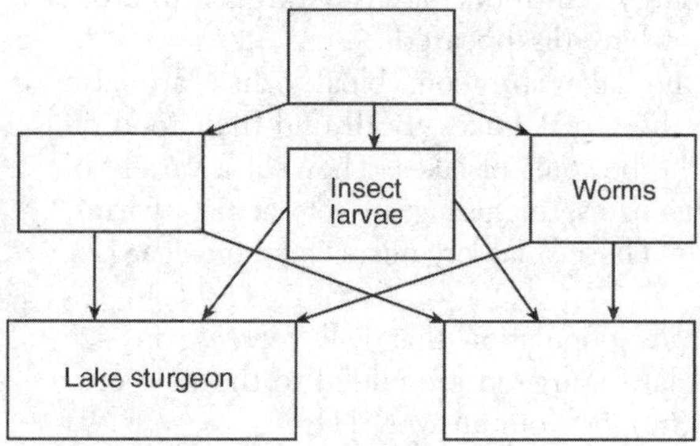

54 Identify which population, other than lake sturgeon, will increase in size after the lake sturgeon are added to the new ecosystems. Support your answer. [1]

____

____

55 State what the arrows in the food web represent. [1]

____

____

PART C

Answer all questions in this part. [17]

Directions **(56–72): Record your answers in the spaces provided.**

Base your answers to questions 56 and 57 on the information and drawing below and on your knowledge of biology. The drawing represents a salamander. Salamanders are small amphibians that live in a variety of environments.

Two species of salamander inhabit an island. The habitat on each side of the island is different. One side tends to be wet; the other side tends to be dry. Researchers want to know if the salamanders will survive equally well on either side of the island. Species *A* lives on the wet side of the island, while Species *B* lives on the dry side of the island. Researchers develop two artificial habitats, one that simulates conditions on the wet side and one that simulates conditions on the dry side.

56 Explain why researchers would put the salamanders in an artificial environment, as opposed to conducting the experiment in their natural habitat. [1]

____

____

57 Researchers put three salamanders of each species in each of the two different artificial environments. Why would other scientists question the validity of the conclusions based on this setup? [1]

____

____

Base your answers to questions 58 through 61 on the information and diagram below and on your knowledge of biology.

Green Roofs

People in Albany and New York City are using "green roofs" to improve the environment. A green roof can be added to many buildings that have large, flat roofs. Green roofs have three parts: a protective layer to separate plant roots from the roof of the building, an absorptive layer to catch and hold rainwater, and a layer of plants. Often, green roofs use *Sedum*, a short, desert plant, because it is efficient at storing water in its leaves and can withstand the colder climate.

A green roof saves energy, reduces carbon dioxide in the atmosphere, and prevents rainwater and melting snow from overloading sewer systems. It can also protect the roof of a building from damage. However, green roofs can be expensive to install, and require care and maintenance.

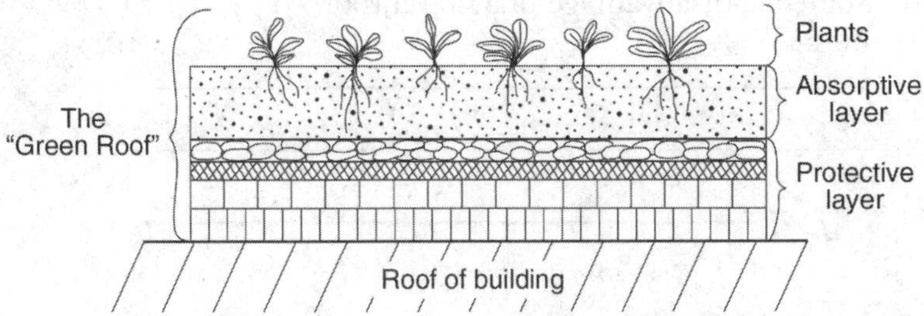

58 Most varieties of *Sedum* are not native plants in Albany or New York City. State *one* reason why it may be dangerous to introduce a new species to an established ecosystem. [1]

____

____

59 State *one* reason why a green roof reduces the amount of carbon dioxide in the atmosphere. [1]

____

____

60 State *one* reason why it is important to reduce the amount of carbon dioxide in the atmosphere. [1]

61 State *one* disadvantage of a green roof. [1]

62 A scientist took samples from a culture of *E. coli* bacteria and placed them in each of 100 petri dishes. Once the bacteria began to grow in the dishes, she exposed 50 of the dishes to x-ray radiation and 50 to natural light. After five days, she examined samples of DNA from the bacteria and recorded any differences she found between the DNA of the two groups. State *one* hypothesis the experiment would test. [1]

Base your answers to questions 63 and 64 on the passage below and on your knowledge of biology.

Super Vaccine Could Eliminate Flu

Every flu season, vaccine makers must bet on which strain of influenza A will pose the greatest threat to the public, and millions of Americans must decide whether to get a shot. In August, virologist Gary Nabel at the National Institutes of Health (NIH) announced progress toward a universal flu vaccine: two shots of it could provide years of protection from every known influenza A virus.

"We use a prime-boost strategy, meaning that we immunize with two vehicles that deliver the vaccine in different ways," Nabel says. In their experimental treatment, he and his colleagues injected mice, ferrets, and monkeys with viral DNA, causing their muscle cells to produce hemagglutinin, a protein found on the surface of all flu viruses. The animals' immune systems then began making antibodies that latch onto the protein and disable the virus. The researchers followed the DNA injection with a traditional seasonal flu shot, which contains dead viruses. This one-two punch protected the test subjects against influenza A viruses that had emerged in 1934 and 2007, and other experiments showed that the antibodies it generated successfully neutralized a wide variety of flu strains. Nabel's colleagues at the NIH are already testing similar approaches in humans.

Source: Rowe, A. "Super Vaccine Could Eliminate Flu." Discover, Jan./Feb. 2011, p. 37.

63 Identify *one* specific difference, other than it is a two-step vaccination, between Nabel's vaccination and a traditional flu vaccine. [1]

64 Explain how injecting dead or weakened viruses into a person can help to fight against future infections from that virus. [1]

Base your answers to questions 65 through 68 on the information below and on your knowledge of biology.

Bald Eagle Facts

- Bald eagles eat primarily fish, carrion (dead animals), smaller birds, and rodents. Their most important non-carrion food is fish, which they catch by swooping down and grabbing fish that are near the surface of the water.
- The number of nesting pairs in the lower 48 United States increased from fewer than 450 in the early 1960s to more than 4,500 adult bald eagle nesting pairs in the 1990s. Today, there are an estimated 9,789 nesting pairs of bald eagles.

- Bald eagles are found in large numbers in certain areas during the winter (known as roosts). These winter roosts are located in areas where prey are plentiful. Winter roosts are protected under federal law, and managed with a buffer zone to reduce human disturbance. As winter ends, the eagles return to their summer nesting/hunting areas.

Bald Eagle Research

In the winter of 2009, volunteers from an Audubon group conducted a survey of roosting bald eagles at four locations in an area in the lower Hudson River Valley. The data below show the average number of eagles sighted and the number of visits made by the volunteers each month. Among the other data collected were percent ice cover and percent cloud cover on the surface of the water. The eagles fly freely between these four sites, depending on a variety of conditions.

Some of the data are shown in the table below.

Bald Eagles Sighted at Four Hudson Valley Locations in 2009

Location	January		February		March	
	Average Number of Eagles	Number of Visits	Average Number of Eagles	Number of Visits	Average Number of Eagles	Number of Visits
Croton Reservoir	22.86	7	47.88	8	9.17	6
George's Island Park	27.00	7	18.38	8	5.00	4
George's Island North	12.29	7	4.43	7	2.20	5
Stony Point	3.57	7	3.63	8	0.00	5

65 State *one* reason why the percent ice cover is important to the ability of eagles to obtain food. [1]

66 What inference can be made about the percent ice cover at Croton Reservoir between January and February 2009? Support your answer. [1]

67 State *one* reason why the number of eagles sighted showed a change at all four sites between February and March. [1]

68 State *one* possible reason why a popular hiking trail in this area is closed during eagle roosting seasons. [1]

Base your answers to questions 69 through 72 on the information below and on your knowledge of biology.

Coral Reef Ecosystems

There are many ecological interactions that maintain the biodiversity present in coral reefs. In addition to coral, microscopic algae, seaweed, sea grasses, sponges and worms, and a variety of fish are among the organisms that live in reef ecosystems. Ocean currents often link different reef systems and move organisms from one reef area to another. This movement is a factor in repopulating a reef that has been damaged by environmental changes.

One environmental change involves an increased growth of seaweed. When the population of seaweed increases, the reef shifts from a coral-dominated ecosystem to a seaweed-dominated ecosystem. This change disrupts the relationships between the organisms that live there.

Studies have shown that, as the density of seaweed in a reef area increases, the number of fish that eat the seaweed in that area decreases. This may be due to the presence of more predators, or the taste of the more mature plants. The fish move to areas where there is less seaweed growth. As this trend continues, the reef areas are taken over by the seaweed. Once this happens, it is very hard to remove the seaweed and restore the reef to a healthy ecosystem.

In addition to this problem, temperature changes are threatening the ocean currents that connect the reef systems. A change in the currents would reduce the movement of fish larvae from one area to another. This contributes to the seaweed problem.

69 State the role of the sea grasses in the reef ecosystem. [1]

___

___

70 Identify *one* abiotic factor that is affecting the stability of the coral reef ecosystems and state how the factor identified is important to the coral reef ecosystems. [1]

Abiotic factor: ___

Effect: ___

___

71 State *one* reason why it is important to maintain the stability of the coral reefs. [1]

___

___

72 State *one* advantage of the fish larvae moving by ocean currents into a damaged reef ecosystem. [1]

___

___

PART D

Answer all questions in this part. [13]

Directions **(73–85): For those questions that are multiple choice, record in the space provided the *number* of the choice that, of those given, best completes the statement or answers the question. For all other questions in this part, follow the directions given and record your answers in the spaces provided.**

Base your answers to questions 73 and 74 on the diagram below, which represents a process that occurs in living cells, and on your knowledge of biology.

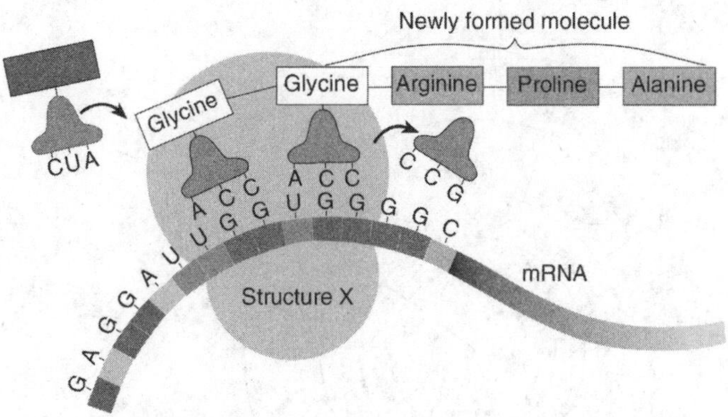

Note: The answer to question 73 should be recorded in the space provided.

73 The process shown in the diagram is

1 cellular respiration
2 cellular reorganization
3 gene recombination
4 protein synthesis

73____

Note: The answer to question 74 should be recorded in the space provided.

74 Structure *X* is a

1 mitochondrion
2 vacuole
3 nucleus
4 ribosome

74_____

Base your answers to questions 75 through 77 on the diagram below and on your knowledge of biology.

Variations in Beaks of Galapagos Islands Finches

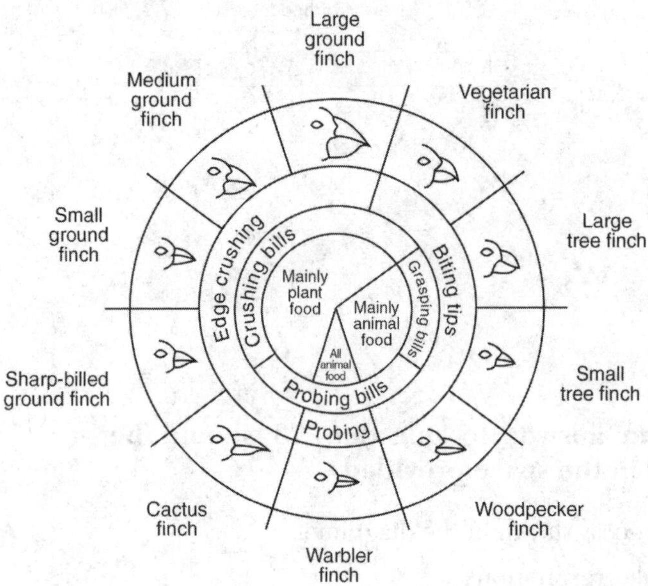

Note: The answer to question 75 should be recorded in the space provided.

75 A finch that picks small insects out from cracks in the bark of trees would most likely have a beak that is

1 sharp and thin
2 sharp and thick
3 rounded and thin
4 rounded and thick

75____

Note: The answer to question 76 should be recorded in the space provided.

76 Which statement is a basic assumption from *The Beaks of Finches* lab?

1 The type of beak indicates the type of food the finch eats.
2 Different birds have different songs.
3 Birds with larger beaks can find mates more easily.
4 Nesting behavior of finches is an inherited trait.

76____

77 State *two* reasons why the large ground finch and sharp-billed ground finch could live on the same island but not compete for food, even though they both eat mainly plant food. [1]

Reason 1: ______________________________

Reason 2: ______________________________

Base your answer to question 78 on the table below and on your knowledge of biology. The table shows which of four enzymes are present in three related plant species.

Comparison of Four Enzymes

Plant Species	Enzyme W	Enzyme X	Enzyme Y	Enzyme Z
Species A	present	present	absent	present
Species B	absent	absent	present	absent
Species C	present	present	absent	present

The tree diagrams below show two possible evolutionary relationships between the three species.

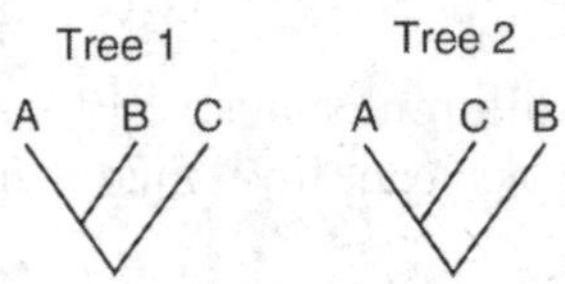

78 In the space below, write the number of the tree diagram that shows the most probable evolutionary relationship between the three species. Support your answer. [1]

Tree diagram: __________

____

____

Base your answers to questions 79 and 80 on the diagram below, which represents the shrinking of a cell in response to an increase in the concentration of a substance outside of the cell.

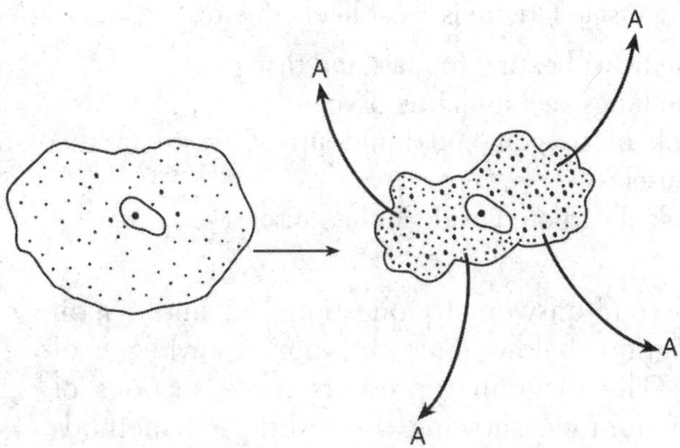

79 Identify substance *A*. [1]

____

80 Identify *one* likely substance in the environment of the cell that caused this response. [1]

____

Note: The answer to question 81 should be recorded in the space provided.

81 A student lifted weights after school and found that his muscles started to burn. He couldn't continue to lift the weights after prolonged exercising. This muscle fatigue is most likely due to

1 the heart beating too fast and tiring out
2 the lungs accumulating oxygen
3 lack of oxygen and build up of waste in the muscles
4 lack of carbon dioxide in the muscles 81____

Base your answers to questions 82 and 83 on the diagram below and on your knowledge of biology. The diagram represents three sections of a cell membrane showing three different methods involved in the transport of various molecules across the membrane.

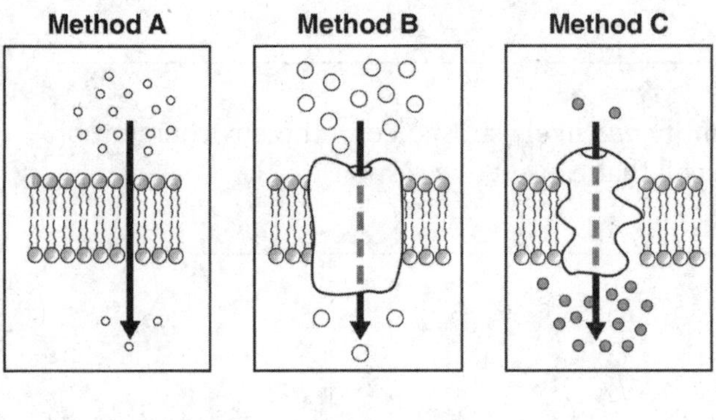

Note: The answer to question 82 should be recorded in the space provided.

82 Methods *A* and *B* are classified as methods of passive transport because they do not require

1 ATP
2 carbon dioxide
3 light
4 DNA

82____

83 Using information from the diagram, state *one* reason why the movement of molecules in method *C* represents active transport. [1]

____

____

____

84 State *one* reason why some species might have similar body structures even if they are *not* closely related. [1]

____

____

85 A student went out to the school track and walked two laps, ran two laps, and then walked two more laps. On the grid below, draw a line that shows what most likely happened to the pulse rate of the student during these activities. [1]

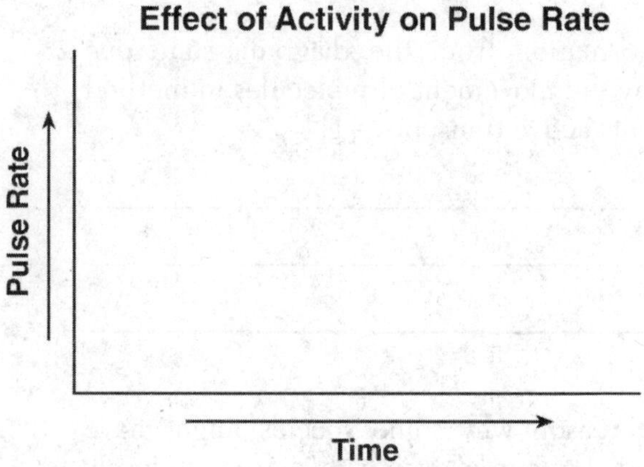

Answers August 2016

Living Environment

Answer Key

PART A

1. 4	**6.** 2	**11.** 1	**16.** 3	**21.** 2	**26.** 1
2. 3	**7.** 1	**12.** 1	**17.** 2	**22.** 4	**27.** 4
3. 3	**8.** 4	**13.** 3	**18.** 4	**23.** 4	**28.** 2
4. 1	**9.** 1	**14.** 3	**19.** 2	**24.** 3	**29.** 2
5. 2	**10.** 3	**15.** 2	**20.** 3	**25.** 4	**30.** 3

PART B–1

31. 1	**34.** 3	**36.** 2	**38.** 2	**40.** 1	**42.** 3
32. 1	**35.** 4	**37.** 1	**39.** 1	**41.** 2	**43.** 4
33. 2					

PART B–2

44. *See* Answers Explained.
45. *See* Answers Explained.
46. *See* Answers Explained.
47. 3
48. *See* Answers Explained.
49. 4
50. 1
51. *See* Answers Explained.
52. *See* Answers Explained.
53. *See* Answers Explained.
54. *See* Answers Explained.
55. *See* Answers Explained.

PART C. *See* **Answers Explained**.

PART D

73. 4
74. 4
75. 1
76. 1
77. *See* Answers Explained.
78. *See* Answers Explained.
79. *See* Answers Explained.
80. *See* Answers Explained.
81. 3
82. 1
83. *See* Answers Explained.
84. *See* Answers Explained.
85. *See* Answers Explained.

Answers Explained

PART A

1. **4** The sequence *oak tree → leaf → guard cell → chloroplast* is the one that represents structures organized from most complex to least complex. An oak tree is a living organism composed of millions of cells. A leaf is an organ of the tree composed of significantly fewer cells than the tree itself. A guard cell is one type of cell in a leaf and is specialized to regulate gas flow into and out of the leaf. A chloroplast is a cell organelle located within the guard cell that is specialized to carry out the photosynthetic reactions. Each of the structures in this sequence is less complex than the one that precedes it.

WRONG CHOICES EXPLAINED

(1), (2), (3) The sequences *chloroplast → guard cell → leaf → oak tree*, *guard cell → chloroplast → leaf → oak tree*, and *oak tree → guard cell → leaf → chloroplast* are *not* the sequences that represent structures organized from most complex to least complex. Each of these sequences has at least one structure out of order compared to the correct answer above.

2. **3** Autotrophs differ from heterotrophs in that only autotrophs *synthesize nutrients using carbon dioxide and water*. Autotrophs (also known as self-feeders) are organisms that carry on biochemical reactions (photosynthesis) that reorganize inorganic compounds (carbon dioxide and water) into organic nutrient molecules (glucose) using the energy of light to power the reactions. Heterotrophs (also known as other-feeders) are incapable of carrying on photosynthesis. Instead, they must obtain nutrients from the bodies of other living things.

WRONG CHOICES EXPLAINED

(1) It is *not* true that autotrophs differ from heterotrophs in that only autotrophs *require carbon dioxide for cellular respiration*. Carbon dioxide is a by-product, not a requirement, of cellular respiration.

(2) It is *not* true that autotrophs differ from heterotrophs in that only autotrophs *release oxygen as a product of cellular respiration*. Oxygen gas is a requirement, not a by-product, of cellular respiration.

(4) It is *not* true that autotrophs differ from heterotrophs in that only autotrophs *break down sugars to assemble other molecules*. Both autotrophs and heterotrophs reorganize the atoms of sugar molecules to synthesize other molecules.

3. **3** One likely effect of the introduction of the pythons is that *alligator populations will decline*. Pythons and alligators are predators that hunt the same prey. So pythons and alligators are in competition for that same food source. As long as that prey is abundant, completion will be minimal and there will be no immediate effect on either pythons or alligators. If the prey becomes scarce, however, pythons and alligators will compete more directly and intensely. The more successful predator organism will gradually reduce the population of the less successful predator organism. It is unclear whether alligators or pythons will prove to be the more successful predator.

[**Note:** There is not enough information provided in the question to reach any conclusive answer as to the most likely outcome of this scenario.]

WRONG CHOICES EXPLAINED

(1) It is *not* true that one likely effect of the introduction of the pythons is that *alligators will have more prey available*. As a result of interspecies competition, alligators (and pythons) will likely have less, not more, prey available.

(2) It is *not* true that one likely effect of the introduction of the pythons is that *pythons will become native to the Everglades*. The python is an invasive species native to Southeast Asia. So by definition, it cannot be native to North America.

(4) It is *not* true that one likely effect of the introduction of the pythons is that *pythons will become an endangered species*. The python appears to be an exceptionally successful invasive species in the Everglades. It is unlikely that it will become an endangered species unless drastic environmental change occurs in the Everglades that limits the population of pythons there.

4. **1** *Selecting and breeding the organisms for specific traits* is the activity that enables humans to produce new genetic combinations in other organisms. This process, known as selective breeding, is a technique used by animal and plant breeders in which organisms with desirable traits are cross-bred with the hope of producing offspring that also display those traits.

WRONG CHOICES EXPLAINED

(2) *Increasing the number of enzymes available to the organisms* is *not* the activity that enables humans to produce new genetic combinations in other organisms. Enzymes are produced in the cells of living organisms as directed by each organism's specific genetic blueprint, not by human activity.

(3) *Growing organisms that reproduce asexually* is *not* the activity that enables humans to produce new genetic combinations in other organisms. Asexual reproduction limits the production of new genetic combinations in organisms that carry it out.

(4) *Decreasing the amount of DNA in the diet of the organisms* is *not* the activity that enables humans to produce new genetic combinations in other organisms. DNA is an organic molecule that is present in all natural

organic foods. DNA in the diet does not affect the production of new genetic combinations.

5. **2** Before the cell divides, an exact copy of each chromosome is made by the process of *replication*. Replication is the exact self-duplication of genetic material (DNA) in the cell prior to cell division. The process of replication ensures that the same number of genes and the same amount of DNA are present in the daughter cells as were present in the parent cell.

WRONG CHOICES EXPLAINED

(1) It is *not* true that before the cell divides, an exact copy of each chromosome is made by the process of *genetic engineering*. Genetic engineering is a laboratory technique in which a gene for a desired trait is snipped from the DNA of a donor cell and inserted into the genome of a recipient cell. Genetic engineering is not the mechanism responsible for copying genetic material in the cell.

(3) It is *not* true that before the cell divides, an exact copy of each chromosome is made by the process of *mutation*. Mutation is a random process by which genetic material is altered by the effects of mutagenic agents. Mutation is not the mechanism responsible for copying genetic material in the cell.

(4) It is *not* true that before the cell divides, an exact copy of each chromosome is made by the process of *recombination*. Sexual reproduction involves the recombination of genes in offspring through the processes of meiosis and fertilization. Recombination is not the mechanism responsible for copying genetic material in the cell.

6. **2** The condors most likely increased in number because humans decided to *pass laws against using lead bullets to kill animals used by condors for food*. Condors are scavengers that feed on the carcasses of deer and other game animals, some of which are killed by lead bullets commonly used by human hunters. Condors have been known to accidentally ingest spent lead bullets along with the meat of these carcasses. Scientific studies of free-flying condors in California have found some birds with spent lead bullets in their digestive tracts and high levels of lead in their blood. The lead in the spent bullets was thought to have been absorbed into the blood of the condors, causing lead poisoning and eventual death. Passing laws requiring the use of lead-free ammunition is one way to help resolve this problem. This is an example of a trade-off between protecting a wild species (condors) and regulating a human activity (hunting).

WRONG CHOICES EXPLAINED

(1) It is *not* true that the condors most likely increased in number because humans decided to *produce lead-resistant condors through asexual reproduction*. Lead is a naturally occurring metal that can produce adverse effects in all living things exposed to it. Other than limiting exposure to the metal, there is

no practical laboratory procedure to produce lead resistance in condors or in any other species. Condors are birds and as such carry on sexual, not asexual, reproduction.

(3) It is *not* true that the condors most likely increased in number because humans decided to *introduce plants that didn't absorb the lead from discharged bullets*. Condors are scavengers that consume the bodies of dead animals, not plants.

(4) It is *not* true that the condors most likely increased in number because humans decided to *produce lead-resistant prey for the condors through genetic engineering*. Lead is a naturally occurring metal that can produce adverse effects in all living things exposed to it. Other than limiting exposure to the metal, there is no practical laboratory procedure to produce lead resistance in condors or in any other species. The laboratory technique of genetic engineering would have no practical effect on lead toxicity in prey animals and therefore would not reduce lead's toxicity in condors.

7. **1** In humans, the maintenance of a stable internal temperature is a direct result of *detection of and reaction to stimuli in the environment*. The nervous system detects the temperature of the blood (internal environment) and sends command signals to effectors, including the sweat glands and muscles. The effectors act to increase or decrease the temperature of the blood as required to maintain a body temperature of 98.6°F (37°C). This mechanism is known as a feedback loop.

WRONG CHOICES EXPLAINED

(2), (3), (4) In humans, the maintenance of a stable internal temperature is *not* a direct result of *digestion of starches and absorption of protein from the internal environment*, *diffusion of water and excretion of glucose to the external environment*, or *transport of ATP and locomotion through the environment*. None of these activities would have any direct effect on the maintenance of body temperature in humans. They are all nonsense distracters.

8. **4** *Biological catalysts* are the molecules needed to cut and copy segments of DNA. These catalysts, also known as restriction enzymes, are used by scientists to snip selected genes from donor DNA and insert the genes into recipient DNA. This process transfers genetic traits from one organism to another organism.

WRONG CHOICES EXPLAINED

(1) *Reproductive hormones* are *not* the molecules needed to cut and copy segments of DNA. Reproductive hormones regulate the reproductive processes of humans and other animals. Reproductive hormones are not used in the cutting or copying of DNA.

(2) *Carbohydrates* are *not* the molecules needed to cut and copy segments of DNA. Carbohydrates are a type of organic compound that includes sugars and starches. Carbohydrates are not used in the cutting or copying of DNA.

(3) *Antibodies* are *not* the molecules needed to cut and copy segments of DNA. Antibodies are proteins produced by the body's immune system in response to a specific antigen. Antibodies are not used in the cutting or copying of DNA.

9. **1** For evolution to occur, there must be *variations within a species*. Favorable variations give some individuals an adaptive advantage over others in a changing environment. Individuals with favorable variations will survive to reproduce offspring carrying genes for the same favorable variation. As a result, the frequency of the genes controlling those favorable variations will tend to increase in the gene pool. This action is the basis for biological evolution.

WRONG CHOICES EXPLAINED

(2) It is *not* true that for evolution to occur, there must be *extinction of the species*. Extinction is the complete and permanent elimination of a species from Earth. Extinct species cannot evolve.

(3) It is *not* true that for evolution to occur, there must be *asexual reproduction*. Asexual reproduction is a type of reproduction that results in stable, slowly changing gene pools. Asexually reproducing species can evolve. However, the degree of genetic variation of such species is less than in sexually reproducing species. As a result, evolution of asexually reproducing species is usually slow by comparison to sexually reproducing species.

(4) It is *not* true that for evolution to occur, there must be *no change in the genes of an organism*. If genes are not altered (mutated), there can be no variation within a species. Species whose genes do not change and for which there is no variation cannot evolve.

10. **3** The cells labeled *A* and *B* are examples of cells known as *white blood cells*. White blood cells are components of the blood tissue of humans and other animals. These cells are specialized to protect the body from foreign invaders, such as pathogenic organisms. Cell *A*, known as a phagocyte, is capable of seeking out and engulfing foreign substances in the blood fluid. Cell *B*, known as a lymphocyte, is capable of detecting the antigens on the surfaces of pathogenic organisms and producing specific antibodies to neutralize those pathogens.

WRONG CHOICES EXPLAINED

(1) The cells labeled *A* and *B* are *not* examples of cells known as *guard cells*. A guard cell is one type of cell in a leaf. It is specialized to regulate gas flow into and out of the leaf. Guard cells neither engulf foreign substances nor produce antibodies.

(2) The cells labeled *A* and *B* are *not* examples of cells known as *reproductive cells*. Reproductive cells, such as sperm cells and egg cells, are specialized to carry out fertilization in sexually reproducing species. Reproductive cells neither engulf foreign substances nor produce antibodies.

(4) The cells labeled *A* and *B* are *not* examples of cells known as *specialized skin cells*. Specialized skin cells, such as hair follicles and sweat glands, are specialized to function in the protection and regulation of the body's internal environment. Skin cells neither engulf foreign substances nor produce antibodies.

11. **1** This type of molecule is composed of a sequence of *amino acids arranged in a specific order*. Proteins are complex organic molecules composed of amino acid molecules linked in a particular sequence as determined by the DNA template in the nucleus of a cell.

WRONG CHOICES EXPLAINED

(2), (3), (4) This type of molecule is *not* composed of a sequence of *simple sugars alternating with starches arranged in a folded pattern*, *large inorganic subunits that form chains that interlock with each other*, or *four bases that make up the folded structure*. No such biochemical structures exist in nature. These are nonsense distracters.

12. **1** *Reproductive technology has medical, agricultural, and ecological applications* is the statement that summarizes these three actions. All three of these actions relate to reproductive processes in humans or other organisms. These actions fall into medical ("diagnose and treat fetal problems" or "insects can be sterilized to prevent mating"), agricultural ("produce plants resistant to drought" or "insects can be sterilized to prevent mating"), and ecological (no specific example provided) applications.

WRONG CHOICES EXPLAINED

(2) *Development is a highly regulated process involving mitosis and differentiation* is *not* the statement that summarizes these three actions. No mention is made in the list concerning cell division, development, or differentiation.

(3) *Reproduction and development are subject to environmental effects* is *not* the statement that summarizes these three actions. No mention is made in the list concerning environmental effects on the reproductive process.

(4) *Human development, birth, and aging should be viewed as a predictable pattern of events* is *not* the statement that summarizes these three actions. No mention is made in the list concerning the predictability of human development, birth, or aging.

13. **3** Natural selection is best described as *differences in survival rates as a result of different inherited characteristics*. Natural selection, sometimes referred to as "survival of the fittest," postulates that certain members of a species are better adapted to their environment than others. These better-

adapted members are able to survive adverse conditions at higher rates. They are therefore more likely to pass on their favorable adaptations to succeeding generations than are less-well-adapted members. These factors act together to increase the proportion of individuals displaying highly adaptive traits.

WRONG CHOICES EXPLAINED

(1), (4) Natural selection is *not* best described as *a change in an organism in response to a need of that organism* or *inheritance of characteristics acquired during the life of an organism*. These concepts closely mirror the discredited theory of "acquired characteristics" forwarded by Jean Lamarck in the early 19th century (1801). Natural selection does not occur as a function of need or acquisition of characteristics.

(2) Natural selection is *not* best described as *a process of nearly constant improvement that leads to an organism that is nearly perfect*. Even highly adapted organisms retain characteristics that make them susceptible to drastic changes in environmental conditions. Natural selection does not occur as a function of perfection.

14. **3** *A deer and a rabbit consume grass in a field* is the statement that best describes a situation where competition occurs in an ecosystem. Interspecies competition occurs when members of two different species (deer and rabbits) utilize the same limited resource (grass) in the same environment (field). As long as the resource is abundant, competition is minimized. When the resource becomes scarce, competition intensifies until one of the two species is eliminated from the environment.

WRONG CHOICES EXPLAINED

(1) *A deer outruns an attacking wolf* is *not* the statement that best describes a situation where competition occurs in an ecosystem. This situation is an example of a predator-prey relationship in a natural environment, not interspecies competition.

(2) *A deer, during the winter, consumes tree bark* is *not* the statement that best describes a situation where competition occurs in an ecosystem. This situation is an example of coping behaviors in response to environmental stress, not interspecies competition.

(4) *A deer and a rabbit are both startled by a hawk flying overhead* is *not* the statement that best describes a situation where competition occurs in an ecosystem. This situation is an example of similar stimulus-response behaviors in the deer and rabbit, not interspecies competition.

15. **2** Her children will not inherit this red hair color because the woman does not have *genes for red hair in her sex cells*. Dyeing one's hair a different color is an example of an acquired characteristic. Acquired characteristics cannot be passed on to future generations because these characteristics do not

usually have a genetic basis in the person acquiring them. Only characteristics controlled by genes that occur in sex cells can be passed on to future generations.

WRONG CHOICES EXPLAINED

(1) It is *not* true that her children will not inherit this red hair color because the woman does not have *genes for red hair in her skin*. It is possible that a mutation for red hair color may be present in some of this person's skin cells. However, genes in skin cells cannot be passed on to future generations.

(3), (4) It is *not* true that her children will not inherit this red hair color because the woman does not have *proteins for red hair in the placenta* or *proteins for red hair in her egg cells*. Proteins for red hair color would be present in the skin cells and hair of a person only if the DNA in those cells contains genes for red hair color and if the genes are switched on. These proteins would not be produced in placental cells or in egg cells because the genes for red hair color would be switched off in those cells.

16. **3** Fossils provide evidence that *many species of organisms that lived long ago are now extinct*. Scientific study of fossils reveals many thousands of species that existed in the past that can no longer be found on Earth. It is inferred that these species succumbed to environmental changes that made their survival impossible and, once extinct, can never be recreated.

WRONG CHOICES EXPLAINED

(1) It is *not* true that fossils provide evidence that *life on Earth millions of years ago was more complex than life is today*. Scientific study of fossils reveals a general pattern of increasing complexity from life's origins on Earth 3.7 billion years ago to the present day.

(2) It is *not* true that fossils provide evidence that *the changes that will occur in species in the future are easy to predict*. Because environmental and genetic changes cannot be predicted, making a sound scientific prediction of how evolutionary change will progress in the future is impossible.

(4) It is *not* true that fossils provide evidence that *most species of organisms that lived long ago are exactly the same today*. Scientific study of fossils reveals that modern-day species of organisms are generally very different compared to species that lived millions of years ago.

17. **2** For the frigatebird, this behavior has most likely resulted in *greater reproductive success*. We can infer from the inclusion of both male and female frigatebirds in this scenario that the behaviors described represent a genetically programmed mating display. Such displays are common in males of many bird species. These displays are known to stimulate nesting and mating responses in the female birds. Such mating displays and the resulting mating/nesting responses help to ensure successful reproduction in, and continuation of, these species.

WRONG CHOICES EXPLAINED

(1), (3) For the frigatebird, this behavior has *not* most likely resulted in *hiding from predators* or *locating new sources of food*. No information is presented in this scenario that would lead a reasonable observer to draw these inferences.

(4) For the frigatebird, this behavior has *not* most likely resulted in *reduced population growth*. These behaviors are likely to result in increased, not decreased, populations of frigatebirds.

18. **4** Plant species *X* could survive if the plants *have a few members of the population that are fungus-resistant*. If fungus-resistant plants mate, their offspring have a high probability of inheriting and displaying the same fungus resistance. Assuming other environmental factors do not intervene, such resistant plants could form a colony that would perpetuate the species in the changing environment.

WRONG CHOICES EXPLAINED

(1) It is *not* true that plant species *X* could survive if the plants *try to mutate quickly and synthesize new proteins*. Mutation is a random event that cannot be controlled at will by species for particular purposes.

(2) It is *not* true that plant species *X* could survive if the plants *are watered often and fertilized with extra nutrients*. Species *X* is adapted to a hot, dry environment. So the species would be unlikely to respond well to being watered and fertilized extensively.

(3) It is *not* true that plant species *X* could survive if the plants *can adapt to the new conditions by mating with the fungus*. It is impossible for members of different species to mate and produce fertile offspring. In addition, fungi reproduce by asexual, not sexual, means.

19. **2** This type of production of offspring is a form of *asexual reproduction*. Asexual reproduction carried on by yeast is a form of mitotic cell division known as budding. Mitotic cell division is characterized by replication (exact self-duplication) of genetic material (DNA) followed by separation of the replicated material into two daughter nuclei/cells. The resulting daughter cells are exact genetic duplicates of each other and of the parent cell. This process helps ensure genetic continuity in the yeast from one generation to the next.

WRONG CHOICES EXPLAINED

(1) This type of production of offspring is *not* a form of *sexual reproduction*. Sexual reproduction involves the fusion of a haploid egg cell and a haploid sperm cell to produce a diploid zygote. This process does not describe budding in yeast cells.

(3), (4) This type of production of offspring is *not* a form of *gene manipulation* or *genetic engineering*. These terms are essentially synonymous for laboratory techniques that involve the use of restriction enzymes to snip genes or gene fragments from the genomes of donor cells and insert the genes into the genomes of recipient cells. These processes do not describe budding in yeast cells.

20. **3** A human cell that contains all of the information necessary for the growth and development of a complete organism is *a zygote*. A zygote, or fertilized egg, is a diploid ($2n$) cell that results from the fusion of a haploid (n) sperm cell and a haploid (n) egg cell (both gametes). The zygote contains the full set of chromosomes (46) and 100% of the genetic information that describes the entire range of characteristics for a unique human being.

WRONG CHOICES EXPLAINED

(1), (2), (4) It is *not* true that a human cell that contains all of the information necessary for the growth and development of a complete organism is *a sperm cell*, *a gamete*, or *an egg cell*. These cells contain the haploid (n) number of chromosomes (23) and only 50% of the genetic information needed to describe the range of characteristics for a unique human being.

21. **2** *During the early stages of pregnancy* is the time when exposure to a potentially harmful substance would be most likely to damage many organs in a developing embryo. Scientific study of human development seems to demonstrate that the embryo is most at risk from exposure to harmful substances (e.g., alcohol, tobacco, and drugs) in the first trimester (months 1–3) of pregnancy. However, these substances are known to cause potential damage to the reproductive process at other stages, including during the formation of gametes, at fertilization, and at the later stages of pregnancy. For that reason, these substances should be avoided.

WRONG CHOICES EXPLAINED

(1), (3), (4) *During the last three months of pregnancy*, *during the formation of the zygote*, and *during meiosis in both males and females* are *not* the times when exposure to a potentially harmful substance would be most likely to damage many organs in a developing embryo. The potential for embryonic damage from exposure to potentially harmful substances is marginally less likely at these stages of the reproductive process, although damage may still occur.

[**Note:** In the opinion of the author, this question goes well beyond the scope of the *Living Environment* curriculum. Although students should know that a growing body of scientific data indicates that exposure to certain harmful substances can be detrimental to embryonic development at almost any stage of the reproductive process, asking them to differentiate the relative danger of exposure during these various stages is not supported by the *Living Environment* Standards, Key Ideas, and Performance Indicators.]

22. **4** Structure *D* is the structure within which the placenta develops. Structure *D* represents the uterus, where the fertilized egg becomes embedded and where embryonic development occurs. The placenta is a structure composed of both maternal and embryonic tissues that supplies the embryo with dissolved food and oxygen and that removes metabolic wastes during development.

WRONG CHOICES EXPLAINED

(1) Structure *A* is *not* the structure within which the placenta develops. Structure *A* represents the oviduct/Fallopian tube, through which the egg travels after ovulation and where fertilization and early development occur.

(2) Structure *B* is *not* the structure within which the placenta develops. Structure *B* represents the ovary, where egg cells mature before being released into the oviduct.

(3) Structure *C* is *not* the structure within which the placenta develops. Structure *C* represents the vagina/birth canal, through which sperm enter the female reproductive tract and through which the baby passes during birth.

23. **4** An energy-rich organic compound needed by organisms is *glucose*. Glucose is an organic molecule (formula $C_6H_{12}O_6$) that is utilized universally by living things on Earth to supply the energy needed for metabolic activities. This energy is held primarily in the carbon-carbon bonds of the molecule and is released through the life process of respiration.

WRONG CHOICES EXPLAINED

(1), (2), (3) It is *not* true that an energy-rich organic compound needed by organisms is *water*, *salt*, or *oxygen*. None of these substances is organic, and none is energy rich in terms of normal metabolic activity.

24. **3** A possible symptom of SCIDS would be an increase in the *number of infections by pathogens*. Because SCIDS is known to attack T-cells and because T-cells are known to function in the immune response in humans, it logically follows that SCIDS would adversely affect the body's response to infectious agents (pathogens) such as viruses and bacteria.

WRONG CHOICES EXPLAINED

(1), (2), (4) It is *not* true that a possible symptom of SCIDS would be an increase in the *number of antigens produced*, *red blood cell count*, or *ability to maintain homeostasis*. No information is provided in the passage that indicates that SCIDS results in an increase in antigen production, red blood cell count, or ability to maintain homeostasis.

25. **4** If this is not done, the *immune system might attack the donated organ*. The human immune system is capable of detecting foreign antigens present in the body and producing specific antibodies designed to neutralize

and destroy those antigens and the cells that carry them. Except in cases of identical twins, donated organs/tissues always contain some foreign antigens. So these organs/tissues must be tested to determine the degree of compatibility in order to minimize the risk of rejection by the recipient's immune system.

WRONG CHOICES EXPLAINED

(1), (2) It is *not* true that if this is not done, the *donated organ might attack the body* or the *donated organ might attack the immune system*. The donated organ lacks an immune system. So it is not capable of detecting foreign antigens or of producing antibodies.

(3) It is *not* true that if this is not done, the *immune system might attack its own body cells*. Under normal circumstances, the human immune system does not attack its own body cells because the antigens present in the body are not foreign antigens.

26. **1** *There were more niches and greater stability in the ecosystem* is the set of observations that represents evidence for this claim. Biodiversity is a measure of the number of different species that inhabit an ecosystem. Biodiverse ecosystems are characterized by a large number of niches (roles) filled by many different species. Biodiverse ecosystems tend to be more stable under environmental stresses than ecosystems that lack biodiversity. More niches and greater stability both indicate that biodiversity increased in that ecosystem.

WRONG CHOICES EXPLAINED

(2) *There were more niches and less energy lost as heat in the ecosystem* is *not* the set of observations that represents evidence for this claim. More niches indicate increasing biodiversity in an ecosystem. Less energy lost as heat is an indication of less metabolic activity (fewer species) and indicates that biodiversity decreased, not increased, in that ecosystem.

(3) *There were fewer niches for decomposers and less stability* is *not* the set of observations that represents evidence for this claim. Fewer niches and less stability in an ecosystem both indicate that biodiversity decreased, not increased, in that ecosystem.

(4) *There were fewer niches for consumers and greater cycling of materials* is *not* the set of observations that represents evidence for this claim. Greater recycling of materials such as carbon and nitrogen is an indication of increasing biodiversity in an ecosystem. Fewer niches indicate that biodiversity decreased, not increased, in that ecosystem.

27. **4** The final consumers in many food webs are *carnivores*. Carnivores (meat eaters) are animals that consume the bodies of other animals for food. Carnivores are considered to be top-order consumers because they are not generally taken as food by any other animal.

WRONG CHOICES EXPLAINED

(1) It is *not* true that the final consumers in many food webs are *autotrophs*. Autotrophs (self-feeders) are producer, not consumer, organisms.

(2) It is *not* true that the final consumers in many food webs are *hosts*. Hosts are organisms that are infected by parasites. Hosts may be plants or animals. Some hosts may also be top-order consumers.

(3) It is *not* true that the final consumers in many food webs are *herbivores*. Herbivores (plant eaters) are animals that consume plants for food. Herbivores are considered to be first-order, not top-order, consumers.

28. **2** These observed changes resulted directly from *ecological succession*. Ecological succession refers to the series of changes that occur to the dominant plant community of an area that has been significantly affected by an environmental change (in this case, farm activity). Over time, several different plant communities inhabit the area, each replacing the last until a stable climax forest community is established.

WRONG CHOICES EXPLAINED

(1) These observed changes did *not* result directly from *evolutionary change*. Biological evolution is a process by which members of a species displaying favorable characteristics survive harsh environmental conditions to reproduce and pass on their favorable characteristics to their offspring. Evolutionary change is not the process described in the question.

(3) These observed changes did *not* result directly from *loss of biodiversity*. Biodiversity is a measure of the number of different species that inhabit an ecosystem. The question describes an ecosystem in which biodiversity is being gained, not lost.

(4) These observed changes did *not* result directly from *selective breeding*. Selective breeding is a technique used by animal and plant breeders in which organisms with desirable traits are cross-bred with the hope of producing offspring that also display those traits. Selective breeding is not the process described in the question.

29. **2** This would interfere most directly with the ability of the lawn ecosystem to *recycle nutrients*. Decomposers, including bacteria and fungi, occupy an ecological niche in which they break down complex organic molecules (such as cellulose) into simple sugars and inorganic compounds. In this role, decomposers play an important role in the recycling of carbon, hydrogen, oxygen, and other materials in the environment. If decomposers cannot live in an ecosystem, such recycling cannot occur effectively, if at all.

WRONG CHOICES EXPLAINED

(1) It is *not* true that this would interfere most directly with the ability of the lawn ecosystem to *recycle energy*. Energy does not recycle in a natural environment. Instead, energy in the natural environment flows in one direction

from light energy from the Sun, to chemical bond energy organic molecules that make up living things, to heat energy that is dissipated into the environment. A lawn is made up of green plants that absorb light and convert it to chemical bond energy via photosynthesis.

(3) It is *not* true that this would interfere most directly with the ability of the lawn ecosystem to *maintain atmospheric pH*. The pH (acidity) of the atmosphere is a function of sulfur and nitrogen oxides dissolved in atmospheric moisture (humidity, fog, rain, and snow). A lawn does not regulate any aspect of atmospheric pH.

(4) It is *not* true that this would interfere most directly with the ability of the lawn ecosystem to *reduce biodiversity*. Biodiversity is a measure of the number of different species that inhabit an ecosystem. By definition, a lawn is a monocrop displaying little or no biodiversity. A lawn does not regulate any aspect of biodiversity.

30. **3** One benefit of this technology is that it *reduces dependence on fossil fuels*. Geothermal technology is one of several alternate energy sources that utilize nonpolluting and renewable sources to produce heat or electrical energy for human uses. The more that these alternate energy sources can be tapped for human use, the less that fossil fuels (coal, gas, and oil) will have to be used to produce the same amount of energy.

WRONG CHOICES EXPLAINED

(1), (2), (4) It is *not* true that one benefit of this technology is that it *contributes to global warming*, *reduces the ozone shield*, or *decreases resources for many species*. Each of these answers represents a negative effect, not a benefit, of any technology on the environment. In fact, competent scientists hold the use of fossil fuels at least partially responsible for each of these negative outcomes.

PART B–1

31. **1** This type of research is often successful in advancing knowledge regarding the functioning of human cells because *cells present in humans often function in similar ways to cells present in other organisms*. Genetic research has shown that complex living things share a large majority of genes in common. As mammals, humans share an even greater percentage of genes in common with other mammals, including mice. It follows, then, that cells that derive their functions from the genes they carry function similarly in humans and other mammals. Because of this fact, research performed using mice and other mammals can provide data useful for human health.

WRONG CHOICES EXPLAINED

(2) It is *not* true that this type of research is often successful in advancing knowledge regarding the functioning of human cells because *cells from different types of organisms function differently when transplanted into humans*. Cells of different organisms cannot be transplanted into humans without triggering the human immune system, which would produce antibodies to counteract the foreign antigens present on those foreign cells.

(3) It is *not* true that this type of research is often successful in advancing knowledge regarding the functioning of human cells because *the cells in all complex organisms contain the same genes and function in similar ways*. Although a majority of genes are held in common among all complex living things, it is not true that these genes function in similar ways in all organisms.

(4) It is *not* true that this type of research is often successful in advancing knowledge regarding the functioning of human cells because *cellular research using mice can always be applied to human cells since all complex organisms produce the same proteins*. Although complex organisms and humans share many genes in common, they do not all produce the same proteins. So research involving mice cannot always be applied to humans.

32. **1** *An indicator* is the substance that would be used to prove that the gas produced by the yeast in the vacuum bottle could change the pH of the liquid in the flask. An acid-base indicator, such as phenolphthalein or bromothymol blue, in the flask would indicate a change of pH (acidity), if it occurred, by means of a visible color change.

WRONG CHOICES EXPLAINED

(2) *A chemical messenger* is *not* the substance that would be used to prove that the gas produced by the yeast in the vacuum bottle could change the pH of the liquid in the flask. Chemical messengers might include neurotransmitters such as acetylcholine. Chemical messengers cannot be used to determine the pH of a mixture.

(3) *An enzyme* is *not* the substance that would be used to prove that the gas produced by the yeast in the vacuum bottle could change the pH of the liquid in the flask. Enzymes, such as pepsin, are organic catalysts that regulate the speed of chemical reactions in the body. Enzymes cannot be used to determine the pH of a mixture.

(4) *A salt solution* is *not* the substance that would be used to prove that the gas produced by the yeast in the vacuum bottle could change the pH of the liquid in the flask. A mixture of water and a salt, such as sodium chloride, constitutes a salt solution. Salt solutions cannot be used to determine the pH of a mixture.

33. **2** *If a conclusion is valid, similar investigations by other scientists should result in the same conclusion* is the statement that best expresses a basic

scientific assumption. The repeatability of experimental results through peer review is a fundamental principle of all scientific investigations.

WRONG CHOICES EXPLAINED

(1) *Interpretation of experimental results has provided explanations for all natural phenomena* is *not* the statement that best expresses a basic scientific assumption. In fact, scientific investigations have provided explanations for only a small fraction of all natural phenomena. Much more remains to be discovered compared to what has already been learned.

(3) *For any conclusion to be valid, the design of the experiment requires that only two groups be compared* is *not* the statement that best expresses a basic scientific assumption. As long as proper scientific methodology is followed, any number of experimental groups may be compared.

(4) *After a scientist formulates a conclusion based on an experiment, no further investigation is necessary* is *not* the statement that best expresses a basic scientific assumption. The repeatability of experimental results though peer review is a fundamental principle of all scientific investigations.

34. **3** *The continued deforestation and removal of forest resources* is the phrase that describes a human activity that could have a negative effect on this cycle. The diagram describes essential aspects of the carbon cycle that occur in nature and that depend on robust interactions among plants, animals, and decomposers. A critical part of this cycle is the role of green plants (such as forest trees) in which carbon dioxide is removed from the atmosphere and incorporated (fixed) into complex carbon compounds such as cellulose. Deforestation and removal of timber resources are human activities that can seriously interfere with this process. They reduce carbon-fixation capacity and remove complex carbon compounds from the environment.

WRONG CHOICES EXPLAINED

(1), (2), (4) *A decrease in the amount of sulfates given off by motor vehicles*, *an increase in recycling programs for plastics and metals*, and *development of programs to conserve wildlife* are *not* the phrases that describe human activities that could have a negative effect on this cycle. Each of these phrases describes a human activity that could have a positive, not a negative, effect on this cycle and/or on environmental quality in general.

35. **4** The resistance gene was present in the aphid population as a result of *a random change in the aphids' DNA sequence*. Mutations are random events in which genetic material is altered by mutagenic agents. Mutations can be passed on to future generations only if they occur in gametic tissues that actually participate in the formation of offspring. If the phenotype of a mutation provides the individual with some adaptive advantage in a changing environment, the mutation may be passed on to future generations through the reproductive process. In this case, it is probable that a random mutation

provided an individual with a phenotype of resistance to insecticide. This gene then probably increased in frequency in the population when most nonresistant aphids were killed by the insecticide but the resistant aphids were not. The resistant aphids survived and mated with other resistant aphids to produce an insecticide-resistant population of aphids.

WRONG CHOICES EXPLAINED

(1) It is *not* true that the resistance gene was present in the aphid population as a result of *the need of the potatoes to become resistant to the insecticide*. Mutation does not occur as a function of need in any species. Insecticides are toxic to insects such as the aphid; they are not toxic to potatoes.

(2) It is *not* true that the resistance gene was present in the aphid population as a result of *changes in the aphids' local habitat by the insecticide*. Although the presence of the insecticide did change the aphids' local habitat by introducing selection pressure, this change did not create the resistance gene in the aphids.

(3) It is *not* true that the resistance gene was present in the aphid population as a result of *a recombination of the proteins in the potato cells*. Changes in the potatoes' biochemistry did not create the resistance gene in the aphids.

36. **2** *The nonresistant aphid population would likely increase* is the statement that would best predict the population in generation 4. Without the selection pressure of the insecticide, more nonresistant aphids would likely survive to mate and produce offspring that were also nonresistant. This would have the effect of increasing the proportion of nonresistant aphids in the population.

WRONG CHOICES EXPLAINED

(1) *The nonresistant aphid would become extinct* is *not* the statement that would best predict the population in generation 4. It is very likely that unless the resistant gene provides an adaptive disadvantage to some other environmental condition, it will remain in the aphid's gene pool and that insecticide resistance will continue to be a part of the phenotype of the aphid population.

(3) *The resistant aphid would mutate to a nonresistant aphid* is *not* the statement that would best predict the population in generation 4. Mutation is a random event in which a gene is altered due to the effect of a mutagenic agent. So it is extremely unlikely that the mutated gene will receive an alteration that would reverse the insecticide-resistant phenotype.

(4) *The plants would be free of insect populations* is *not* the statement that would best predict the population in generation 4. It is likely that in the absence of the insecticide, the aphids would continue to form populations on the plants.

37. **1** One *negative* consequence of using an insecticide is that it *selects for insecticide-resistant organisms*. By selecting for those individuals that carry the gene for insecticide resistance, the insecticide helps to ensure that

insecticide-resistant aphids will survive at a higher rate than nonresistant aphids. This action has the consequence that the gene for insecticide resistance will increase in frequency in the aphid gene pool, making future generations more and more resistant to the insecticide and compounding the problem of aphid infestations.

WRONG CHOICES EXPLAINED

(2) It is *not* true that one *negative* consequence of using an insecticide is that it *keeps a balance of organic compounds*. This answer has no meaning in modern science. This is a nonsense distracter.

(3) It is *not* true that one *negative* consequence of using an insecticide is that it *encourages biodiversity in plants*. The insecticide would have no known effect on the biodiversity of plants in this scenario.

(4) It is *not* true that one *negative* consequence of using an insecticide is that it *gives the nonresistant aphids a survival advantage*. In fact, the nonresistant aphids will be killed by the insecticide, providing them with a distinct disadvantage compared to resistant aphids.

38. **2** *Single-celled organisms carry out life functions that are essential for survival* is the concept that is best illustrated by the diagram. The diagram depicts a single-celled organism as it simultaneously carries out multiple life functions utilizing its specialized organelles and enzymes for this purpose. Organelles shown include the cell membrane, nucleus, endoplasmic reticulum, vacuole, and mitochondrion. Life functions illustrated include respiration, absorption, and excretion.

WRONG CHOICES EXPLAINED

(1) *The life functions performed by single-celled organisms are different from the life functions performed by complex organisms* is *not* the concept that is best illustrated by the diagram. In fact, the life functions carried out by single-celled organisms are nearly exactly the same as those carried out by complex multicellular organisms, including humans.

(3) *Since single-celled organisms lack organs, they can survive only in moist environments* is *not* the concept that is best illustrated by the diagram. Most single-celled organisms live in moist environments because they lack a means to prevent water loss in dry environments. In complex organisms, the organ known as the skin functions to reduce water loss.

[**Note:** This is a possible correct answer to the question as written, but it is not the best answer.]

(4) *Single-celled organisms contain one organelle that performs all the life functions* is *not* the concept that is best illustrated by the diagram. In fact, single-celled organisms contain several types of organelles, each type specialized to perform specific life functions on behalf of the cell.

39. **1** *Scientific explanations are built by combining evidence that can be observed with what people already know* is the statement that is the best interpretation of the material presented in the chart. The chart presents information about "early discoveries" made by scientists who had incomplete information about their observations of the natural world. Subsequently, "later discoveries" were made by other scientists that shed light on the earlier discoveries and provided more complete explanations of the observed phenomena. This is a central tenet of modern science that drives scientific knowledge forward from one generation to the next.

WRONG CHOICES EXPLAINED

(2), (4) *Inquiry involves making judgments about the reliability of the source and relevance of the information* and *hypotheses are valuable even if they turn out to not be true, because they may lead to further investigation* are *not* the statements that are the best interpretation of the material presented in the chart. These are important tasks that should be part of any scientific investigation in order to ensure freedom from bias. However, these concepts are not reflected in the information provided in the chart.

(3) *Science provides information, but values are also essential to making ethical decisions* is *not* the statement that is the best interpretation of the material presented in the chart. This is an important consideration that should be made before scientific and technological discoveries are implemented. However, this concept is not reflected in the information provided in the chart.

40. **1** Row *1* in the table correctly identifies the functions of these structures. Body system *X*, which is described as functioning to break down carbohydrates and release sugar into the blood, is the human digestive system. Organ *Y*, which is described in the table as functioning to make and release hormones *A* and *B* that regulate sugar concentration in the blood, is the pancreas, which is part of the human regulatory (endocrine) system.

WRONG CHOICES EXPLAINED

(2) Rows *2*, *3*, and *4* in the table do *not* correctly identify the functions of these structures. Each of these rows contains information contrary to the correct answer as explained above.

41. **2** When body system *X* releases too much sugar into the blood, the body can maintain homeostasis by making *more hormone B, only*. Hormone *B* is insulin, a hormone produced by the pancreas (organ *Y*) that regulates blood sugar concentrations by removing excess sugar from the blood. This excess sugar is stored in the liver and muscles as the complex carbohydrate glycogen.

WRONG CHOICES EXPLAINED

(1) It is *not* true that when body system *X* releases too much sugar into the blood, the body can maintain homeostasis by making *more hormone A*,

only. Hormone *A* is glucagon, a hormone produced by the pancreas (organ *Y*) that regulates blood sugar concentrations by converting glycogen to sugar and releasing it into the blood. This is the opposite reaction called for in the question.

(3) It is *not* true that when body system *X* releases too much sugar into the blood, the body can maintain homeostasis by making *more hormone A and more hormone B*. Releasing more of both hormones would result in a counter-active reaction, with glucagon (hormone *A*) increasing and insulin (hormone *B*) decreasing blood sugar concentrations simultaneously.

(4) It is *not* true that when body system *X* releases too much sugar into the blood, the body can maintain homeostasis by making *no hormone A and no hormone B*. Releasing neither of these hormones would result in an unregulated situation in which blood sugar concentrations would rise to dangerously high levels whenever the digestive process released sugar into the bloodstream.

42. **3** To restore homeostasis and compensate for the lack of hormone *B*, one useful action would be to *reduce the carbohydrates in body system X*. By limiting carbohydrate consumption, fewer complex carbohydrates would be digested into sugar and less sugar would enter the bloodstream. If properly planned and carried out, this method could control the concentration of sugar in the blood at homeostatic levels.

WRONG CHOICES EXPLAINED

(1) To restore homeostasis and compensate for the lack of hormone *B*, one useful action would *not* be to *increase the production of hormone A*. Increasing the production of hormone *A* (glucagon) would have the effect of releasing even more sugar into the blood from the liver and muscles. This would have the opposite effect of the one needed in this case and would further disrupt homeostasis.

(2) To restore homeostasis and compensate for the lack of hormone *B*, one useful action would *not* be to *remove organ Y from the body surgically*. Removing organ *Y* (pancreas) from the body would stop the production of both hormone *A* (glucagon) and hormone *B* (insulin). This would result in an unregulated situation in which blood sugar concentrations would rise and fall to dangerously high and low levels depending on the digestive process and further disrupt homeostasis.

(4) To restore homeostasis and compensate for the lack of hormone *B*, one useful action would *not* be to *reduce the synthesis of enzymes in organ Y*. Enzymes in organ *Y* (pancreas) are responsible for the regulation of all metabolic activities in this organ. Reducing enzyme production would slow or stop these metabolic activities and further disrupt homeostasis.

43. **4** If body system *X* temporarily stops releasing sugar into the blood, a likely response of the body would be to *start to increase synthesis of hormone A*. Hormone *A* (glucagon) is produced by the pancreas (organ *Y*). Glucagon

regulates blood sugar concentrations by converting glycogen to sugar and releasing it into the blood. Increased levels of hormone *A* would compensate for the lack of sugar being released from body system *X* (digestive system).

WRONG CHOICES EXPLAINED

(1) If body system *X* temporarily stops releasing sugar into the blood, a likely response of the body would *not* be to *stop using enzymes in body system X*. Body system *X* (digestive system) depends on the production of enzymes to carry out its digestive function. Stopping enzyme production in this system would completely block the digestive process.

(2) If body system *X* temporarily stops releasing sugar into the blood, a likely response of the body would *not* be to *stop organ Y from producing hormone A*. Stopping organ *Y* (pancreas) from producing hormone *A* (glucagon) would cause sugar to be retained in the liver instead of being released into the blood. This would have the opposite effect of the one needed in this case.

(3) If body system *X* temporarily stops releasing sugar into the blood, a likely response of the body would *not* be to *start to increase synthesis of hormone B*. Increasing production of hormone *B* (insulin) would have the effect of storing even more sugar as glycogen and would further reduce blood sugar levels. This would have the opposite effect of the one needed in this case.

PART B–2

44. One credit is allowed for correctly marking an appropriate scale, without any breaks in the data, on each labeled axis. [1]

45. One credit is allowed for correctly plotting the data on the grid, connecting the points, and surrounding each point with a small circle. [1]

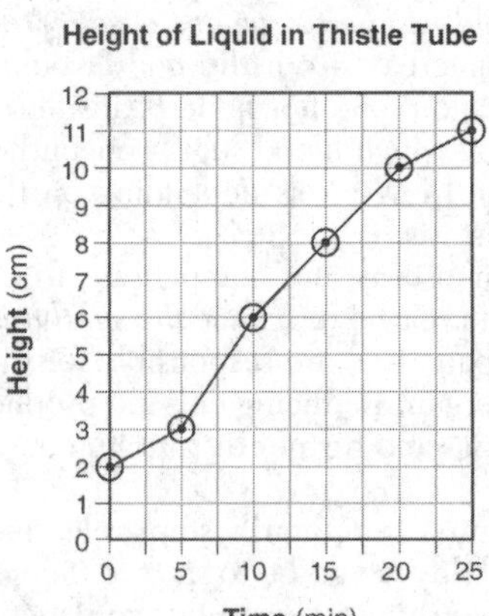

46. One credit is allowed for correctly explaining why the height of the solution in the thistle tube increased during the 25-minute period. Acceptable responses include but are not limited to: [1]

- *Water diffused into the thistle tube.*
- *Water moved from an area of high concentration to an area of low concentration.*
- *Osmosis took place.*
- *Water diffused through the membrane.*
- *Water molecules migrated through the membrane with the concentration gradient.*

47. **3** The most likely reason for this observation is that the *amber-colored solution moved into the thistle tube*. First, we must assume that the amber-colored liquid is iodine (Lugol's) solution, an indicator that is known to react with starch molecules to produce a distinct blue-black coloration. The molecules that make up Lugol's solution are small and can readily pass through the pores of a semipermeable membrane from the beaker to the thistle tube. Starch molecules are too large to pass through this same semipermeable membrane and so remain inside the thistle tube. These facts explain why the contents of the beaker remain amber-colored while the contents of the thistle tube turn blue-black.

WRONG CHOICES EXPLAINED

(1) The most likely reason for this observation is *not* that *starch molecules moved out of the thistle tube*. Starch molecules are too large to pass through this same semipermeable membrane and so remain inside the thistle tube.

(2), (4) The most likely reason for this observation is *not* that *water molecules moved into the thistle tube* or that *water molecules moved out of the thistle tube*. Water molecules can readily pass in either direction through a semipermeable membrane, but its net movement is always with the concentration gradient. In this experiment, water serves only as a suspension medium and solvent, not as a reactant.

48. One credit is allowed for correctly explaining why the chemical released from the injured fish may *not* cause an alarm response in other fish species. Acceptable responses include but are not limited to: [1]

- *Other species lack the receptors for the chemical.*
- *The alarm chemicals are specific to the species.*
- *Other species lack the ability to sense/recognize the chemical.*
- *Animal behaviors are genetic traits that are unique to the species displaying them.*

49. **4** The process represented in the diagram is *genetic engineering*. Genetic engineering is a laboratory technique in which a gene for a desired trait is snipped from the DNA of a donor cell and inserted into the genome of a recipient cell. This is the process illustrated in the diagram.

WRONG CHOICES EXPLAINED

(1) It is *not* true that the process represented in the diagram is *DNA replication*. Replication is the exact self-duplication of genetic material (DNA) in the cell prior to cell division. DNA replication is not illustrated in the diagram.

(2) It is *not* true that the process represented in the diagram is *natural selection*. Natural selection, sometimes referred to as survival of the fittest, postulates that certain members of a species are better adapted to their environment than others. Natural selection is not illustrated in the diagram.

(3) It is *not* true that the process represented in the diagram is *gel electrophoresis*. In this laboratory technique, electrically charged fragments of DNA are placed into a gel and then drawn through the gel by an electric current until they separate based on size. Gel electrophoresis is not illustrated in the diagram.

50. **1** The original gene for the production of a human hormone was most likely removed from a *chromosome*. Chromosomes are cell organelles that contain DNA molecules that carry the genetic codes needed for the production of specific proteins in the cell. The snipped gene for a human hormone was taken from a human chromosome.

WRONG CHOICES EXPLAINED

(2), (3), (4) The original gene for the production of a human hormone was *not* most likely removed from a *ribosome*, a *mitochondrion*, or a *cell membrane*. None of these organelles contains a gene for the production of a human hormone. The mitochondrion contains DNA that codes for the production of enzymes involved in respiration.

51. One credit is allowed for correctly stating *one* possible reason why a gene for the production of a human hormone would be placed in bacterial DNA. Acceptable responses include but are not limited to: [1]

- *The bacteria will produce the human hormone quickly.*
- *This is done in order to increase the production of insulin or some other hormone.*
- *The human hormone produced can be given to people who need it, and the number of allergic reactions will be reduced.*
- *By inserting the gene, the bacteria can be made to produce the human hormone.*

52. One credit is allowed for correctly identifying *one* population that will *decrease* in size after the lake sturgeon are added to the new ecosystem and for supporting the answer. Acceptable responses include but are not limited to: [1]

- *Populations of insect larvae/clams/worms will decrease in size because they will be eaten by both lake sturgeon and bloater fish.*
- *The bloater fish may temporarily decrease in number due to competition with the newly introduced lake sturgeon.*
- *If the lake sturgeon are more successful competitors, then they will gradually eliminate bloater fish from their habitat.*
- *If the bloater fish are more successful in the environment, then at some time after its reintroduction, the lake sturgeon may decrease in number due to competition with the bloater fish.*

53. One credit is allowed for correctly indicating the position in the food web where each organism listed would be placed and for writing the name of each in the appropriate box. Acceptable responses include: [1]

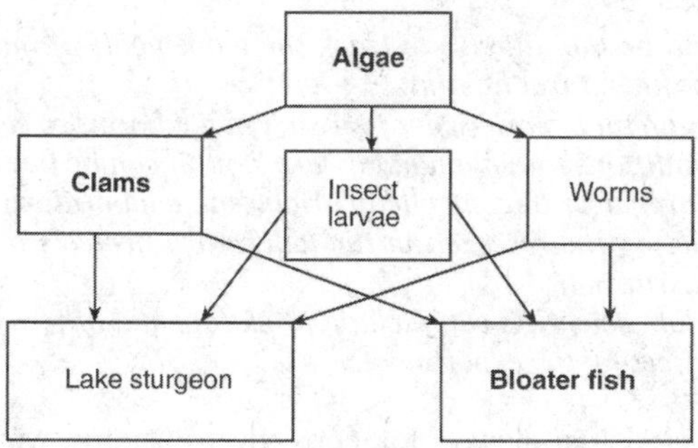

54. One credit is allowed for correctly identifying *one* population, other than lake sturgeon, that will increase in size after the lake sturgeon are added to the new ecosystems and for supporting the answer. Acceptable responses include but are not limited to: [1]

- *Algae, because there will be fewer organisms that feed on them.*
- *More algae will survive and reproduce more because populations of insect larvae, clams, and worms that eat algae will decrease in size because they will be eaten by both lake sturgeon and bloater fish.*

55. One credit is allowed for correctly stating what the arrows in the food web represent. Acceptable responses include but are not limited to: [1]

- *The arrows show the direction of energy movement from one organism to another.*
- *The arrows show the flow of food energy through the ecosystem.*
- *The arrows indicate the movement of nutrient energy through the food web.*
- *Each arrow represents what each organism is eaten by or what each organism it eats.*

PART C

56. One credit is allowed for correctly explaining why researchers would put the salamander in an artificial environment, as opposed to conducting the experiment in their natural habitat. Acceptable responses include but are not limited to: [1]

- *It would be too difficult to track the movements of individual salamanders in the natural habitat.*
- *The salamanders are easier to watch in a laboratory setting.*
- *The conditions the salamanders are kept in can be better controlled.*
- *There are fewer uncontrolled variables in a laboratory setup.*
- *Conducting the research in the laboratory protects the natural habitat from disruption.*
- *In the lab, scientists can exclude predators and other adverse conditions from affecting the experiment.*

57. One credit is allowed for correctly explaining why other scientists would question the validity of the conclusions based on this setup. Acceptable responses include but are not limited to: [1]

- *The sample size is too small to provide adequate data.*
- *In such a small experimental group, it would be easy to have some abnormal salamanders.*
- *No hypothesis is stated and no experimental method is outlined.*
- *The artificial environment lacks many factors present in the natural habitat.*
- *The lab environments might not accurately duplicate the natural habitats.*
- *No statement is made concerning what is being measured and how data will be collected and analyzed.*
- *It is not clear on what basis experimental inferences/conclusions will be made.*

58. One credit is allowed for correctly stating *one* reason why it may be dangerous to introduce a new species to an established ecosystem. Acceptable responses include but are not limited to: [1]

- *New species might compete with native species for limited resources.*
- *Native species can have their space, light, and water taken away by invasive plants.*
- *New plant species could have nonnative insects/pathogens on them.*
- *Invasive plants can crowd out native plants, which also eliminate animals that use the native plants for food/habitat.*

59. One credit is allowed for correctly stating *one* reason why a green roof reduces the amount of carbon dioxide in the atmosphere. Acceptable responses include but are not limited to: [1]

- *Green plants take in carbon dioxide to perform photosynthesis.*
- *Plants use* CO_2 *to produce food.*
- *Plants take in carbon dioxide.*
- *Green roofs increase the area of photosynthesizing plants in an urban area that doesn't have other green spaces.*

60. One credit is allowed for correctly stating *one* reason why it is important to reduce the amount of carbon dioxide in the atmosphere. Acceptable responses include but are not limited to: [1]

- *Carbon dioxide in the atmosphere adds to the greenhouse effect.*
- *Excessive* CO_2 *in the air traps solar heat and raises the temperature of Earth.*
- *Carbon dioxide has been associated with global warming/global climate change.*
- *High concentrations of* CO_2 *in the atmosphere have been linked to melting of the polar ice caps and rising sea levels.*

61. One credit is allowed for correctly stating *one disadvantage* of a green roof. Acceptable responses include but are not limited to: [1]

- *Green roofs are expensive to install.*
- *Green roofs need more maintenance than traditional roofs.*
- *Such roofs might be more susceptible to leakage and rot.*
- *Green roofs might be too heavy for the building structure to support.*
- *The damp absorptive layer might attract burrowing rodents or insects that could damage the roof.*

62. One credit is allowed for correctly stating *one* hypothesis the experiment would test. Acceptable responses include but are not limited to: [1]

- *Bacteria exposed to x-ray radiation will show a greater number of mutations than bacteria exposed to natural light.*
- *Bacteria exposed to x-rays will show less DNA damage than bacteria exposed to natural light.*
- *If bacteria are exposed to x-rays, they will show more mutations than bacteria not exposed to x-rays.*
- *The DNA of bacteria exposed to x-rays will change more than that of bacteria not exposed to x-rays.*
- *X-ray exposure will affect bacterial DNA by increasing the frequency of mutation.*

63. One credit is allowed for correctly identifying *one* specific difference, other than it is a two-step vaccination, between Nabel's vaccination and a traditional flu vaccine. Acceptable responses include but are not limited to: [1]

- *Nabel's vaccination protects from every known influenza A virus.*
- *The new NIH vaccine directly injects viral DNA.*
- *This new approach may add years of protection compared to traditional vaccines.*
- *It makes muscle cells produce hemagglutinin, which triggers the immune response against flu.*
- *It causes the body to make a protein found on the surface of all flu viruses.*

64. One credit is allowed for correctly explaining how injecting dead or weakened viruses into a person can help fight against future infections from that virus. Acceptable responses include but are not limited to: [1]

- *The body responds by making antibodies that protect against future infections.*
- *There are antigens on the surface of the dead virus, which then trigger the body to produce antibodies specific to that antigen.*
- *The body makes memory cells that can fight the virus in the future.*

65. One credit is allowed for correctly stating *one* reason why the percent ice cover is important to the ability of eagles to obtain food. Acceptable responses include but are not limited to: [1]

- *The access to fish would be limited by the amount of ice cover.*
- *Eagles would have to search for food other than fish.*
- *The ice cover might limit the number of organisms that are available for the eagles to eat.*
- *During periods of ice cover, fish tend to stay near the bottom of the river where the temperature is stable at 4°C, so eagles must find other food sources.*

[**Note:** Because no data are provided concerning percent of ice cover and other environmental factors affecting eagle survival, this question can be answered only in the form of a hypothesis.]

66. One credit is allowed for correctly stating what inference can be made about the percent ice cover at Croton Reservoir between January and February 2009 and for supporting the answer. Acceptable responses include but are not limited to: [1]

- *The ice cover was probably less in February because there were more eagles sighted.*
- *Since fewer eagles were reported in January, there might have been more ice on the reservoir.*

[**Note:** Because no data are provided concerning percent of ice cover and other environmental factors affecting eagle survival, this question can be answered only in the form of a hypothesis.]

67. One credit is allowed for correctly stating *one* reason why the number of eagles sighted showed a change at all four sites between February and March. Acceptable responses include but are not limited to: [1]

- *Winter was ending, so the eagles returned to their summer nesting areas.*
- *As winter ended, the eagles migrated away.*
- *The warming weather and longer days stimulated the eagles to begin their annual migration north.*
- *The eagles are free flying and so have left their winter roosts for their summer nesting areas.*

[**Note:** No data are provided concerning percent of ice cover, late-winter storms, and other environmental factors affecting eagle migration. So responses including these factors are speculative and should not be acceptable. In addition, the data on number of eagles in all months at all sites are averaged. So the smaller number of visits in March becomes less important as a factor affecting the data.]

68. One credit is allowed for correctly stating *one* possible reason why a popular hiking trail in this area is closed during eagle roosting seasons. Acceptable responses include but are not limited to: [1]

- *Winter roosts are protected under federal law and managed with a buffer zone to reduce human interference.*
- *Humans might disturb the eagles.*
- *People using the hiking trails in these areas might make noise, leave trash, or otherwise disrupt the roosting behaviors of the eagles.*

69. One credit is allowed for correctly stating the role of the sea grasses in the reef ecosystem. Acceptable responses include but are not limited to: [1]

- *Their role is to convert inorganic materials into organic matter/food.*
- *They capture radiant energy.*
- *They serve as the producer in this food web/ecosystem.*
- *They serve as food for fish.*
- *They produce oxygen/absorb carbon dioxide.*
- *They disrupt the coral-dominated ecosystem.*
- *They change the coral reef food web by eliminating seaweed-eating fish.*
- *They reduce the amount of sunlight reaching the sea floor near the reefs.*

70. One credit is allowed for correctly identifying *one* abiotic factor that is affecting the stability of the coral reef ecosystems and for stating how the factor identified is important to the coral reef ecosystems. Acceptable responses include but are not limited to: [1]

- *Temperature—causes change in ocean currents*
- *Water temperature—can affect the survival of coral polyps*
- *Light—used by plants for photosynthesis*
- *Ocean currents—carry organisms from one reef to another*
- *Carbon dioxide—used by plants for photosynthesis*
- *Dissolved oxygen—used by all organisms for respiration*
- *Salinity—provides proper salt/water concentration for the survival of marine life*
- *Acidity—can affect the efficiency of enzyme-catalyzed reactions/survival of reef organisms*

71. One credit is allowed for correctly stating *one* reason why it is important to maintain the stability of coral reefs. Acceptable responses include but are not limited to: [1]

- *Reefs should be protected in order to maintain the food webs that exist there.*
- *This is important because many organisms depend on them for food/shelter/nesting sites.*
- *By keeping coral reefs healthy, we contribute to biodiversity in the marine environment.*
- *Healthy reefs contribute to healthy oceans and to the fish we depend on for food.*
- *Maintaining healthy coral reefs helps to prevent the extinction of reef species.*
- *Reef ecosystems are indicators of the overall health of the global ecosystem that we depend on for human survival.*

72. One credit is allowed for correctly stating *one* advantage of the fish larvae moving by ocean currents into a damaged reef ecosystem. Acceptable responses include but are not limited to: [1]

- *The fish larvae could repair/repopulate the damaged reefs.*
- *The reef might become more stable.*
- *They might keep the seaweed under control/eat the seaweed.*
- *The reef would provide the fish larvae with food/shelter.*
- *The fish larvae would have less competition/more resources.*

PART D

73. **4** The process shown in the diagram is *protein synthesis*. The diagram illustrates a molecule of messenger RNA (mRNA) attached to Structure *X* (a ribosome). Smaller bell-shaped molecules of transfer RNA (tRNA) carrying specific amino acids (glycine, arginine, proline, alanine) are shown linking to the mRNA codon in a predetermined sequence to form a chain of amino acids ("newly formed molecule"), also known as a protein.

WRONG CHOICES EXPLAINED

(1) It is *not* true that the process shown in the diagram is *cellular respiration*. An illustration of cellular respiration would show molecules of glucose and oxygen reacting to produce water, carbon dioxide, and cellular energy as ATP.

(2) It is *not* true that the process shown in the diagram is *cellular reorganization*. An illustration of cellular reorganization would show a cell about to divide as it gathers its chromosomes and morphs its organelles prior to spindle formation and mitosis.

(3) It is *not* true that the process shown in the diagram is *gene recombination*. An illustration of gene recombination would show the reuniting of segregated chromosome pairs during fertilization, a process that brings alleles together in a new genetic combination.

74. **4** Structure *X* is a *ribosome*. The ribosome is a cell organelle that functions as the site for protein synthesis in the cell.

WRONG CHOICES EXPLAINED

(1) Structure *X* is *not* a *mitochondrion*. The mitochondrion is a cell organelle that functions as the site for aerobic respiration in the cell.

(2) Structure *X* is *not* a *vacuole*. The vacuole is a cell organelle that functions as the site for food and/or waste storage in the cell.

(3) Structure *X* is *not* a *nucleus*. The nucleus is a cell organelle that functions as the site for genetic regulation in the cell.

75. **1** A finch that picks small insects out from cracks in the bark of trees would most likely have a beak that is *sharp and thin*. An examination of the diagram reveals that finches eating all or mainly animal food (including insects) tend to have more pointed bills than finches that eat mainly plant food. In addition, insect-eating finches whose bills are described as "probing" (warbler finch and woodpecker finch) have the sharpest and thinnest bills of those illustrated.

WRONG CHOICES EXPLAINED

(2), (3), (4) A finch that picks small insects out from cracks in the bark of trees would *not* most likely have a beak that is *sharp and thick*, *rounded and thin*, or *rounded and thick*. Each of these combinations of characteristics contains at least one characteristic (i.e., thick, rounded) that would not be efficient at probing for insects in the cracks in the bark of trees.

76. **1** *The type of beak indicates the type of food the finch eats* is the statement that represents a basic assumption from *The Beaks of Finches* lab. In this lab exercise, students are asked to pick up small objects using different tools with different shapes and gripping surfaces and then record data concerning their observations. The lab is designed to help students understand the mechanisms (natural selection) that favored some variations (beak shapes) over others in the evolution of the Galapagos finches.

WRONG CHOICES EXPLAINED

(2), (3), (4) *Different birds have different songs*, *birds with larger beaks can find mates more easily*, and *nesting behavior of finches is an inherited trait* are *not* the statements that represent a basic assumption from *The Beaks of Finches* lab. None of these statements relates its concept to the actual activities of this lab. Exploring these concepts would require laboratory investigations using different tools and techniques than those used in *The Beaks of Finches* lab.

77. One credit is allowed for correctly stating *two* reasons why the large ground finch and sharp-billed ground finch could live on the same island but *not* compete for food, even though they both eat mainly plant food. Acceptable responses include but are not limited to: [1]

- *They have different beak structures.*
- *They might eat different types of plants.*
- *They might eat at different times of day or night.*
- *They might live in different areas of the island.*
- *They might feed at different elevations on the same island.*
- *They are adapted to eat different kinds of seeds.*
- *They would compete only if they both required the same limited food resources.*

78. One credit is allowed for correctly writing the number "2" to designate the tree diagram that shows the most probable evolutionary relationships among the three species and for supporting the answer. Acceptable responses include but are not limited to: [1]

- *Tree 2—Species B has enzyme Y, which is not present in either species A or C.*
- *2—Enzymes W and X are present in species A and C but not in B.*
- *2—A and C have enzymes W, X, and Z, but B doesn't.*
- *Tree 2—A and C have three enzymes in common: W, X, and Z.*
- *2—A and C have more enzymes in common.*

79. One credit is allowed for correctly identifying substance A. Acceptable responses include: [1]

- *Water*
- H_2O
- *H-O-H*

80. One credit is allowed for correctly identifying *one* likely substance in the environment of the cell that caused this response. Acceptable responses include but are not limited to: [1]

- *Salt*
- *Sugar*
- *Seawater*
- *Sodium chloride (or any other soluble salt)*

81. **3** This muscle fatigue is most likely due to *lack of oxygen and build up of waste in the muscles*. During periods of strenuous exercise, muscle cells quickly use up available dissolved oxygen required for aerobic respiration. As a result, the cells resort instead to a less-efficient form of anaerobic respiration known as lactic acid fermentation or glycolysis. This process results in the production of a small amount of ATP energy and the metabolic waste lactic acid. Lactic acid stimulates a pain response in the muscles where it forms. This combination of factors is what we describe as muscle fatigue.

WRONG CHOICES EXPLAINED

(1) This muscle fatigue is *not* most likely due to *the heart beating too fast and tiring out*. Although an increased heart rate is typical during exercise, it is not the cause of muscle fatigue.

(2) This muscle fatigue is *not* most likely due to *the lungs accumulating oxygen*. The lungs function to absorb oxygen from the air and transfer the oxygen to the blood. This activity is not the cause of muscle fatigue.

(4) This muscle fatigue is *not* most likely due to *lack of carbon dioxide in the muscles*. Carbon dioxide is a waste product of aerobic respiration but not of lactic acid fermentation. Carbon dioxide is not the cause of muscle fatigue.

82. **1** Methods *A* and *B* are classified as methods of passive transport because they do not require *ATP*. In the illustrations of both of these methods, a higher concentration of molecules is shown outside the cell compared to inside the cell. In these cases, the molecules move across the membrane by the process of diffusion. Diffusion does not require the expenditure of ATP energy in order to occur.

WRONG CHOICES EXPLAINED

(2), (3), (4) It is *not* true that methods *A* and *B* are classified as methods of passive transport because they do not require *carbon dioxide*, *light*, or *DNA*. None of these factors has any direct effect on the diffusion of molecules from a region of higher relative concentration to an area of lower relative concentration of those molecules.

83. One credit is allowed for correctly stating *one* reason why the movement of molecules in method *C* represents active transport. Acceptable responses include but are not limited to: [1]

- *Method C shows the transport of molecules from an area of lower concentration to an area of higher concentration.*
- *The molecules are moving against the concentration gradient.*

84. One credit is allowed for correctly stating *one* reason why some species might have similar body structures even if they are *not* closely related. Acceptable responses include but are not limited to: [1]

- *They may have evolved in the same or similar environments, and these structures and traits were advantageous adaptations.*
- *They have two separate mutations that produce the same appearance.*
- *These structures might help them use similar food/adapt to their environment.*
- *The species inhabit similar environments.*
- *They occupy similar niches.*
- *Natural selection may favor the body structures for survival in a particular environment.*

85. One credit is allowed for correctly drawing a line that shows what most likely happened to the pulse rate of the student during these activities. Acceptable responses include but are not limited to: [1]

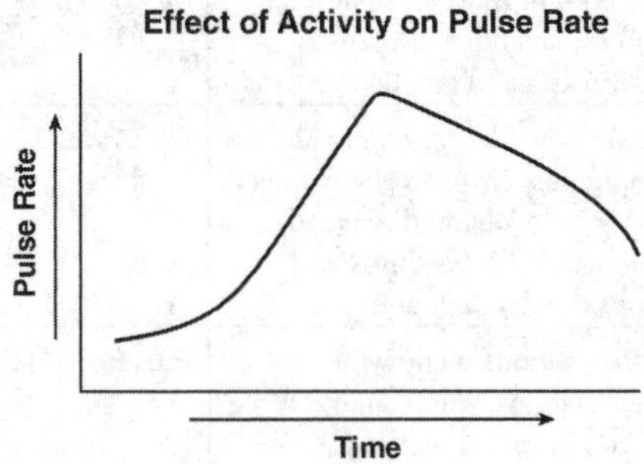

Note: Allow credit for a graph that shows a gradual increase followed by a gradual decrease.

STANDARDS/KEY IDEAS	AUGUST 2016 QUESTION NUMBERS	NUMBER OF CORRECT RESPONSES
STANDARD 1		
Key Idea 1: The central purpose of scientific inquiry is to develop explanations of natural phenomena in a continuing and creative process.	39, 47	
Key Idea 2: Beyond the use of reasoning and consensus, scientific inquiry involves the testing of proposed explanations involving the use of conventional techniques and procedures and usually requiring considerable ingenuity.	56, 57, 62	
Key Idea 3: The observations made while testing proposed explanations, when analyzed using conventional and invented methods, provide new insights into natural phenomena.	33, 36, 38, 44, 45	
Laboratory Checklist	32	
STANDARD 4		
Key Idea 1: Living things are both similar to and different from each other and from nonliving things.	1, 14, 31, 40, 46, 48, 52, 53, 54, 55, 69	
Key Idea 2: Organisms inherit genetic information in a variety of ways that result in continuity of structure and function between parents and offspring.	4, 5, 8, 11, 35, 49, 50, 51	
Key Idea 3: Individual organisms and species change over time.	9, 13, 15, 16, 17, 18, 37	
Key Idea 4: The continuity of life is sustained through reproduction and development.	12, 19, 20, 21, 22	
Key Idea 5: Organisms maintain a dynamic equilibrium that sustains life.	2, 7, 10, 23, 24, 25, 41, 42, 43, 63, 64	
Key Idea 6: Plants and animals depend on each other and their physical environment.	26, 27, 28, 65, 66, 67, 70, 71, 72	
Key Idea 7: Human decisions and activities have a profound impact on the physical and living environment.	3, 6, 29, 30, 34, 58, 59, 60, 61, 68	

STANDARDS/KEY IDEAS	AUGUST 2016 QUESTION NUMBERS	NUMBER OF CORRECT RESPONSES
REQUIRED LABORATORIES		
Lab 1: "Relationships and Biodiversity"	73, 74, 78, 84	
Lab 2: "Making Connections"	81, 85	
Lab 3: "The Beaks of Finches"	75, 76, 77	
Lab 5: "Diffusion Through a Membrane"	79, 80, 82, 83	

Examination June 2017

Living Environment

PART A

Answer all questions in this part. [30]

***Directions* (1–30): For *each* statement or question, record in the space provided the *number* of the word or expression that, of those given, best completes the statement or answers the question.**

1 An example of recycling is

1 using a paper cup instead of a mug
2 turning off the lights when leaving a room
3 using plastic from a soda bottle to make a certain type of clothing
4 using two paper towels rather than five to clean up a spill 1____

2 One characteristic of a stable ecosystem is

1 a high number of predators
2 an interdependence of organisms
3 a lack of biodiversity
4 an increase in human interference 2____

3 According to scientists, ocean waves could be a source of energy. Devices are being designed to capture the energy from waves and supply electricity to coastal areas. A direct benefit of utilizing this technology to produce energy would be the

1 destruction of habitats near the devices
2 decreased use of nonrenewable resources
3 release of gases needed for photosynthesis
4 increased use of finite resources 3_____

4 The emerald ash borer is an insect that is thought to have been accidentally brought to the United States from China in shipping containers. It attacks ash trees, eventually killing the trees, destroying an important part of the ecosystem. The presence of the emerald ash borer in the United States can be used as an example of how humans have

1 purposely introduced an insect to correct a problem in an ecosystem
2 used an insect to remove one insect species and replace it with another
3 worked to increase the biodiversity in a particular area
4 altered the equilibrium in an ecosystem by introducing a new species 4_____

5 Which graph best shows the changes in global human population and natural resource use over the past 500 years?

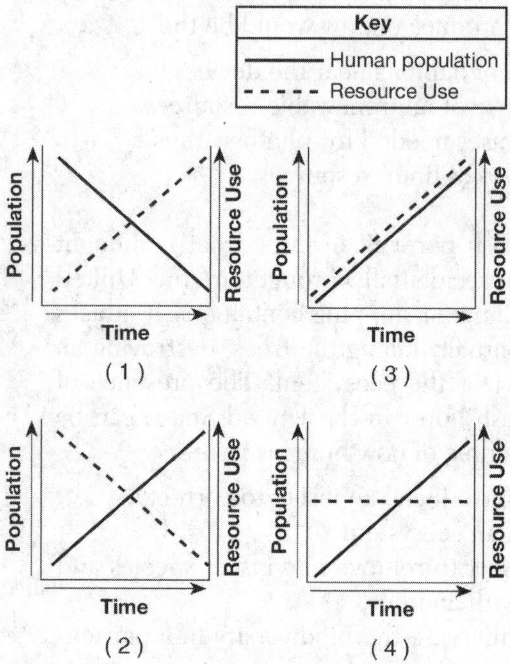

5_____

6 The burning of fossil fuels has harmed the environment by

1 decreasing acid rain in the northeast United States
2 adding carbon dioxide to the atmosphere
3 increasing biodiversity in the lakes and ponds of the Adirondacks
4 depleting the ozone shield directly over western New York State

6_____

7 Which method of collecting data would provide the most accurate information about how an ecosystem is being affected by human development?

1 The people in the neighborhood record the number of birds they see in the area both before and after the construction of a new building.
2 Deer in the area are identified before construction so that scientists can see where they go after the building is finished.
3 More trees are planted in the area around the construction site to allow the animals a new place to live.
4 The population size is recorded for each species present in the construction area before and after the building is completed. 7____

8 Natural selection produces changes most quickly in

1 species with short reproductive cycles
2 individual pathogens killed by antibiotics
3 complex multicellular organisms
4 individuals that produce a small number of offspring 8____

9 As a result of habitat destruction, the size of the Florida panther population has been drastically reduced. It is estimated that there are only 100 to 160 Florida panthers in the wild. Which statement best explains why the Florida panther population may not continue to evolve?

1 There is no longer a chance of mutations occurring in the population.
2 There is a lack of competition for limited environmental resources.
3 There is no longer a chance of a trait providing a reproductive advantage to the population.
4 There is a lack of genetic variation for selection to act upon. 9____

10 Which level of the pyramid below is correctly paired with the type of organism that would most likely be found at that level in an ecosystem?

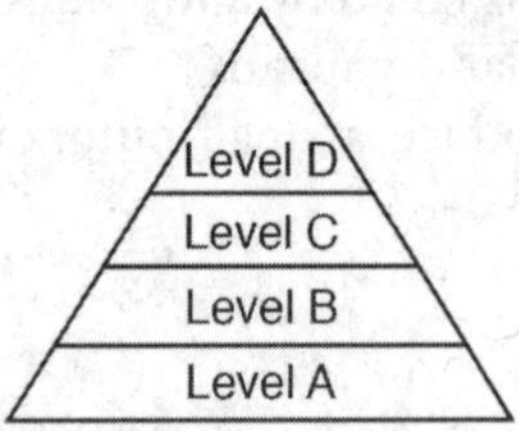

1 Level *A* – producers
2 Level *B* – carnivores
3 Level *C* – herbivores
4 Level *D* – decomposers 10____

11 Which statement best describes what is most likely to occur if an animal population grows larger than the carrying capacity of its environment?

1 The birth rate will increase.
2 Both the birth rate and death rate will decrease.
3 The death rate will increase.
4 Neither the birth rate nor the death rate will decrease. 11_____

12 Many biotic factors affect individuals in a population. An example of an organism being directly affected by a biotic factor is

1 a squirrel cannot find a mate
2 a flood washes away a maple tree
3 a plant is in a dark room
4 a chipmunk finds a rock pile to use for a home 12_____

13 When people receive organ transplants, they often need to take medications that decrease immune responses because

1 transplanted organs contain antigens that can trigger white blood cell activity
2 hormones present in replacement organs prevent the synthesis of antibiotics
3 transplanted organs produce their own antibiotics
4 antigens present in these organs attack antibodies already present in the blood 13_____

14 Which structures regulate water loss and gas exchange in the leaves of plants?

1 vacuoles
2 chloroplasts
3 guard cells
4 mitochondria 14_____

15 Which life process carried out by a green plant is represented in the diagram below?

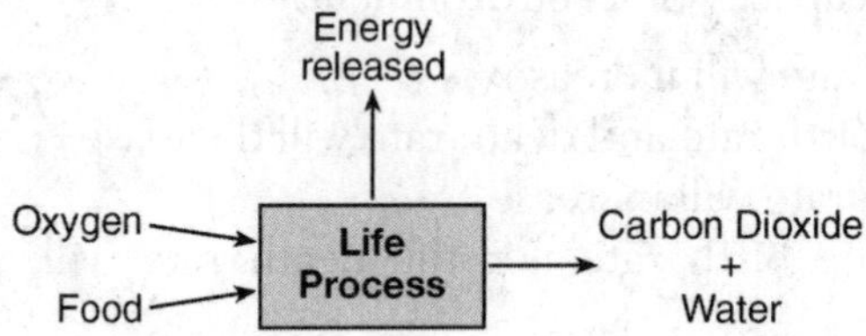

1 respiration
2 photosynthesis
3 digestion
4 replication

15_____

16 Scientists are developing a malaria vaccine that would most likely contain

1 weakened drugs used to treat the symptoms of malaria
2 white blood cells from the malaria-infected individuals
3 a weakened form of the malaria-causing organism
4 antibodies made from the malaria-causing organism

16_____

17 Proteins, starch, and DNA are similar in that they are all

1 organic compounds
2 parts of genes
3 made of amino acids
4 made of simple sugars

17_____

18 In response to an increasing blood glucose level, the human body will normally

1 store the glucose in cell nuclei
2 release a hormone that lowers the blood glucose
3 produce a hormone that destroys the glucose
4 use the excess glucose to make proteins 18_____

19 A kitten was born with black fur and green eyes. The fur and eye color of its parents are shown in the chart below.

Cat	Fur	Eye Color
Father	striped	green
Mother	black	yellow
Kitten	black	green

Which statement helps explain why the kitten has black fur?

1 Chromosomes present on the genes code for the characteristics of its fur.
2 Genetic mutations always cause the fur color and eye color to change.
3 Offspring receive genetic information from both parents.
4 Gene expression is changed in every generation, resulting in evolution. 19_____

20 Scientists have been investigating a way to recreate extinct species such as the saber-toothed cat illustrated below.

Source: https://IGS.Indiana.edu

Which technique would use DNA from an extinct species to recreate an organism of the species?

1 natural selection
2 differentiation
3 cloning
4 selective breeding

20 _____

21 The sequence that best illustrates the flow of energy through an ecosystem is

1 sunlight → plant → wolf → rabbit
2 plant → sunlight → rabbit → wolf
3 sunlight → plant → rabbit → wolf
4 wolf → rabbit → plant → sunlight

21 _____

22 Which cellular change in an organism could be inherited by the next generation?

1 a change in the ribosomes in the pancreas of a squirrel
2 the deletion of a single DNA base in a sperm cell of a trout
3 a decrease in the size of a vacuole in a rose leaf cell
4 the transfer of a piece of a chromosome in the skin cell of a raccoon

22 _____

23 A chemical was added to hand sanitizers and dish detergents to kill bacteria. Certain species of bacteria are no longer killed by this chemical. One likely reason for the decreased effectiveness of this chemical is that these bacteria have

1 slower metabolic rates
2 a mutation for resistance
3 been selectively bred for survival
4 an adaptation to a different niche 23____

24 The hemlock wooly adelgid is an invasive insect species that is destroying native hemlock trees in New York State. These insects can upset natural ecosystems because they

1 provide food for native bird species
2 can carry diseases that can be spread to pets
3 increase biodiversity in New York State forests
4 disrupt habitats that native species depend upon 24____

25 A shark and a dolphin have similarly shaped bodies and fins. However, these two organisms are not closely related: The shark is a fish, and the dolphin is a mammal. Some species may have similar body structures even if they are not related because they evolved in

1 similar environments and specific traits increased their chances of survival
2 similar environments and were exposed to factors that caused exactly the same mutations
3 different environments, but tried to adapt in the same ways so they could survive
4 different environments, but ate similar foods that affected their growth and development 25____

26 A dog gave birth to the three puppies shown in the photograph below. One of the puppies has darker fur on its face than the other two.

Source: http://germanshepherdsatsdikennels.blogspot.com

Which two biological processes account for this difference between the puppies?

1 meiosis and recombination
2 meiosis and cloning
3 mitosis and differentiation
4 mitosis and cloning

26 _____

27 The diagram below represents a cell that produces digestive enzymes.

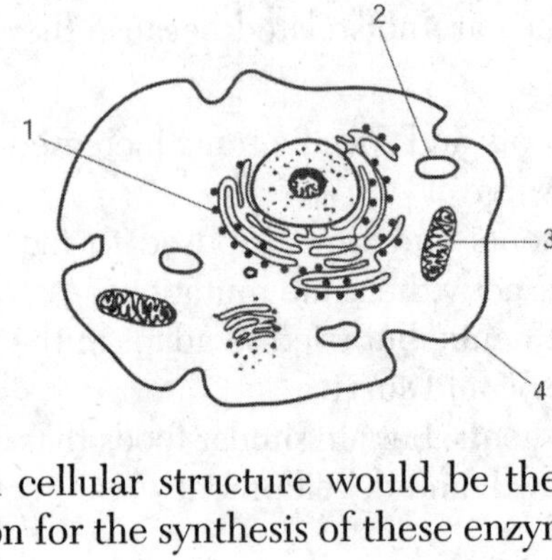

Which cellular structure would be the most likely location for the synthesis of these enzymes?

1 1
2 2
3 3
4 4

27 _____

28 Energy drinks have become increasingly popular. Some of these drinks contain large amounts of caffeine, which is known to increase heart rates in most individuals. This effect on the heart rate can be dangerous because it can lead to

1 a disruption in the absorption of starch
2 an increase in blood volume
3 a decrease in oxygen levels
4 an imbalance in homeostasis

28_____

29 The kidney is an organ that collects wastes and excess water from the blood and sends them to the bladder where they are stored before being removed from the body. Which two systems work together to perform this function?

1 immune and respiratory
2 circulatory and excretory
3 skeletal and nervous
4 digestive and circulatory

29_____

30 Scientists have recently discovered a community of bacteria and clams living under an ice shelf in Antarctica. These organisms live under 600 feet of ice, in the absence of sunlight, and in temperatures considered too cold for most living organisms. The location where these organisms live is unusual because

1 only biotic factors control the size of the populations
2 bacteria and clams are found in the same area
3 abiotic factors are found in their environment
4 green plants make energy-rich compounds available

30_____

PART B–1

Answer all questions in this part. [13]

Directions **(31–43): For *each* statement or question, record in the space provided the *number* of the word or expression that, of those given, best completes the statement or answers the question.**

31 The diagram below represents an important biological concept.

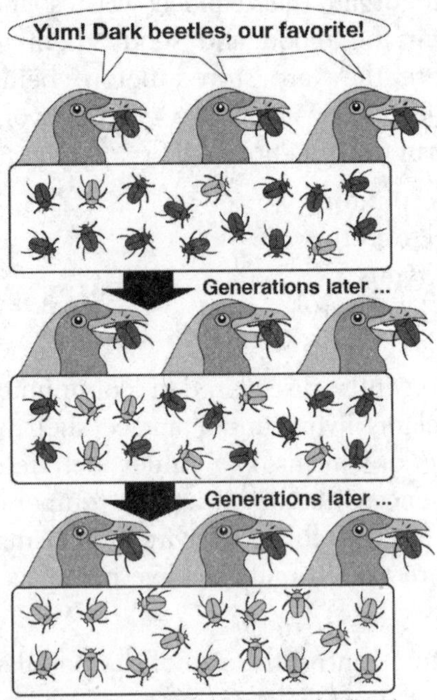

Adapted from: http://evolution.berkeley.edu/evolibrary/

The concept being represented is

1 overproduction
2 natural selection
3 homeostasis
4 ecological succession

31____

32 The graph below shows levels of a form of ultraviolet radiation (UV-B) and ozone thickness in Australia during December 1987 and January 1988.

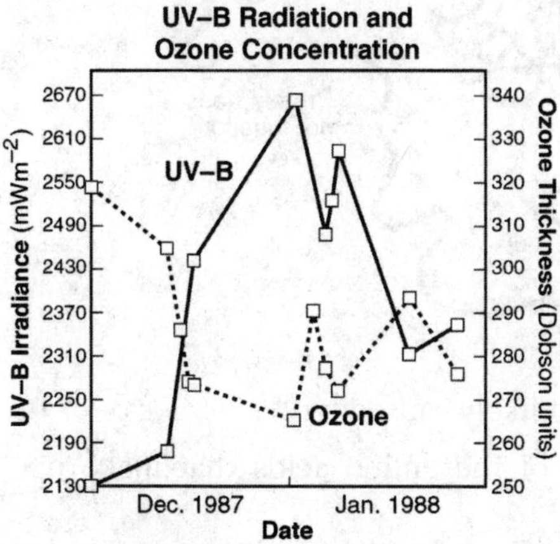

Adapted from: C. R. Roy, et. al., *Nature* 347:235, 1990

Which statement best describes the apparent relationship between ozone and UV-B?

1 When ozone levels are at 2550 Dobson units, the UV-B levels are at 250 Dobson units.
2 The increase in UV-B reduces the destruction of the ozone layer.
3 When the ozone layer is thinner, more UV-B gets through it.
4 If the ozone layer is thicker, UV-B levels on the ground increase. 32_____

33 In the diagram below, *X* represents a process that causes a protein to unfold and stop functioning.

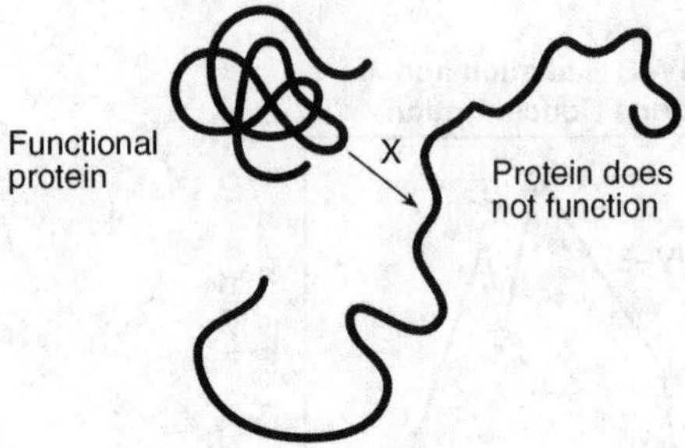

Process *X* is most likely caused by

1 the digestion of the amino acids that make up the proteins
2 the synthesis of a protein with different simple sugars
3 removal of the gene that codes for the production of the protein
4 an internal factor in the body, such as a temperature increase

33____

34 Which statement is an example of how the external environment can influence gene expression?

1 Some flowering plants that inherit a gene for white flowers and a gene for red flowers will produce pink flowers.
2 Some animals that inherit genes for brown fur will grow white fur if the outside temperature falls below a certain level.
3 In some breeds of cat, certain fur-color genes are found only in females.
4 A pea plant is short-stemmed only if it inherits the genes for the trait from both parents.

34____

35 Which statement best illustrates a concept represented in the diagram below?

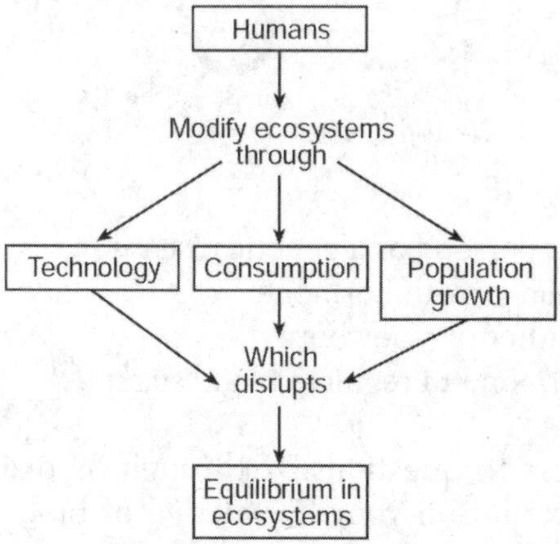

1 Tsunamis triggered by oceanic earthquakes cause widespread flooding that can lead to large scale environmental destruction.
2 Annual hunting laws determine the number of deer that can be hunted to ensure population stability.
3 More individuals are purchasing hybrid cars that use less gasoline and produce less carbon dioxide.
4 The increased use of electronics has led to increased mining for precious metals and minerals in developing countries. 35_____

36 The process shown below is used to

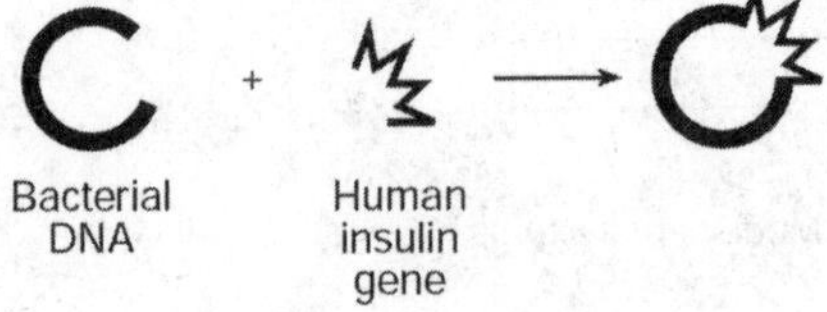

1 determine if a person has a genetic disease
2 produce human growth hormone
3 identify the father of a newborn
4 produce a hormone to regulate blood sugar 36____

Base your answers to questions 37 through 39 on the diagrams below and on your knowledge of biology. The diagrams represent a single-celled organism and a multicellular organism.

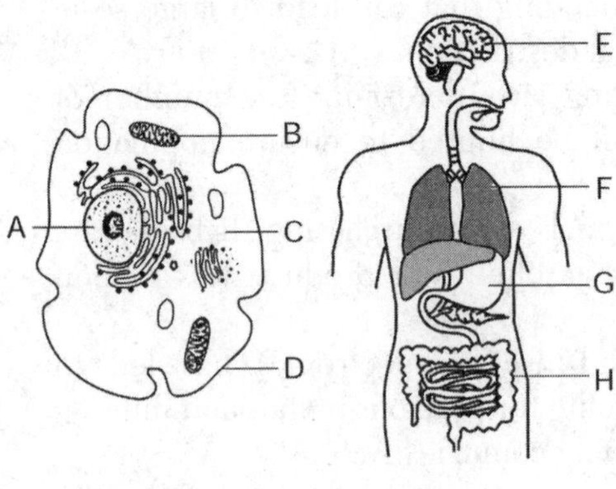

37 Which statement correctly identifies the levels of organization for the structures indicated?

1 *A* and *B* are tissues; *E* and *G* are organs.
2 *A* and *B* are organs; *E* and *G* are systems.
3 *A* and *B* are tissues; *E* and *G* are organelles.
4 *A* and *B* are organelles; *E* and *G* are organs. 37____

38 Cells from structure *E* and cells from structure *G* are similar in that they

1 have the same structure and function
2 contain the same genetic material
3 are identical in structure, but different in function
4 contain only the genetic information needed for their specific job

38_____

39 Rotenone is an insecticide that is toxic to humans as well as to insects. Rotenone interferes with the process of ATP production in the cell. Which row in the chart below correctly identifies the structure where ATP is produced and the reason it is affected by rotenone?

Row	Structure	Reason Affected
(1)	A	It would be unable to store enzymes for ATP production.
(2)	B	Production of ATP would occur less efficiently.
(3)	C	The raw materials used for ATP production would be altered.
(4)	D	Absorption of the ATP would increase here.

39_____

Base your answers to questions 40 and 41 on the information and diagram below and on your knowledge of biology.

A student used a microscope like the one represented below to observe cell division in onion cells.

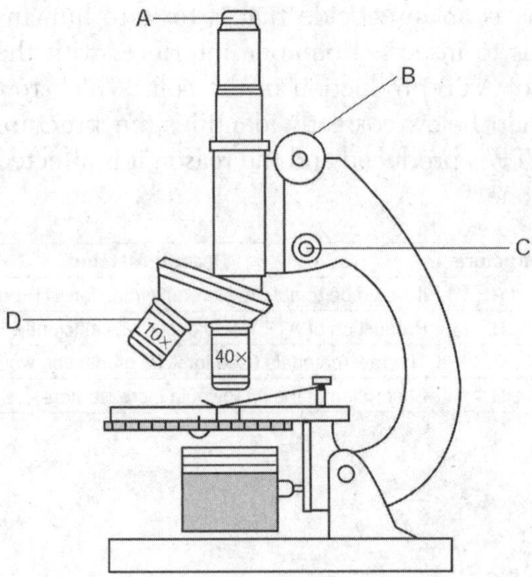

40 The part of the microscope that should be adjusted in order to better view the onion cells while using high power is

1 *A* 3 *C*
2 *B* 4 *D* 40____

41 The student noticed that as the new cells formed, they contained rod-shaped chromosomes. It is necessary for onion cells to contain chromosomes because chromosomes

1 are composed of genes that contain the instructions for an organism's traits
2 are made of carbohydrates and are needed as an energy source
3 direct the production of inorganic molecules within the cell
4 are composed of lipids that contain stored nutrients for the new cell 41____

42 The diagram below indicates a few of the many varieties of domestic dogs thought to have originated from wolves that were domesticated thousands of years ago.

(Not drawn to scale)

Adapted from: http://evolution.berkeley.edu/evolibrary/article

The many varieties of domesticated dogs were most likely produced as a result of

1 mutating the body cells of the dogs
2 selective breeding over many generations
3 genetic engineering with specific enzymes
4 cloning dogs with desirable traits 42 ____

43 The graph below represents some changes in the number of individuals in a particular population in a stable ecosystem over a period of time.

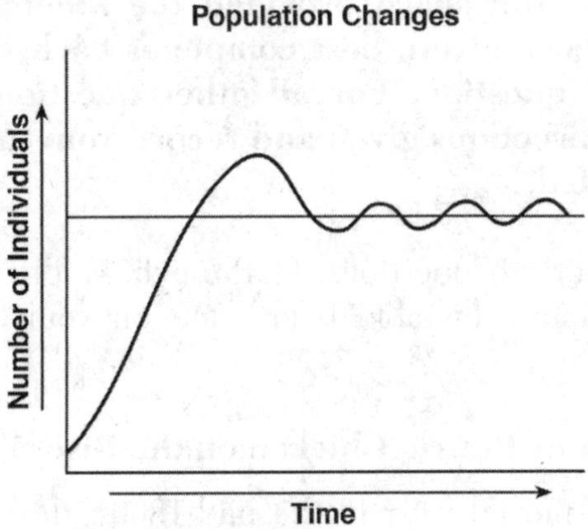

Which statement best describes the trend shown in this graph?

1 Ecosystem conditions will eventually cause a population to become extinct.
2 In a stable ecosystem, the number of individuals in a population is usually maintained within a certain range.
3 The interactions between a population and various factors in an environment are always predictable.
4 In order for any ecosystem to maintain a balance, populations must be reduced to half their original number.

43____

PART B–2

Answer all questions in this part. [12]

***Directions* (44–55): For those questions that are multiple choice, record in the space provided the *number* of the choice that, of those given, best completes each statement or answers each question. For all other questions in this part, follow the directions given and record your answers in the spaces provided.**

Base your answers to questions 44 through 47 on the information, and data table below and on your knowledge of biology.

Illinois Greater Prairie Chicken on the Rise

As pioneers moved west in the mid-1800s, the greater prairie chicken population in Illinois was estimated to number in the millions. Since then, their population has drastically declined.

Evidence of the rapidly declining population was obtained from studying the number of eggs that hatched over several years. In Jasper County, Illinois, the number of prairie chickens fell from 2,000 to less than 50 in under 35 years. Researchers compared the DNA from feather samples from the living Illinois chickens to the DNA from feather samples dating from the year 1930 found in a museum. It was found that the living Illinois chicken population had a very low level of genetic diversity.

In 1992, researchers attempted to increase genetic variation by transporting more than 500 healthy prairie chickens into Illinois from the states of Minnesota, Kansas, and Nebraska. The data table below shows the changes in the percent of eggs that hatched from samples taken in different years. Researchers documented that this increase in the percent of eggs that hatched was not influenced by environmental events.

Prairie Chicken Egg Hatching

Years	Percent of Eggs Hatched
1970–1974	89
1975–1979	88
1980–1984	83
1985–1989	78
1990	38
1993–1996	94

Adapted from: http://www.sciencedaily.com/releases/1998/11/981130045644.htm

***Directions* (44–46): Using the information given, construct a bar graph on the grid following the directions below.**

44 Label the y-axis on the line provided. [1]

45 Mark an appropriate scale, without any breaks in the data, on the y-axis. [1]

46 Construct vertical bars to represent the data. Shade in *each* bar. [1]

Hatching Rate

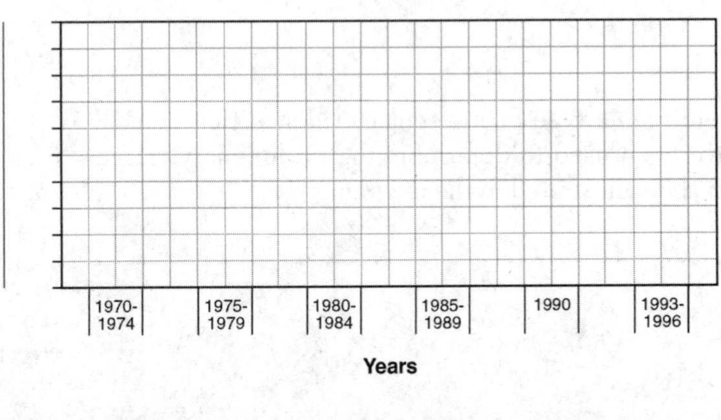

Note: The answer to question 47 should be recorded in the space provided.

47 The scientists transported prairie chickens from three different states into the state of Illinois in order to

1 decrease egg hatching rate
2 increase genetic diversity
3 increase egg fertilization
4 develop different feather colors 47_____

Base your answers to questions 48 and 49 on the diagram below and on your knowledge of biology. The diagram shows the evolutionary history of several plant species.

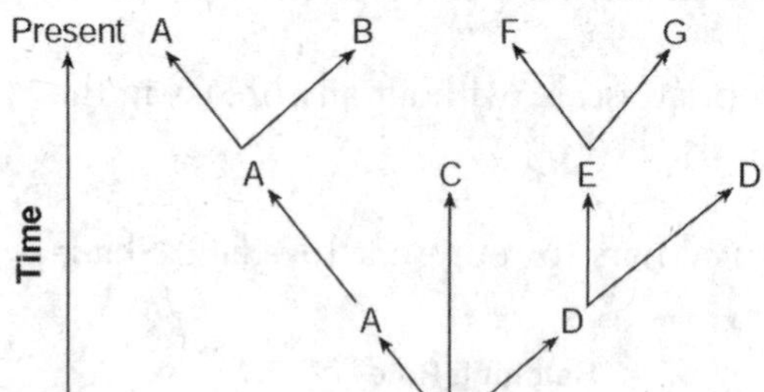

48 Identify *one* type of molecular evidence that could have been used to determine the evolutionary relationships illustrated in the diagram. [1]

____

Note: The answer to question 49 should be recorded in the space provided.

49 Which biological technique could be used to obtain some structural evidence that species *A* and *B* are closely related?

1 glucose testing
2 cloning
3 genetic engineering
4 dissection

49____

Base your answers to questions 50 and 51 on the diagram below and on your knowledge of biology. The diagram illustrates the steps in a process that occurs in the cells of many organisms.

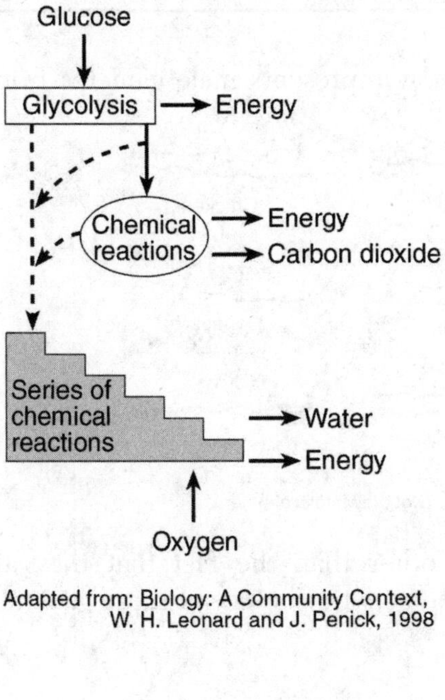

Adapted from: Biology: A Community Context,
W. H. Leonard and J. Penick, 1998

Note: The answer to question 50 should be recorded in the space provided.

50 Based on the diagram, the process of glycolysis most likely

1 begins the breakdown of glucose
2 produces oxygen for organisms to use
3 stores energy in molecules of water and carbon dioxide
4 recycles glucose within the cells of simple organisms 50_____

51 Identify *one* specific molecule used to store the energy being released during this process. [1]

Molecule: ______________________________

____

52 The diagram below represents male gametes from different animals.

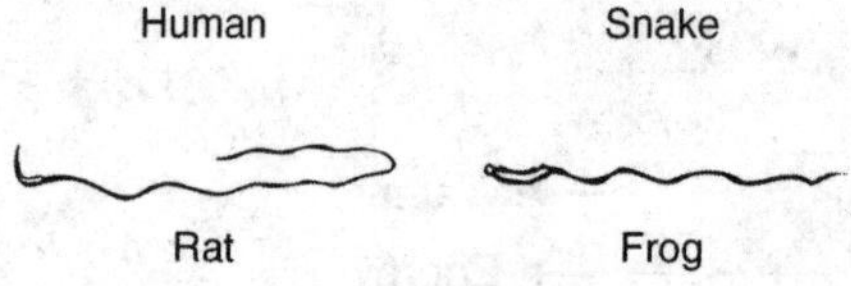

(Not drawn to scale)

State *one* way, other than the fact that they all contain DNA, the genetic content of these gametes is similar. [1]

____

____

53 Part of a food web is represented below. It includes organisms located in a stream near farm fields.

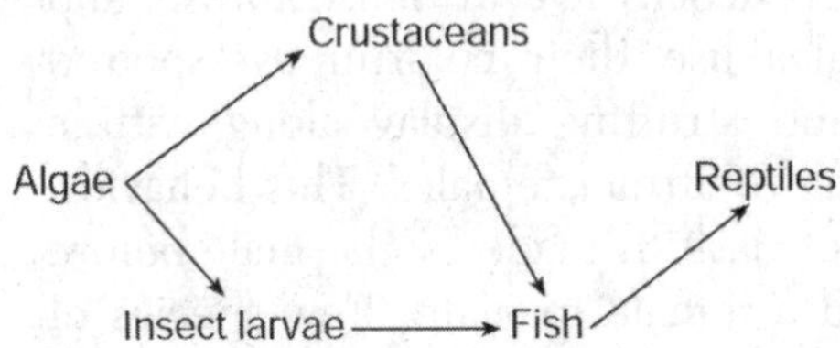

Select *one* type of organism, other than the crustaceans, from the food web. State how the population of organisms you selected might be affected if the population of crustaceans in this food web were reduced due to the use of chemicals harmful to crustaceans in the fields near the stream. Support your answer. [1]

Organism: ______________________________

____

____

Base your answer to question 54–55 on the information below and on your knowledge of biology.

Indian blue peacocks live in dense forests and scrubland. Males use their colorful eye-spotted tail feathers and strutting display along with a loud mating call to attract females. This behavior, known as a hoot-dash, is made by the male before dashing toward a female to mate. Few species of other animals use loud courtship calls as the call attracts predators and uses energy.

54–55 Discuss the use of the hoot-dash by male peacocks. In your answer, be sure to:

- state *one* advantage of the hoot-dash [1]
- state *one disadvantage* of the hoot-dash [1]

____

____

____

____

PART C

Answer all questions in this part. [17]

***Directions* (56–72): Record your answers in the spaces provided.**

Base your answers to questions 56 and 57 on the information below and on your knowledge of biology.

The placenta secretes progesterone and estrogen during pregnancy. Progesterone is responsible for the following functions:

- maintains the lining of the uterus
- inhibits (interferes with) contractions of the uterus
- inhibits the production and release of eggs

56 Describe *one* likely result if the placenta became damaged and could *not* maintain progesterone levels. Support your answer. [1]

____

____

57 Explain how the release of additional eggs is prevented during pregnancy. [1]

____

____

Base your answers to questions 58 and 59 on the information below and on your knowledge of biology.

"Cancer is a disease of genes gone wrong. When certain genes mutate, they make cells behave in odd ways. The cells divide swiftly, they hide from the immune system that could kill them and they gain the nourishment they need to develop into tumors ...."

Source: Carl Zimmer, *NY Times*, February 6, 2014

58 Explain why the body of a person infected with HIV, the virus that causes AIDS, would have a different immune response to the presence of cancer cells than a person *not* infected with HIV. [1]

____

____

59 Explain why certain chemicals and radiation sources are risk factors for cancer. [1]

____

____

60 Scientists have learned that when a pregnant woman smokes, one of the chemicals absorbed, nicotine, can narrow the diameter of her blood vessels that lead to the placenta. Explain why narrowing the diameter of these blood vessels can result in low birth weight babies. [1]

____

____

Base your answers to questions 61 through 63 on the information below and on your knowledge of biology.

In an experiment to test the effectiveness of a new vaccine, 50 rats received an injection of equal doses of the vaccine and 50 other rats received an injection of equal doses of a weak salt solution. Two months later, all of the rats received injections that contained equal doses of live, disease-causing organisms.

The experimental results are shown in the chart below.

Effectiveness of a New Vaccine

Injection: 50 Rats Received	Number of Rats That Developed the Disease	Number of Rats That Did Not Develop the Disease
vaccine	7	43
weak salt solution	48	2

61 Was the vaccine effective in preventing the disease? Use the information in the data table to support your answer. [1]

____

____

62 State *one* possible reason why two of the rats did *not* get sick even though they did *not* receive the vaccine. [1]

____

____

63 Do the results of this experiment indicate that the vaccine is ready for human testing? Support your answer with information from the table. [1]

____

____

Base your answers to questions 64 through 66 on the information below and on your knowledge of biology.

Saving Florida Oranges

A disease that affects orange trees has led to the destruction of numerous orange trees in Florida. Orange growers have unsuccessfully tried to stop the spread of the disease by cutting down infected trees and using a variety of pesticides on the insects that spread the disease. The growers fear that if nothing further is done, entire crops could be wiped out in the near future. In hopes of saving Florida's orange industry, scientists are attempting to alter the DNA of orange trees by inserting DNA, that codes for disease resistance, from a different plant species.

64 State the name of the specific technique that is used to alter the DNA of orange trees. [1]

____

65 Explain why growing oranges with disease resistance is better for the environment than using pesticides to control the spread of the disease. [1]

____

____

66 Identify a trait, other than disease resistance, that the orange trees could have that would be beneficial to the growers. [1]

____

67 The diagram below represents two energy pyramids. Each pyramid represents the productivity of one acre of land.

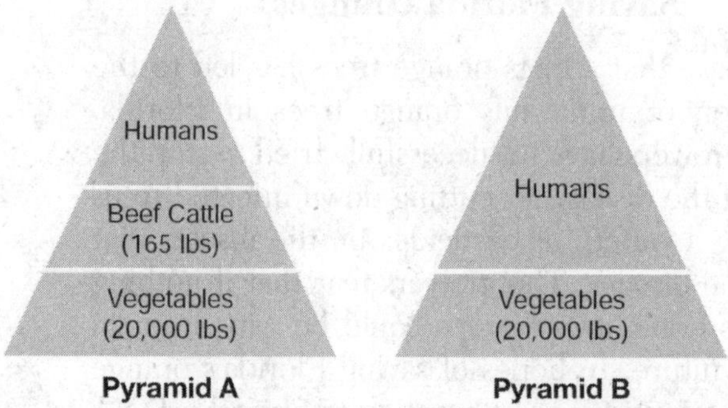

Based on the concept of energy transfer, explain why one acre of land can produce more vegetables for human consumption than beef for human consumption. [1]

____

____

Base your answers to questions 68 through 72 on the information below and on your knowledge of biology.

Nature Will Have to Clean Up Hawaii Molasses Leak That Killed Thousands of Fish

A massive spill of thick molasses has turned Honolulu Harbor into a watery wasteland where thousands of fish have been suffocated—a disaster that officials say Mother Nature will have to clean up.

"There's nothing alive there at all," diver Roger White told NBC affiliate KHNL after making a seven-minute video of dead sea life blanketing the bottom of the harbor ....

..."Unlike with an oil spill, it's a sugar product so it will dissipate on its own," Matson spokesman Jeff Hull told NBC News on Thursday. "There's not an active cleanup."

"The molasses is not toxic but it's heavier than water so it's spreading around on the sea floor, displacing the oxygen-rich water down there, and the fish are suffocating," said Keith Korsmeyer, a professor of biology at Hawaii Pacific University.

The die-off also could lure predators like sharks, barracuda and eels to the harbor and neighboring Keehi Lagoon, experts warned....

...Korsmeyer said marine life would probably repopulate the harbor, after the low oxygen water moves out, but that could take months or even years....

Source: http://www.nbcnews.com/news/other/nature-will-have-clean-hawaii-molasses-leak-killed-thousands-fish-f8C11137030

68 Explain how the molasses spill caused many of the animals to die off. [1]

____

____

69 State *one* reason why the die-off could attract sharks, barracuda, and eels to the harbor and lagoon. [1]

____

____

70 Identify *one* group of organisms responsible for the recycling of dead sea life that is *not* mentioned in the article. [1]

____

71 Explain why it is important to preserve the biodiversity of the Honolulu Harbor ecosystem. [1]

____

____

72 Predict what will most likely happen to this ecosystem in 20 years if no other disasters occur. [1]

____

____

PART D

Answer all questions in this part. [13]

Directions **(73–85): For those questions that are multiple choice, record in the space provided the *number* of the choice that, of those given, best completes the statement or answers the question. For all other questions in this part, follow the directions given and record your answers in the spaces provided.**

Note: The answer to question 73 should be recorded in the space provided.

73 On various Galapagos islands, finch species have different diets: seeds, insects, flowers, the blood of seabirds, and leaves. This is evidence that each species has a different

1 mating behavior
2 nesting material
3 niche
4 predator

73____

Note: The answer to question 74 should be recorded in the space provided.

74 A person's pulse rate generally goes down while sleeping. One reason for this decrease is

1 the body is producing more carbon dioxide when a person is sleeping
2 a person is not using energy while sleeping
3 a person requires less oxygen when sleeping
4 the body is using more nutrients while asleep

74____

Base your answers to questions 75 through 77 on the information below and on your knowledge of biology.

Unique populations of giant tortoises live in the Galapagos Islands. This is the same group of islands where Darwin studied his famous finches. It is thought that the original ancestors of today's giant tortoises came from the mainland of South America. These animals left and drifted in the ocean to the islands where they began to live, reproduce, and evolve. Each of the islands has a different habitat. Each species varies in shell shape and body structure. Information about two of these species of giant tortoises and their habitats is found in the table below.

Giant Tortoises of the Galapagos Islands

Tortoise Shell Type	Habitat and Body Description
Dome-shaped 	• The island receives a lot of rain, and there is an abundant amount of available food (plant material). • The tortoises have shorter necks. They do not have to reach for food.
Saddle-backed 	• The land on the island is dry, so there is limited available plant food. • The tortoises have long necks and legs, which allow them to reach for scarce food.

Adapted from: BenchPrep.com/blog/AP Biology-evolution-part-1/

Note: The answer to question 75 should be recorded in the space provided.

75 Which statement best explains the differences observed in the tortoises on each island?

1 Each tortoise adapted to its environment during its lifetime.
2 Sudden mutations changed the appearance of all of the tortoises.
3 The tortoises grew different structures based on the available food.
4 Different adaptations gave some tortoises a better chance of survival. 75____

Note: The answer to question 76 should be recorded in the space provided.

76 When the saddle-backed tortoises extend their long necks out of the shell, they are unprotected from attack. Scientists hypothesize that during the evolution of this tortoise, they had few predators. Which statement best supports this hypothesis?

1 Competition between the predators and the tortoises would have resulted in a greater number of long-necked tortoises.
2 The number of predators was greater than the number of tortoises.
3 Saddle-backed tortoises were not a part of the food chain.
4 Predators would have killed the tortoises with long necks, leaving more tortoises with shorter necks. 76____

77 If a group of saddle-backed tortoises were brought to an island inhabited by dome-shaped tortoises, could both species survive? Circle yes *or* no and support your answer. [1]

Circle one: Yes *or* No

____

____

____

78 During the laboratory activity *The Beaks of Finches*, you obtained food under two conditions: with competition and with no competition. State *one* way the results obtained from these two conditions differed when you did this activity. [1]

____

____

79 An investigation was carried out to determine the effects of exercise on the human body. Identify *one* body system, other than the circulatory system, that becomes more active as a result of exercise. State *one* change the system you identified will undergo as a result of becoming more active. [1]

Body system: ____________________________

Change: ________________________________

____

Base your answers to questions 80 through 82 on the information below and on your knowledge of biology.

A student designed an experiment to investigate a claim that athletes would have lower heart rates than nonathletes during exercise. After the students classified themselves as an athlete or a nonathlete, their resting pulses were determined. Then all the students performed the same exercise for four minutes and their heart rates were determined by recording the pulse rate in beats per minute. The students continued to measure their pulse rates for an additional four minutes. The average heart rate per minute for each group was determined. The data were recorded, as shown on the table below.

Average Heart Rate Response to Exercise (beats per minute)

	Time (minutes)	**Athlete Students**	**Nonathlete Students**
Resting Pulse	0	68	72
Exercising Period	1	76	78
	2	82	90
	3	95	115
	4	110	130
After Exercise	5	100	125
	6	95	120
	7	85	100
	8	68	95

80 State *one* appropriate hypothesis for this experiment. [1]

____

____

Note: The answer to question 81 should be recorded in the space provided.

81 Which statement is best supported by the data in the chart?

1 After exercise, the nonathletic students had a lower heart rate than the athletic students.
2 After exercise, the heart rates of the athletic students returned to resting pulse in four minutes.
3 During exercise, both groups of students had the same increase above their resting pulse.
4 During exercise, the athletic students had a higher heart rate than the nonathletic students. 81____

Note: The answer to question 82 should be recorded in the space provided.

82 To improve the validity of the conclusion reached in this experiment, the students should repeat the experiment

1 disregarding any data that don't fit the hypothesis
2 with a larger number of athletes and non-athletes
3 comparing the heart rates and breathing rates of males and females
4 with the athletes doing different exercises than the nonathletes 82____

Base your answer to question 83 on the information below and on your knowledge of biology.

A student added glucose indicator to a beaker of an unknown liquid. Starch indicator was added to a different beaker containing an equal amount of the same unknown liquid. The color of the indicator solutions before they were added to the beakers and the color of the contents of the beakers after adding the indicator solutions are recorded in the chart below.

Beaker	Solution	Color of Indicator Solution Before Adding to Beaker	Color of Contents of Beaker After Adding Indicator Solution
1	unknown liquid + glucose indicator	blue	blue (after heating)
2	unknown liquid + starch indicator	amber	blue-black

83 State *one* conclusion the student would make about the unknown liquid based on the results. Support your answer with information from the data table. [1]

____

____

84 While getting ready to perform the *Making Connections* lab, a teacher did not have enough of the old wooden clothespins she was handing out to the students to squeeze in the lab. The teacher opened a bag of new plastic clothespins and handed them out to the students who had not received a wooden clothespin.

Explain why using new clothespins for some students and not others was an error in the experimental procedure. [1]

____

____

Base your answer to question 85 on the information and diagram below and on your knowledge of biology.

An unknown sample of DNA found at a crime scene was compared to DNA samples taken from three individuals. The results of the technique used to compare the samples are represented below.

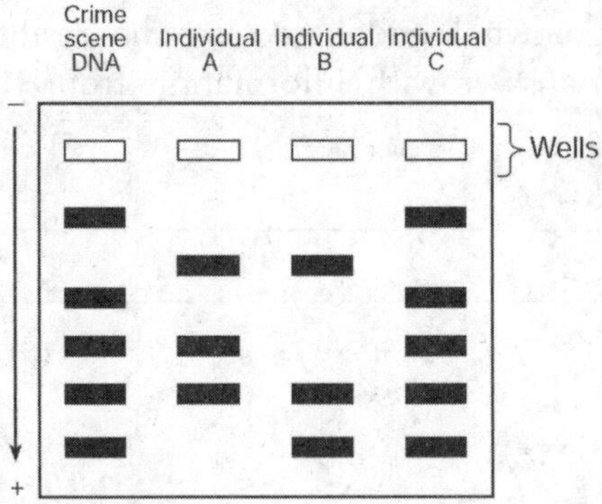

85 What factor causes the DNA fragments to move in this technique? [1]

____

Answers June 2017

Living Environment

Answer Key

PART A

1. 3	**6.** 2	**11.** 3	**16.** 3	**21.** 3	**26.** 1
2. 2	**7.** 4	**12.** 1	**17.** 1	**22.** 2	**27.** 1
3. 2	**8.** 1	**13.** 1	**18.** 2	**23.** 2	**28.** 4
4. 4	**9.** 4	**14.** 3	**19.** 3	**24.** 4	**29.** 2
5. 3	**10.** 1	**15.** 1	**20.** 3	**25.** 1	**30.** *

PART B–1

31. 2	**34.** 2	**36.** 4	**38.** 2	**40.** 3	**42.** 2
32. 3	**35.** 4	**37.** 4	**39.** 2	**41.** 1	**43.** 2
33. 4					

PART B–2

44. *See* Answers Explained.
45. *See* Answers Explained.
46. *See* Answers Explained.
47. 2
48. *See* Answers Explained.
49. 4
50. 1
51. *See* Answers Explained.
52. *See* Answers Explained.
53. *See* Answers Explained.
54. *See* Answers Explained.
55. *See* Answers Explained.

PART C. *See* **Answers Explained**.

PART D

73. 3
74. 3
75. 4
76. 4
77. *See* Answers Explained.
78. *See* Answers Explained.
79. *See* Answers Explained.
80. *See* Answers Explained.
81. 2
82. 2
83. *See* Answers Explained.
84. *See* Answers Explained.
85. *See* Answers Explained.

*Due to a typographical error, Question 30 does not have a correct answer choice. All students should be awarded credit for Question 30 regardless of the answer, if any, they record on the answer sheet for this question.

Answers Explained

PART A

1. **3** *Using plastic from a soda bottle to make a certain type of clothing* is an example of recycling. The soda bottle plastic is subjected to physical and chemical actions that result in a fabric-like material (known as fleece) that can be crafted into clothing and other products.

WRONG CHOICES EXPLAINED

(1) *Using a paper cup instead of a mug* is *not* an example of recycling. This action will use a renewable resource (paper) that will likely be discarded after use rather than being converted into a new paper product.

(2), (4) *Turning off the lights when leaving a room* and *using two paper towels rather than five to clean up a spill* are *not* examples of recycling. These actions are examples of conservation of resources (energy, paper). However, they do not result directly in the conversion of those resources into new products.

2. **2** *An interdependence of organisms* is one characteristic of a stable ecosystem. Typical stable ecosystems are characterized by many thousands of plants, animals, fungi, and bacteria, each filling its own environmental niche and interacting with other species in the environment.

WRONG CHOICES EXPLAINED

(1) *A high number of predators* is *not* one characteristic of a stable ecosystem. Typical stable ecosystems are characterized by a relatively low number of predators compared with other types of organisms, not a high number.

(3) *A lack of biodiversity* is *not* one characteristic of a stable ecosystem. Typical stable ecosystems are characterized by high biodiversity, not a lack of biodiversity.

(4) *An increase in human interference* is *not* one characteristic of a stable ecosystem. Typical stable ecosystems are characterized by a decrease, not an increase, in human interference.

3. **2** *Decreased use of nonrenewable resources* is a direct benefit of utilizing this technology to produce energy. Capturing the energy contained in ocean waves will mean that less energy will be needed from power plants fueled by coal, oil, and natural gas, which are all nonrenewable resources.

WRONG CHOICES EXPLAINED

(1) *Destruction of habitats near the devices* is *not* a direct benefit of utilizing this technology to produce energy. The destruction of habitats essential to the survival of native species cannot be described as a benefit.

(3) *Release of gases needed for photosynthesis* is *not* a direct benefit of utilizing this technology to produce energy. This technology presumably will not produce carbon dioxide, the gas needed for photosynthesis, into the environment.

(4) *Increased use of finite resources* is *not* a direct benefit of utilizing this technology to produce energy. This technology will decrease, not increase, the use of finite resources such as coal, oil, and natural gas.

4. **4** The presence of the emerald ash borer in the United States can be used as an example of how humans have *altered the equilibrium in an ecosystem by introducing a new species*. Such organisms are known as invasive species. They disrupt the natural balance of an ecosystem by competing with or physically harming native species.

WRONG CHOICES EXPLAINED

(1), (2), (3) It is *not* true that the presence of the emerald ash borer in the United States can be used as an example of how humans have *purposely introduced an insect to correct a problem in an ecosystem*, *used an insect to remove one insect species and replace it with another,* or *worked to increase the biodiversity in a particular area*. The question clearly states that the emerald ash borer was accidentally introduced to the United States from China. All of these responses imply an intentional introduction of the ash borer for some purpose.

5. **3** Graph *3* best shows the changes in the global human population and natural resource use over the past 500 years. This graph indicates that both human population and the use of natural resources have increased over time. This depiction matches actual historical events that have occurred during this timeframe. As the human population has increased, more and more natural resources have been used to support it.

WRONG CHOICES EXPLAINED

(1), (2), (4) Graphs *1*, *2*, and *4* do *not* best show the changes in the global human population and natural resource use over the past 500 years. Each of these graphs illustrates either decreasing human population or deceasing use of natural resources. Both of these measures have increased, not decreased, during this period.

6. **2** The burning of fossil fuels has harmed the environment by *adding carbon dioxide to the atmosphere*. Excessive carbon dioxide added to the atmosphere has created a greenhouse effect that traps solar heat and causes

Earth's temperature to rise. The effects of this global warming include climate change, habitat destruction, and species dislocation.

WRONG CHOICES EXPLAINED

(1) It is *not* true that the burning of fossil fuels has harmed the environment by *decreasing acid rain in the northeast United States*. Carbon dioxide dissolves in rainwater to produce a weak acid that falls as acid rain and causes negative environmental effects on the ground and in bodies of water. The burning of fossil fuels that produces carbon dioxide has increased, not decreased, the incidence of acid rain.

(3) It is *not* true that the burning of fossil fuels has harmed the environment by *increasing biodiversity in the lakes and ponds of the Adirondacks*. Many native fish populations have been eliminated from Adirondack lakes and ponds as waters have become increasingly acidic due to acid rain. The burning of fossil fuels that produces carbon dioxide has decreased, not increased, the biodiversity of lakes and ponds.

(4) It is *not* true that the burning of fossil fuels has harmed the environment by *depleting the ozone shield directly over western New York State*. The ozone shield is known to have been damaged by the release of chloroflourocarbons (CFCs) that serve as aerosol propellants and refrigerants. CFCs are not produced by the burning of fossil fuels. The burning of fossil fuels that produces carbon dioxide has no known effect on the ozone shield.

7. **4** *The population size is recorded for each species present in the construction area before and after the building is completed* is the method of collecting data that would provide the most accurate information about how an ecosystem is being affected by human development. In this method, the greatest number of different species will be studied both before and after the construction. Comparison of the gathered data will allow scientists to determine the effects of the development on the entire ecosystem.

WRONG CHOICES EXPLAINED

(1), (2) *The people in the neighborhood record the number of birds they see in the area both before and after the construction of a new building* and *deer in the area are identified before construction so that scientists can see where they go after the building is finished* are *not* the methods of collecting data that would provide the most accurate information about how an ecosystem is being affected by human development. In these methods, very limited numbers of species are studied. Therefore, little useful data will be gathered about the effects of the development on the entire ecosystem.

(3) *More trees are planted in the area around the construction site to allow the animals a new place to live* is *not* the method of collecting data that would provide the most accurate information about how an ecosystem is being affected by human development. This action does not represent a method of

data collection. Therefore, it will not result in any useful information about the effects of development on this ecosystem.

8. **1** Natural selection produces changes most quickly in *species with short reproductive cycles*. Genetic data, including mutations, are passed on through the reproductive process. Therefore, it is reasonable to assume that species with short reproductive cycles will produce new variations and respond to environmental changes more rapidly than species with long reproductive cycles.

WRONG CHOICES EXPLAINED

(2) It is *not* true that natural selection produces changes most quickly in *individual pathogens killed by antibiotics*. Killed organisms are not capable of reproduction, so they cannot change by natural selection.

(3) It is *not* true that natural selection produces changes most quickly in *complex multicellular organisms*. Complex multicellular organisms normally display many genetic traits, so genetic changes that lead to new variations are not as readily apparent in such species. In addition, complex multicellular organisms tend to have longer reproductive cycles than simpler organisms, so natural selection works more slowly.

(4) It is *not* true that natural selection produces changes most quickly in *individuals that produce a small number of offspring*. Natural selection is an evolutionary mechanism that works at the species level, not the individual level. Species that produce many offspring tend to change by natural selection more rapidly than species that produce a small number of offspring.

9. **4** *There is a lack of genetic variation for selection to act upon* is the statement that best explains why the Florida panther population may *not* continue to evolve. Evolution of any species depends on genetic variation, which is most readily found in species with extensive gene pools carried in many individuals. A population size of only 100 to 160 individuals is generally insufficient to support natural selection.

WRONG CHOICES EXPLAINED

(1) *There is no longer a chance of mutations occurring in the population* is *not* the statement that best explains why the Florida panther population may *not* continue to evolve. Mutations are random events in which genetic material is altered though the effects of mutagenic agents. Mutations can occur in any individual in a population at any time and be passed on through the reproductive process.

(2) *There is a lack of competition for limited environmental resources* is *not* the statement that best explains why the Florida panther population may *not* continue to evolve. Any time an environmental resource becomes limited, competition intensifies among individuals of a species requiring that resource.

(3) *There is no longer a chance of a trait providing a reproductive advantage to the population* is *not* the statement that best explains why the Florida panther population may *not* continue to evolve. In such a small population, it is likely that all, or nearly all, members of the population will share identical traits that provide reproductive advantage.

10. **1** *Level A—producers* is the level of the pyramid that is correctly paired with the type of organism that would most likely be found at that level in an ecosystem. In any stable ecosystem, producer organisms (green plants) are the most abundant types of organisms. By convention, producers are shown at the base of a food pyramid because they serve as the energy foundation that supports all the upper levels.

WRONG CHOICES EXPLAINED

(2) *Level B—carnivores* is *not* the level of the pyramid that is correctly paired with the type of organism that would most likely be found at that level in an ecosystem. By convention, herbivores are shown at Level *B* because they directly depend on the producers for their nutritional energy.

(3) *Level C—herbivores* is *not* the level of the pyramid that is correctly paired with the type of organism that would most likely be found at that level in an ecosystem. By convention, first-order carnivores are shown at Level *C* because they directly depend on the herbivores for their nutritional energy.

(4) *Level D—decomposers* is *not* the level of the pyramid that is correctly paired with the type of organism that would most likely be found at that level in an ecosystem. By convention, top-order carnivores are shown at Level *D* because they directly depend on the herbivores and first-order carnivores for their nutritional energy.

11. **3** *The death rate will increase* is the statement that best describes what is most likely to occur if an animal population grows larger than the carrying capacity of its environment. The carrying capacity of an environment is the maximum number of individuals of a species that can be supported given limited food and other resources. When the carrying capacity is exceeded, there will be a growing scarcity of the limited resource until individuals of the species begin to die from lack of that resource. These deaths will tend to reduce the species population and help to bring it back into balance, although that may require several generations to accomplish.

WRONG CHOICES EXPLAINED

(1) *The birth rate will increase* is *not* the statement that best describes what is most likely to occur if an animal population grows larger than the carrying capacity of its environment. During periods of extreme deprivation, such as would occur in this circumstance, the fertility rate would tend to decrease and infant mortality would tend to increase. The overall birth rate would tend to decrease, not increase.

(2), (4) *Both the birth rate and death rate will decrease* and *neither the birth rate nor the death rate will decrease* are *not* the statements that best describe what is most likely to occur if an animal population grows larger than the carrying capacity of its environment. In this circumstance, the birth rate would tend to decrease while the death rate would tend to increase.

12. **1** *A squirrel cannot find a mate* is an example of an organism being directly affected by a biotic factor. Biotic factors are those associated with organisms and the life activities of those organisms. Reproduction is a major life activity; therefore, the inability to find a mate represents a biotic factor in the squirrel's environment.

WRONG CHOICES EXPLAINED

(2), (3), (4) *A flood washes away a maple tree*, *a plant is in a dark room*, and *a chipmunk finds a rock pile to use for a home* are *not* examples of organisms being directly affected by biotic factors. In each of these responses, an abiotic (nonliving) factor is at work and affecting the organisms. These abiotic factors are a flood, a darkened room, and a rock pile, respectively.

13. **1** When people receive organ transplants, they often need to take medications that decrease immune responses because *transplanted organs contain antigens that can trigger white blood cell activity*. These white blood cells respond to the foreign antigens by attaching to and destroying the cells of the transplanted organ. Without immunosuppressant medications, this immune response can result in the death and rejection of the transplanted organ.

WRONG CHOICES EXPLAINED

(2), (3) It is *not* true that when people receive organ transplants, they often need to take medications that decrease immune responses because *hormones present in replacement organs prevent the synthesis of antibiotics* or because *transplanted organs produce their own antibiotics*. Antibiotics are substances that are synthesized naturally in certain fungi and, more recently, in pharmaceutical laboratories. Antibiotics are not synthesized by transplanted organs.

(4) It is *not* true that when people receive organ transplants, they often need to take medications that decrease immune responses because *antigens present in these organs attack antibodies already present in the blood*. Antibodies attack antigens; antigens do not attack antibodies.

14. **3** *Guard cells* are the structures that regulate water loss and gas exchange in the leaves of plants. The guard cells surround small openings (stomates) on the leaf surface. When the plant has an adequate supply of moisture, the guard cells absorb water and become turgid, opening the

stomates. This allows water vapor and oxygen to diffuse out of the leaf and allows carbon dioxide to diffuse into the leaf.

WRONG CHOICES EXPLAINED

(1) *Vacuoles* are *not* the structures that regulate water loss and gas exchange in the leaves of plants. Vacuoles are cell organelles that store food and wastes inside a living cell.

(2) *Chloroplasts* are *not* the structures that regulate water loss and gas exchange in the leaves of plants. Chloroplasts are cell organelles that contain the structures and enzymes necessary for photosynthesis inside a living cell.

(4) *Mitochondria* are *not* the structures that regulate water loss and gas exchange in the leaves of plants. Mitochondria are cell organelles that contain the structures and enzymes necessary for respiration inside a living cell.

15. **1** *Respiration* is the life process carried out by a green plant that is represented in the diagram. In the diagram, oxygen and food are taken in and carbon dioxide and water are released. Cellular energy in the form of ATP is the main product released in this reaction. Respiration is carried out by the vast majority of living things on Earth, not just green plants.

WRONG CHOICES EXPLAINED

(2) *Photosynthesis* is *not* the life process carried out by a green plant that is represented in the diagram. A diagram representing photosynthesis would show carbon dioxide and water as raw materials and oxygen and food as products. It would also show energy entering the reaction in the form of sunlight.

(3) *Digestion* is *not* the life process carried out by a green plant that is represented in the diagram. A diagram representing digestion would show complex food as raw materials and soluble food molecules as products of a reaction catalyzed by hydrolytic enzymes.

(4) *Replication* is *not* the life process carried out by a green plant that is represented in the diagram. A diagram representing replication would show DNA unwinding to accept free complementary nucleotides in its open nucleotide bonds. The resulting DNA strands would be identical to the original strand.

16. **3** Scientists are developing a malaria vaccine that would most likely contain *a weakened form of the malaria-causing organism*. In order to be effective at stimulating the immune response is its recipients, a vaccine must contain antigens of the pathogens that cause the disease. This may be accomplished most effectively by killing or weakening the pathogens and introducing these killed or weakened pathogens into the bloodstream of the recipient. The recipient will then normally produce specific antibodies to protect him or her from live, strong pathogens.

WRONG CHOICES EXPLAINED

(1), (2), (4) It is *not* true that scientists are developing a malaria vaccine that would most likely contain *weakened drugs used to treat the symptoms of malaria*, *white blood cells from the malaria-infected individuals*, or *antibodies made from the malaria-causing organism*. None of these techniques is used to formulate vaccines against malaria or any other known disease.

17. **1** Proteins, starch, and DNA are similar in that they are all *organic compounds*. Each of these substances is built around a chain of carbon atoms to which atoms of hydrogen, oxygen, and other elements are attached.

WRONG CHOICES EXPLAINED

(2) It is *not* true that proteins, starch, and DNA are similar in that they are all *parts of genes*. Of the substances mentioned, only DNA is part of genes.

(3) It is *not* true that proteins, starch, and DNA are similar in that they are all *made of amino acids*. Of the substances mentioned, only proteins are made of amino acids.

(4) It is *not* true that proteins, starch, and DNA are similar in that they are all *made of simple sugars*. Of the substances mentioned, only starch is made of simple sugars.

18. **2** In response to an increasing blood glucose level, the human body will normally *release a hormone that lowers the blood glucose*. The hormone in question is insulin. Its principal action is to stimulate the conversion of simple glucose dissolved in the blood into complex glycogen, which is then stored in the liver and muscles until needed.

WRONG CHOICES EXPLAINED

(1) It is *not* true that in response to an increasing blood glucose level, the human body will normally *store the glucose in cell nuclei*. Glucose cannot be stored in the cell unless it is converted to glycogen.

(3) It is *not* true that in response to an increasing blood glucose level, the human body will normally *produce a hormone that destroys the glucose*. No known human process destroys glucose, although it is broken down into carbon dioxide and water during the respiratory reactions.

(4) It is *not* true that in response to an increasing blood glucose level, the human body will normally *use the excess glucose to make proteins*. Proteins are constructed from amino acid molecules, not glucose molecules.

19. **3** *Offspring receive genetic information from both parents* is the statement that helps explain why the kitten has black fur. The chart indicates that the mother of the kitten has black fur and the father has striped fur. Although the kitten received genetic information regulating its fur color from both parents, it is likely that the black fur color is dominant over the striped fur color and thus is the visible phenotype in the kitten.

WRONG CHOICES EXPLAINED

(1) *Chromosomes present on the genes code for the characteristics of its fur* is *not* the statement that helps explain why the kitten has black fur. Chromosomes are not present on genes. Genes present on chromosomes code for genetic characteristics.

(2) *Genetic mutations always cause the fur color and eye color to change* is *not* the statement that helps explain why the kitten has black fur. Mutations are random events that may result in genetic change only if they occur in gametes that participate in sexual reproduction. Mutations do not always cause any particular characteristic to change.

(4) *Gene expression is changed in every generation, resulting in evolution* is *not* the statement that helps explain why the kitten has black fur. The continuity of species depends on the relatively unchanging nature of gene expression.

20. **3** *Cloning* is the technique that would use DNA from an extinct species to recreate an organism of the species. Although theoretical at this point, it might become technically feasible in the future to extract the DNA of an extinct animal and use it to replace the DNA of the nucleus of a fertilized egg of a similar extant species. If successful, that egg could develop into a viable offspring that could then be raised to adulthood.

WRONG CHOICES EXPLAINED

(1) *Natural selection* is *not* the technique that would use DNA from an extinct species to recreate an organism of the species. Natural selection is not a laboratory technique but a natural process by which favorable genetic traits are selected for survival in a changing environment.

(2) *Differentiation* is *not* the technique that would use DNA from an extinct species to recreate an organism of the species. Differentiation is not a laboratory technique but a natural process by which the cells of an embryo specialize into functional tissues, organs, and systems.

(4) *Selective breeding* is *not* the technique that would use DNA from an extinct species to recreate an organism of the species. Selective breeding is a technique used by animal and plant breeders in which organisms with desirable traits are cross-bred with the hope of producing offspring that also display those traits.

21. **3** *Sunlight → plant → rabbit → wolf* is the sequence that best illustrates the flow of energy through an ecosystem. Energy always enters an ecosystem in the form of sunlight that is absorbed by green plants. Plants capture this radiant energy and convert it to the chemical bond energy of glucose. When an herbivore such as a rabbit consumes plants, this energy is released into the cells of the rabbit as ATP energy. When the rabbit is consumed by a predator such as a wolf, the energy is passed on to the wolf, which also derives ATP energy from this nutritional process.

WRONG CHOICES EXPLAINED

(1), (2), (4) *Sunlight → plant → wolf → rabbit, plant → sunlight → rabbit → wolf*, and *wolf → rabbit → plant → sunlight* are *not* the sequences that best illustrate the flow of energy through an ecosystem. Each of these sequences contains one or more items out of correct order. See the correct answer above.

22. **2** *The deletion of a single DNA base in a sperm cell of a trout* is the cellular change in an organism that could be inherited by the next generation. Because the change is genetic and because it occurred in a gamete of the trout, it may be passed along to the next generation through the reproductive process. This will occur only if the sperm cell containing the change successfully fertilizes an egg cell.

WRONG CHOICES EXPLAINED

(1), (3), (4) It is *not* true that *a change in the ribosomes in the pancreas of a squirrel, a decrease in the size of a vacuole in a rose leaf cell*, or *the transfer of a piece of a chromosome in the skin cell of a raccoon* are cellular changes in an organism that could be inherited by the next generation. All of these changes are described as occurring in somatic (body) tissues, not in gametic (reproductive) tissues. These changes cannot be passed on to the next generation because there is no way for them to become encoded in sperm cells or in egg cells of these species.

23. **2** One likely reason for the decreased effectiveness of this chemical is that these bacteria have *a mutation for resistance*. Such a mutation would provide resistance to the chemical to individual bacteria that inherit the mutation, giving them an adaptive advantage over other bacteria in the presence of the chemical. As the nonresistant bacteria are killed off by the chemical, the resistant bacteria will continue to reproduce and thrive.

WRONG CHOICES EXPLAINED

(1) It is *not* true that one likely reason for the decreased effectiveness of this chemical is that these bacteria have *slower metabolic rates*. Each bacterial species has a unique metabolic rate that is affected by the temperature and the presence of appropriate resources. Metabolic rate does not explain this phenomenon.

(3) It is *not* true that one likely reason for the decreased effectiveness of this chemical is that these bacteria have *been selectively bred for survival*. Bacteria reproduce asexually. So they cannot be "bred," only "cultured." It is unlikely that scientists would attempt to culture resistant bacteria.

(4) It is *not* true that one likely reason for the decreased effectiveness of this chemical is that these bacteria have *an adaptation to a different niche*. The niches (or ecological roles) of bacteria may vary species by species

according to mutations the bacteria receive and the selection pressures placed upon them by natural events or human intervention. In a sense, the resistant bacteria described in the question have adapted to a different niche due to the selection pressures provided by the chemical in the hand sanitizers and dish detergents.

[NOTE: Due to the wording of this question, choice (4) may be regarded as a correct answer, although it is not the best answer.]

24. **4** These insects can upset natural ecosystems because they *disrupt habitats that native species depend upon*. Invasive species can alter the natural balance of an ecosystem by competing with or physically harming native species. As more and more native hemlock trees are destroyed by the invasive hemlock wooly adelgid, the environmental niche that the hemlocks held in their Adirondack habitat will go unfilled or will be filled by other species.

WRONG CHOICES EXPLAINED

(1), (2) It is *not* true that these insects can upset natural ecosystems because they *provide food for native bird species* or *can carry diseases that can be spread to pets*. The hemlock wooly adelgid may, in fact, serve as food for native birds or carry diseases that infect household pets. However, no information is provided in the question that would allow a determination to be made concerning these potential roles.

(3) It is *not* true that these insects can upset natural ecosystems because they *increase biodiversity in New York State forests*. The destructive actions of the hemlock wooly adelgid decrease, not increase, biodiversity in the Adirondack forest environment.

25. **1** Some species may have similar body structures even if they are not related because they evolved in *similar environments and specific traits increased their chances of survival*. It is likely that sharks and dolphins both evolved in a marine environment. Sharks are extremely ancient animals that have evolved from earlier fish ancestors. Marine mammals such as the dolphin are thought to have evolved from land-based ancestors that inhabited the ocean edge. These ancestors likely found it an advantage to hunt for food in the water itself. Early dolphins that were able to swim faster could catch more food, could avoid predators, and could be more successful reproducers. Adaptations such as fin-shaped limbs, sleek body lines, powerful lower-body muscles, and breathing holes on top of the head proved successful and provided adaptive advantages to the early dolphins that displayed them. Successful adaptations were passed on to succeeding generations, while unsuccessful adaptations were not. These physical traits evolved slowly over millions of years and were caused by the forces of genetic mutation, reproduction, and natural selection.

WRONG CHOICES EXPLAINED

(2) It is *not* true that some species may have similar body structures even if they are not related because they evolved in *similar environments and were exposed to factors that caused exactly the same mutations*. It is likely that sharks and dolphins both evolved in a marine environment. However, mutation is a random event that produces changes in the genetic material of cells. It is extremely unlikely that two different species would receive exactly the same mutations.

(3), (4) It is *not* true that some species may have similar body structures even if they are not related because they evolved in *different environments, but tried to adapt in the same ways so they could survive* or *different environments, but ate similar foods that affected their growth and development*. It is likely that sharks and dolphins both evolved in a marine environment. Explanations that imply that choice, need, or acquisition were forces at work in this or any evolutionary process are incorrect.

26. **1** *Meiosis and recombination* are the two biological processes that account for this difference between the puppies. Meiosis is a process in which homologous chromosome pairs and the genes they carry are separated into haploid gametes during gametogenesis. Recombination is a process in which these chromosome pairs are reestablished in the zygote during fertilization. The resulting new gene combinations are unique for each individual offspring. These two processes together are responsible for creating the variability within sexually reproducing species.

WRONG CHOICES EXPLAINED

(2) *Meiosis and cloning* are *not* the two biological processes that account for this difference between the puppies. Cloning is a laboratory technique that, if successful, results in an exact genetic duplicate of the cloned organism. Cloning does not account for variation within a species.

(3) *Mitosis and differentiation* are *not* the two biological processes that account for this difference between the puppies. Mitosis is a process in which homologous chromosome pairs of a parent nucleus are replicated exactly and separated into two genetically identical daughter nuclei. Differentiation is a process in which embryonic cells specialize to form specific tissues, organs, and systems of the adult organism. Neither of these processes accounts for variation within a species.

(4) *Mitosis and cloning* are *not* the two biological processes that account for this difference between the puppies. Mitosis is a process in which homologous chromosome pairs of a parent nucleus are replicated exactly and separated into two genetically identical daughter nuclei. Cloning is a laboratory technique that, if successful, results in an exact genetic duplicate of the cloned organism. Neither of these processes accounts for variation within a species.

27. **1** Cellular structure *1* would be the most likely location for the synthesis of these enzymes. Structure *1* illustrates a ribosome, an organelle known to be the site of protein synthesis in the cell. Digestive enzymes are types of proteins. So they would be synthesized at this location in the cell.

WRONG CHOICES EXPLAINED

(2) Cellular structure 2 would *not* be the most likely location for the synthesis of these enzymes. Structure 2 illustrates a vacuole, an organelle known to be the site of material storage in the cell, not protein synthesis.

(3) Cellular structure 3 would *not* be the most likely location for the synthesis of these enzymes. Structure 3 illustrates a mitochondrion, an organelle known to be the site of aerobic respiration in the cell, not protein synthesis.

(4) Cellular structure 4 would *not* be the most likely location for the synthesis of these enzymes. Structure 4 illustrates a cell membrane, an organelle known to be the site of material transport into and out of the cell, not protein synthesis.

28. **4** This effect on the heart rate can be dangerous because it can lead to *an imbalance in homeostasis*. Homeostasis is the chemical and physical balance that makes it possible for living things to react to changes in their environments. When this balance is disrupted, essential biological processes can fail. Some people have become ill or have even died as a result of rapid heat rates caused by caffeine overdoses.

WRONG CHOICES EXPLAINED

(1) It is *not* true that this effect on the heart rate can be dangerous because it can lead to *a disruption in the absorption of starch*. Starch is a complex carbohydrate that is too large to be absorbed into living cells.

(2), (3) It is *not* true that this effect on the heart rate can be dangerous because it can lead to *an increase in blood volume* or *a decrease in oxygen levels*. These are not documented direct effects of caffeine on the human body.

29. **2** The *circulatory and excretory* systems work together to perform this function. The circulatory system is responsible for absorbing wastes from body tissues and transporting them (via the blood) to excretory organs for elimination. The excretory system serves to collect (kidney), store (bladder), and expel metabolic wastes from the body. The mention of specific organs in the question makes the correct response to this question clear.

WRONG CHOICES EXPLAINED

(1) The *immune and respiratory* systems *do not* work together to perform this function. The immune system functions in the protection of the body from disease caused by foreign pathogens. The respiratory system provides a moist surface for the absorption of oxygen gas and its transfer to the blood.

The blood, kidney, and bladder are not organs associated with the immune or respiratory systems.

(3) The *skeletal and nervous* systems *do not* work together to perform this function. The skeletal system provides a mechanical system of supports, articulations, and protections to give the body form and enable locomotion. The nervous system is made up of tissues that sense the environment around the body and enable it to respond to environmental stimuli. The blood, kidney, and bladder are not organs associated with the skeletal or nervous systems.

(4) The *digestive and circulatory* systems *do not* work together to perform this function. The digestive system functions to take in, digest, and egest food materials in the body. The blood, kidney, and bladder are not organs associated with the digestive system.

30. [**no correct answer***] Scientists have recently discovered a community of bacteria and clams living under an ice shelf in Antarctica. These organisms live under 600 feet of ice, in the absence of sunlight, and in temperatures considered too cold for most living organisms. The location where these organisms live is unusual because

(1) only biotic factors control the size of the populations
(2) bacteria and clams are found in the same area
(3) abiotic factors are found in their environment
(4) green plants make energy-rich compounds available

[***Note:** This question was officially eliminated from the Regents examination before the examination date, and students were given credit for any answer they entered on the answer sheet.]

PART B–1

31. **2** *Natural selection* is the concept being represented in the diagram. In the diagram. A population of beetles of varying colors is illustrated. In the original population sample, about 80% of the beetles have dark bodies. Birds are shown consuming only dark-colored beetles from this population, leaving the light-colored beetles to live, reproduce, and pass on their genes for light coloration to succeeding generations. The proportion of dark-colored beetles declines over time until virtually none are present in the most recent population sample shown. In this diagram, the birds represent selection pressure on the beetles in that the birds eliminate the dark-colored beetles and allow the light-colored beetles to become the more successful variation in the species population.

WRONG CHOICES EXPLAINED

(1) *Overproduction* is *not* the concept being represented in the diagram. Overproduction is a concept described in the theory of natural selection that states Darwin's general observation that all natural species tend to produce

far more offspring than can actually survive to adulthood. No information is given in the diagram concerning the number of beetle offspring that are produced.

(3) *Homeostasis* is *not* the concept being represented in the diagram. Homeostasis is the chemical and physical balance that makes it possible for living things to react to changes in their environments. No information is given in the diagram concerning the factors affecting the homeostatic balance of the beetles.

(4) *Ecological succession* is *not* the concept being represented in the diagram. Ecological succession refers to the series of changes that occur to the dominant plant community of an area that has been significantly affected by an environmental change. No information is given in the diagram concerning this type of change in the beetle's environment.

32. **3** *When the ozone layer is thinner, more UV-B gets through it* is the statement that best describes the apparent relationship between ozone and UV-B. A general examination of the graphical data shows that whenever the thickness of the ozone layer decreased, the UV-B irradiation increased. When the ozone layer thickness was restored partially, UV-B irradiation declined.

WRONG CHOICES EXPLAINED

(1) *When ozone levels are at 2550 Dobson units, the UV-B levels are at 250 Dobson units* is *not* the statement that best describes the apparent relationship between ozone and UV-B. In the graph, ozone layer thickness is measured in Dobson units, whose range is indicated on the right side vertical axis as varying between 250 and 340. UV-B irradiance is measured in milliwatts per square centimeter (mWm^{-2}), not Dobson units, and its range is indicated on the left side vertical axis as varying between 2130 and 2670. The data represented in this response do not match any of the parameters presented in the graph.

(2) *The increase in UV-B reduces the destruction of the ozone layer* is *not* the statement that best describes the apparent relationship between ozone and UV-B. Ozone in the upper atmosphere is known to be damaged by the presence of chloroflourocarbons (CFCs) such as aerosol propellants and refrigerants released by humans. The ozone layer is not reduced by naturally occurring UV-B.

(4) *If the ozone layer is thicker, UV-B levels on the ground increase* is *not* the statement that best describes the apparent relationship between ozone and UV-B. The graph shows that as the ozone layer thickens, UV-B irradiance decreases, not increases.

33. **4** Process *X* is most likely caused by *an internal factor in the body, such as a temperature increase*. As is true of most organic molecules, proteins are affected by the environmental conditions in their surroundings. When

proteins are subjected to extreme heat, acidity, chemical action, or physical trauma, they can become distorted though a process known as denaturation. Once proteins lose their physical shape through denaturation, they cannot refold themselves or catalyze their substrates effectively.

WRONG CHOICES EXPLAINED

(1) Process *X* is *not* most likely caused by *the digestion of the amino acids that make up the proteins*. Amino acids are the end products of protein digestion. Amino acids cannot be further digested.

(2) Process *X* is *not* most likely caused by *the synthesis of a protein with different simple sugars*. Proteins are synthesized through the chemical linkage of many different amino acids, not simple sugars. The specific arrangement of amino acids in a protein helps determine the protein's shape and function in the body.

(3) Process *X* is *not* most likely caused by *removal of the gene that codes for the production of the protein*. The gene that codes for this protein is present in the DNA of every cell in the body, so its removal would prove impossible in practice. However, if this gene could be removed, the result would be the complete disappearance of the protein in that organism.

34. **2** *Some animals that inherit genes for brown fur will grow white fur if the outside temperature falls below a certain level* is the statement that is an example of how the external environment can influence gene expression. The statement describes a situation in which a fur color–controlling genotype (inherited trait) produces different phenotypes (displayed traits) depending on environmental temperature exposure during development. A real-life example of an animal that displays this temperature control of gene expression is the Himalayan rabbit, whose fur is all white in rabbits reared above 30°C but is black on the extremities in rabbits reared at or below 20°C.

WRONG CHOICES EXPLAINED

(1) *Some flowering plants that inherit a gene for white flowers and a gene for red flowers will produce pink flowers* is *not* the statement that is an example of how the external environment can influence gene expression. This is an example of incomplete dominance, not environmental influence on gene expression.

(3) *In some breeds of cat, certain fur-color genes are found only in females* is *not* the statement that is an example of how the external environment can influence gene expression. This is an example of sex linkage, not environmental influence on gene expression.

(4) *A pea plant is short-stemmed only if it inherits the genes for the trait from both parents* is *not* the statement that is an example of how the external environment can influence gene expression. This is an example of dominance/recessiveness, not environmental influence on gene expression.

35. **4** *The increased use of electronics has led to increased mining for precious metals and minerals in developing countries* is the statement that best illustrates a concept represented in the diagram. The diagram illustrates a series of human actions, including population growth, technology, and consumption, that adversely affect the natural environment. Modern computer technology, with its increasing dependence on and use of scarce metals and rare-earth elements, has led to significant degradation of environmental quality in certain developing nations that are often ill-equipped to regulate the destructive mining activities and the profit motive of big business.

WRONG CHOICES EXPLAINED

(1) *Tsunamis triggered by oceanic earthquakes cause widespread flooding that can lead to large scale environmental destruction* is *not* the statement that best illustrates a concept represented in the diagram. Earthquakes and their resulting tsunamis are natural events not related to human influence on the environment.

(2), (3) *Annual hunting laws determine the number of deer that can be hunted to ensure population stability* and *more individuals are purchasing hybrid cars that use less gasoline and produce less carbon dioxide* are *not* the statements that best illustrate a concept represented in the diagram. These statements are examples of positive actions that have been taken by humans to improve the quality of the natural environment.

36. **4** The process shown is used to *produce a hormone to regulate blood sugar*. The diagram illustrates a strand of bacterial DNA being linked with a human gene for the production of insulin. Once linked in a single bacterium, that bacterium is capable of producing human insulin. When this bacterium is grown in cultures to produce colonies containing millions of bacterial cells, all of the resulting bacterial cells are capable of insulin production. This insulin is drawn off and purified for use in clinical hormone treatments of human diabetics.

WRONG CHOICES EXPLAINED

(1) The process shown is *not* used to *determine if a person has a genetic disease*. Such a process might involve testing the DNA of the individual concerned and looking for genes known to be responsible for certain genetic disorders.

(2) The process shown is *not* used to *produce human growth hormone*. Such a process might involve using the same procedure but inserting the gene for human growth hormone instead of the gene for insulin.

(3) The process shown is *not* used to *identify the father of a newborn*. Such a process might involve comparing the DNA of the father with that of the newborn to determine the degree of genetic similarity between them.

37. **4** *A and B are organelles; E and G are organs* is the statement that correctly identifies the levels of organization for the structures indicated. The diagram illustrates sections of a single cell (left) and of the human body (right). All structures indicated in the diagram of the single cell must be organelles (nucleus, mitochondrion) since they are substructures of the cell. The structures indicated in the diagram of the human body are organs (brain, stomach), which are composed of many cells and numerous tissues.

WRONG CHOICES EXPLAINED

(1), (2), (3) *A and B are tissues; E and G are organs* and *A and B are organs; E and G are systems* and *A and B are tissues; E and G are organelles* are *not* the statements that correctly identify the levels of organization for the structures indicated. Tissues and organs are composed of many cells. So they cannot be found within the confines of a single cell.

38. **2** Cells from structure *E* and cells from structure *G* are similar in that they *contain the same genetic material*. All somatic (body) cells of the body contain identical genetic material due to the process of mitosis of the original zygote formed at fertilization.

WRONG CHOICES EXPLAINED

(1), (3) It is *not* true that cells from structure *E* and cells from structure *G* are similar in that they *have the same structure and function* or in that they *are identical in structure, but different in function*. Due to the process of differentiation, in which tissues take on specialized roles in the body, cells of different organs differ in both structure and function.

(4) It is *not* true that cells from structure *E* and cells from structure *G* are similar in that they *contain only the genetic information needed for their specific job*. All somatic (body) cells of the body contain identical genetic material due to the process of mitosis of the original zygote formed at fertilization.

39. **2** Row (2) in the chart correctly identifies the structure where ATP is produced and the reason it is affected by rotenone. This row in the chart correctly identifies structure *B* (mitochondrion) in the diagram as the site of ATP production in the cell. ATP is an energy-rich substance produced as a result of the process of aerobic respiration that occurs in the mitochondrion.

WRONG CHOICES EXPLAINED

(1) Row (1) in the chart does *not* correctly identify the structure where ATP is produced and the reason it is affected by rotenone. Structure *A* indicates the cell nucleus, which contains the genetic material of the cell.

(3) Row (3) in the chart does *not* correctly identify the structure where ATP is produced and the reason it is affected by rotenone. Structure *C* indicates the cytoplasm, which acts to suspend the organelles.

(4) Row (4) in the chart does *not* correctly identify the structure where ATP is produced and the reason it is affected by rotenone. Structure *D* indicates the cell membrane, which bounds the cell and regulates passage of materials into and out of the cell.

40. **3** The part of the microscope that should be adjusted in order to better view the onion cells while using high power is *C*. Part *C* is the fine adjustment knob, which is used to make very small changes in the focus of the microscope.

WRONG CHOICES EXPLAINED

(1) It is *not* true that the part of the microscope that should be adjusted in order to better view the onion cells while using high power is *A*. Part *A* is the ocular lens, which is located close to the eye and is used by the student to view the specimen on the slide.

(2) It is *not* true that the part of the microscope that should be adjusted in order to better view the onion cells while using high power is *B*. Part *B* is the coarse adjustment knob, which is used to make relatively large changes in the focus of the microscope. It is never used when viewing a slide under high power because the slide could crack if the stage is moved into the high power objective lens.

(4) It is *not* true that the part of the microscope that should be adjusted in order to better view the onion cells while using high power is *D*. Part *D* is the low power objective lens, which is used to view relatively large specimens.

41. **1** It is necessary for onion cells to contain chromosomes because chromosomes *are composed of genes that contain the instructions for an organism's traits*. Every somatic (body) cell of an organism contains all the genetic information necessary to create and maintain that organism for a lifetime.

WRONG CHOICES EXPLAINED

(2), (4) It *not* true that it is necessary for onion cells to contain chromosomes because chromosomes *are made of carbohydrates and are needed as an energy source* or *are composed of lipids that contain stored nutrients for the new cell*. Chromosomes and the genes they contain are composed primarily of nucleic acid, not carbohydrates or lipids.

(3) It *not* true that it is necessary for onion cells to contain chromosomes because chromosomes *direct the production of inorganic molecules within the cell*. The genes on chromosomes code for the production of enzymes that catalyze biochemical reactions in the cell. Some of these reactions result in the production of inorganic molecules, including carbon dioxide, water, and oxygen gas. However, it cannot be said that chromosomes direct these reactions. Rather, certain genes on certain chromosomes direct these reactions.

42. **2** The many varieties of domesticated dogs were most likely produced as a result of *selective breeding over many generations*. Selective breeding is a technique used by animal and plant breeders in which organisms with desirable traits are cross-bred with the hope of producing offspring that also display those traits. The large variety of domesticated dogs is a prime example of the results of this technique.

WRONG CHOICES EXPLAINED

(1) The many varieties of domesticated dogs were *not* most likely produced as a result of *mutating the body cells of the dogs*. Mutations are random events in which genetic material is altered though the effects of mutagenic agents. Mutating genes in body (somatic) cells cannot result in offspring with those alterations because body cells do not participate in sexual reproduction.

(3) The many varieties of domesticated dogs were *not* most likely produced as a result of *genetic engineering with specific enzymes*. Genetic engineering is a laboratory technique in which a gene for a desired trait is snipped from the DNA of a donor cell and inserted into the genome of a recipient cell. This technique deals with single genes or small groups of genes in a single cell. Genetic engineering would be insufficient to create the large variety of different traits in a breed of domesticated dog.

(4) The many varieties of domesticated dogs were *not* most likely produced as a result of *cloning dogs with desirable traits*. Cloning is a laboratory technique that, if successful, results in an exact genetic duplicate of the cloned organism. This technique cannot be used to produce many different varieties of dog since it results in only exact genetic duplicates of the donor organisms.

43. **2** *In a stable ecosystem, the number of individuals in a population is usually maintained within a certain range* is the statement that best describes the trend shown in the graph. The carrying capacity (horizontal line in the graph) of an environment is the maximum number of individuals of a species that can be supported given limited food and other resources. When the carrying capacity is exceeded, there will be a growing scarcity of the limited resource until individuals of the species begin to die from lack of that resource. These deaths will tend to reduce the species population (curved line in the graph) and help to bring it back into balance. The population size of a species will tend to vary around the carrying capacity of the environment, as illustrated in the graph.

WRONG CHOICES EXPLAINED

(1) *Ecosystem conditions will eventually cause a population to become extinct* is *not* the statement that best describes the trend shown in the graph. Such a graph would show the population declining steadily until it intersected

with the horizontal axis. An examination of the graph does not indicate this event.

(3) *The interactions between a population and various factors in an environment are always predictable* is *not* the statement that best describes the trend shown in the graph. The response has nothing to do with the factors represented in the graph (number of individuals vs. time). It is a nonsense distracter.

(4) *In order for any ecosystem to maintain a balance, populations must be reduced to half their original number* is *not* the statement that best describes the trend shown in the graph. Such a graph would show the population declining steadily to about half its highest value. An examination of the graph does not indicate this event.

PART B–2

44. One credit is allowed for correctly labeling the y-axis on the line provided. See the graph below. [1]

45. One credit is allowed for marking an appropriate scale, without any breaks in the data, on the y-axis. [1]

46. One credit is allowed for correctly constructing vertical bars to represent the data and for shading *each* bar. [1]

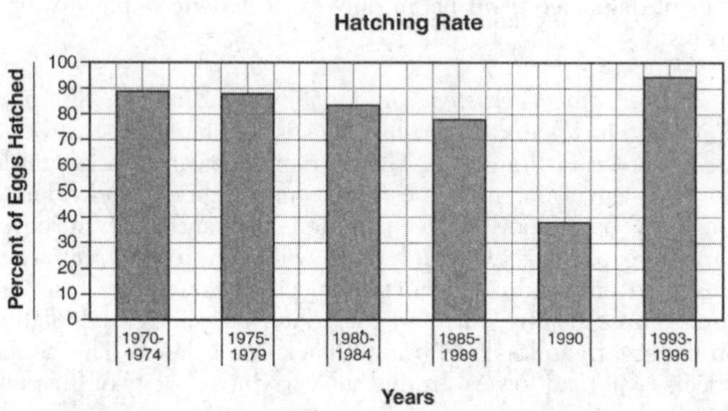

47. **2** The scientists transported prairie chickens from three different states into the state of Illinois in order to *increase genetic diversity*. This information is provided in the passage in paragraph 3, line 1. By introducing prairie chickens from different gene pools, the scientists hoped to diversify the variety of genetic traits in the Illinois prairie chicken population.

WRONG CHOICES EXPLAINED

(1) It is *not* true that the scientists transported prairie chickens from three different states into the state of Illinois in order to *decrease egg hatching rate*. The hatching rate was used by scientists as a measure of the prairie chicken's survival in Illinois. Scientists expected that the hatching rate would increase, not decrease, as a result of this program.

(3), (4) It is *not* true that the scientists transported prairie chickens from three different states into the state of Illinois in order to *increase egg fertilization* or to *develop different feather colors*. These factors are not mentioned in the passage.

48. One credit is allowed for correctly identifying *one* type of molecular evidence that could have been used to determine the evolutionary relationships illustrated in the diagram. Acceptable responses include but are not limited to: [1]

- *Amino acid sequences*
- *DNA sequences*
- *Gene sequences*
- *The results of a gel electrophoresis test*
- *The results of a paper chromatography test*
- *Analysis of pigments present*
- *Analysis of proteins/enzymes present*

49. **4** *Dissection* is the biological technique that could be used to obtain some structural evidence that species *A* and *B* are closely related. Studying the anatomical similarities between species provides the researcher with valuable information when the relationship of those species is called into question. Dissection is a technique that allows the researcher to make such structural/anatomical observations in the laboratory.

WRONG CHOICES EXPLAINED

(1) *Glucose testing* is *not* the biological technique that could be used to obtain some structural evidence that species *A* and *B* are closely related. Glucose is an organic compound with the chemical formula $C_6H_{12}O_6$. Glucose does not vary in its composition among different plant species.

(2) *Cloning* is *not* the biological technique that could be used to obtain some structural evidence that species *A* and *B* are closely related. Cloning is a laboratory technique that, if successful, results in an exact genetic duplicate of the cloned organism. Cloning provides no direct evidence of structural similarities/differences among plant species.

(3) *Genetic engineering* is *not* the biological technique that could be used to obtain some structural evidence that species *A* and *B* are closely related. Genetic engineering is a laboratory technique in which a gene for a desired trait is snipped from the DNA of a donor cell and inserted into the genome

of a recipient cell. Genetic engineering provides no direct evidence of structural similarities/differences among plant species.

50. **1** Based on the diagram, the process of glycolysis most likely *begins the breakdown of glucose*. The diagram shows that glucose enters the reaction but does not emerge on the other end. The diagram also shows that various chemical components found in glucose (energy, carbon, oxygen) are stripped out of the molecular structure of the glucose. This process converts glucose into other compounds that are acted upon by other respiratory reactions.

WRONG CHOICES EXPLAINED

(2) It is *not* true that, based on the diagram, the process of glycolysis most likely *produces oxygen for organisms to use*. The diagram shows oxygen being taken into, not released from, these reactions.

(3) It is *not* true that, based on the diagram, the process of glycolysis most likely *stores energy in molecules of water and carbon dioxide*. Water and carbon dioxide are energy-poor molecules that are released as glucose is gradually broken down during these reactions.

(4) It is *not* true that, based on the diagram, the process of glycolysis most likely *recycles glucose within the cells of simple organisms*. The diagram shows glucose being broken down in these reactions, not recycled.

51. One credit is allowed for correctly identifying *one* specific molecule used to store energy being released during this process. Acceptable responses include: [1]

- *ATP/ADP*
- *NADH/NAD*

52. One credit is allowed for correctly stating *one* way, other than the fact that they all contain DNA, the genetic content of these gametes is similar. Acceptable responses include but are not limited to: [1]

- *All of the male gametes in the diagram contain half the chromosome number (haploid, monoploid, n) compared with a body cell of the same species.*
- *They all contain half of the genetic information of their particular species.*
- *They had similar DNA sequences compared with a common ancestor.*
- *The DNA is composed of the same four subunits (A, T, G, C).*

53. One credit is allowed for selecting *one* type of organism (other than crustaceans) from the food web, correctly stating how the population of organisms you selected might be affected if the population of crustaceans in this food web were reduced due to the use of chemicals harmful

to crustaceans in the fields near the stream, and for supporting your answer. Acceptable responses include but are not limited to: [1]

- *Algae—they may increase because there are fewer crustaceans to eat them.*
- *Insect larvae—they will decrease in population because fish will eat more of them.*
- *Insect larvae—they will increase in in numbers because there is more algae for them to eat.*
- *Fish—they will have to eat more insect larvae because the crustaceans will be gone.*
- *Reptiles—they will have fewer fish to eat because the fish will starve with no crustaceans to eat.*

54–55. Two credits are allowed for correctly discussing the use of the hoot-dash by male peacocks. In your answer be sure to:

- State *one* advantage of the hoot-dash [1]
- State *one disadvantage* of the hoot-dash [1]

Acceptable responses include but are not limited to: [2]

- *An advantage of the hoot-dash is that it helps ensure successful reproduction in this species. A disadvantage of this behavior is that it consumes a great deal of energy that has to be made up through the nutritional process.*
- *The hoot-dash is a mating display used by male peacocks to find a mate. This is an advantage of the behavior. One problem with this behavior is that predators of the peacock may be attracted by the loud call. This becomes a disadvantage if the predator is successful.*

PART C

56. One credit is allowed for correctly describing *one* likely result if the placenta became damaged and could not maintain progesterone levels and for supporting the answer. Acceptable responses include but are not limited to: [1]

- *If adequate progesterone levels are not maintained, the lining of the uterus will not be maintained.*
- *The fetus might be born prematurely because the lining of the uterus will not be maintained and contractions of the uterus will not be inhibited.*
- *The woman will go into labor because uterine contractions will begin to occur.*
- *Production of eggs will not be inhibited.*
- *The fetus may die if born prematurely.*

57. One credit is allowed for correctly explaining how the release of additional eggs is prevented during pregnancy. Acceptable responses include but are not limited to: [1]

- *The presence of progesterone secreted by the placenta inhibits egg production.*
- *Progesterone is a human reproductive hormone that has the effect of inhibiting the maturation and release of egg cells from the ovaries during pregnancy.*
- *The sustained production of progesterone inhibits the production of FSH and LH in the pituitary gland, which has the effect of inhibiting ovulation.*

58. One credit is allowed for correctly explaining why the body of a person infected with HIV, the virus that causes AIDS, would have a different immune response to the presence of cancer cells than a person *not* infected with HIV. Acceptable responses include but are not limited to: [1]

- *HIV/AIDS attacks the immune system directly, and the body is unable to deal with the invaders.*
- *AIDS damages the immune system so it does not respond efficiently.*
- *AIDS damages the immune system so it cannot produce enough antibodies to fight infectious disease or cancer.*

59. One credit is allowed for correctly explaining why certain chemicals and radiation sources are risk factors for cancer. Acceptable responses include but are not limited to: [1]

- *Mutations are sometimes a cause of cancer, and exposure to radiation increases the occurrence of mutations.*
- *Radiation such as UV can cause cells to mutate and cause skin cancer.*
- *Certain chemicals may increase the rate of mutations.*
- *Radiation can alter the genetic information in cells.*
- *Radiation weakens the immune system.*

60. One credit is allowed for correctly explaining why narrowing the diameter of the blood vessels that lead to the placenta can result in low birth weight babies. Acceptable responses include but are not limited to: [1]

- *The amount of food that can pass through these vessels to the placenta and to the embryo will be less than if the vessels are normal in diameter.*
- *Less food and oxygen will reach the embryo.*
- *The developing embryo will not receive enough food for normal development since not as much blood will flow through the placenta.*

61. One credit is allowed for correctly stating if the vaccine was effective in preventing the disease and for supporting the answer. Acceptable responses include but are not limited to: [1]

- *Yes. A much smaller number of rats receiving the vaccine developed the disease.*
- *No. Since seven out of fifty got the disease, the vaccine may not be effective.*
- *Yes, since 86% of vaccinated rats did not get the disease.*

62. One credit is allowed for correctly stating *one* possible reason why two of the rats did *not* get sick even though they did *not* receive the vaccine. Acceptable responses include but are not limited to: [1]

- *The rats may have had a natural immunity to the disease.*
- *Those rats that did not get sick had a variation that protected them.*
- *They may already have had the disease and are immune.*
- *Their immune system fought off the virus.*

63. One credit is allowed for correctly stating if the results of this experiment indicate that the vaccine is ready for human testing and for supporting the answer. Acceptable responses include but are not limited to: [1]

- *No. Seven rats getting sick is too many out of a total of 50.*
- *No, because a larger sample should be tested before testing on humans.*
- *No. Only 100 rats were tested, so a larger sample size should be used before testing on humans.*
- *Yes, because most rats did not develop the disease.*

64. One credit is allowed for correctly stating the name of the specific technique that is used to alter the DNA of orange trees. Acceptable responses include but are not limited to: [1]

- *Genetic engineering*
- *Gene manipulation*
- *Gene splicing*
- *Gene modification*

65. One credit is allowed for correctly explaining why growing oranges with disease resistance is better for the environment than using pesticides to control the spread of the disease. Acceptable responses include but are not limited to: [1]

- *Use of pesticides can have a negative impact on other species in the area.*
- *Pesticides can pollute the groves.*
- *Pesticides could be harmful to people.*
- *Pesticides could kill beneficial insects such as bees.*
- *The insect vectors of the disease could become resistant to pesticides.*

66. One credit is allowed for correctly identifying a trait, other than disease resistance, that the orange trees could have that would be beneficial to the growers. Acceptable responses include but are not limited to: [1]

- *Drought resistance*
- *Insect resistance*
- *Freeze tolerance*
- *Faster growth rate*
- *Larger fruit*
- *Juicier/better tasting/more nutritious fruit.*

67. One credit is allowed for correctly explaining why one acre of land can produce more vegetables for human consumption than beef for human consumption. Acceptable responses include but are not limited to: [1]

- *Energy is lost as heat at each feeding level, and pyramid B has fewer levels.*
- *Consumers that feed directly on plants have more energy available to them than consumers feeding on other consumers since energy is lost at each level.*
- *Raising cattle on the land is not as productive because energy must be transferred from plants to the cattle and then to people. Much energy will be lost as heat.*

68. One credit is allowed for correctly explaining how the molasses spill caused many of the animals to die off. Acceptable responses include but are not limited to: [1]

- *The molasses displaced oxygen-rich water, and the fish suffocated.*
- *As fish died, the predators had less food available to them.*
- *Molasses turned the harbor into a water wasteland where the fish died from lack of oxygen.*

69. One credit is allowed for correctly stating *one* reason why the die-off could attract sharks, barracuda, and eels to the harbor and lagoon. Acceptable responses include but are not limited to: [1]

- *They would come into the area to feed on the dead and dying animals.*
- *They were attracted by the scent of dead organisms.*

70. One credit is allowed for correctly identifying *one* group of organisms responsible for the recycling of dead sea life that is *not* mentioned in the article. Acceptable responses include but are not limited to: [1]

- *Fungi*
- *Decomposers*
- *Bacteria*
- *Zooplankton*
- *Scavengers*

71. One credit is allowed for correctly explaining why it is important to preserve the biodiversity of the Honolulu Harbor ecosystem. Acceptable responses include but are not limited to: [1]

- *Biodiversity increases the chances that at least some organisms will survive in the event of environmental change.*
- *Biodiversity increases the stability of the entire ecosystem.*
- *Biodiversity increases the variety of genetic material present in an ecosystem.*
- *Humans rely on the ecosystem for fishing, food, and tourism.*

72. One credit is allowed for correctly predicting what will most likely happen to this ecosystem in 20 years if no other disasters occur. Acceptable responses include but are not limited to: [1]

- *Marine life will repopulate the harbor.*
- *The ecosystem will recover on its own.*
- *The ecosystem might not recover due to the degree of destruction that occurred.*
- *It will recover due to ecological succession and will reestablish a self-sustaining climax community.*

PART D

73. **3** This is evidence that each species has a different *niche*. An ecological niche is the role that an organism plays in its environment. Niches represent the organism's nutritional, symbiotic, reproductive, behavioral, and biochemical characteristics. The specific mix of roles is unique for each species, although different species may share some roles in common.

WRONG CHOICES EXPLAINED

(1), (2), (4) This is *not* evidence that each species has a different *mating behavior*, *nesting material*, or *predator*. The question references different feeding niches of Galapagos finches, not their mating and nesting behaviors or their position in the food web.

74. **3** One reason for this decrease is *a person requires less oxygen when sleeping*. During sleep, the body's metabolic rate decreases in order to allow the body to renew itself and repair damaged tissues. Since less energy is needed during sleep, less oxygen is needed to fuel the respiratory reactions. As a result, the body's vital systems, including its pulse rate, slow down. This slows circulation and reduces the rate at which oxygen is supplied to the body's organs.

WRONG CHOICES EXPLAINED

(1) It is *not* true that one reason for this decrease is *the body is producing more carbon dioxide when a person is sleeping*. Since the respiratory process slows during sleep, so does the production of carbon dioxide, which is a waste product of respiration.

(2) It is *not* true that one reason for this decrease is *a person is not using energy while sleeping*. Energy is always needed to maintain homeostasis in the body. If energy production ceases, the body quickly dies.

(4) It is *not* true that one reason for this decrease is *the body is using more nutrients while asleep*. As with other metabolic activities, the nutritional process also slows during sleep. So fewer, not more, nutrients are used.

75. **4** *Different adaptations gave some tortoises a better chance of survival* is the statement that best explains the differences observed in the tortoises on each island. Darwin's theory of natural selection, sometimes referred to as "survival of the fittest," postulates that certain members of a species are better adapted to their environment than others. These better-adapted members are able to survive adverse conditions at higher rates and are therefore more likely to pass on their favorable adaptations to succeeding generations than are less well-adapted members. These factors act together to increase the proportion of individuals displaying highly adaptive traits.

WRONG CHOICES EXPLAINED

(1) *Each tortoise adapted to its environment during its lifetime* is *not* the statement that best explains the differences observed in the tortoises on each island. The physical traits of Galapagos tortoises are determined by the genetic codes present in a tortoise's cells. They are passed from generation to generation via the reproductive process.

(2) *Sudden mutations changed the appearance of all of the tortoises* is *not* the statement that best explains the differences observed in the tortoises on each island. Mutations are random events in which genetic material is altered though the effects of mutagenic agents. Only mutations that occur in gametic (reproductive) tissues can be passed on to future generations.

(3) *The tortoises grew different structures based on the available food* is *not* the statement that best explains the differences observed in the tortoises on each island. There is no competent scientific evidence that supports this concept. Genetic traits in the tortoises do not develop based on exposure to available food or other resources needed. Rather, selection pressures in the environments favored certain variations already present in the gene pool of the tortoise for perpetuation in the population.

76. **4** *Predators would have killed the tortoises with long necks, leaving more tortoises with shorter necks* is the statement that best supports the hypothesis. The Galapagos saddle-backed tortoise is well-protected by a massive carapace (shell) and has developed other physical and behavioral adaptations that allow it to protect itself from predation (except human predation). It may or may not be true that the Galapagos Islands were free of predators during the long period of time during which the saddle-backed tortoises were evolving. However, it is unlikely that the saddle-back's evolutionary path would have been very different with or without the presence of predators in their desert island environment.

WRONG CHOICES EXPLAINED

(1) *Competition between the predators and the tortoises would have resulted in a greater number of long-necked tortoises* is *not* the statement that best supports the hypothesis. Competition between species in an ecosystem is most intense when it occurs between species that use the same limited resources, including food. It is highly unlikely that the saddle-backed tortoise consumes the same food resources as those consumed by predators in this island desert ecosystem.

(2) *The number of predators was greater than the number of tortoises* is *not* the statement that best supports the hypothesis. In a naturally balanced ecosystem, the biomass of herbivores (such as the saddle-backed tortoise) always exceeds the biomass of predators.

(3) *Saddle-backed tortoises were not a part of the food chain* is *not* the statement that best supports the hypothesis. As successful herbivores, saddle-

backed tortoises are considered a key component of the food chain/web in this island desert ecosystem.

77. One credit is allowed for indicating if both species could survive if a group of saddle-backed tortoises were brought to an island inhabited by dome-backed tortoises and for supporting the answer. Acceptable responses include but are not limited to: [1]

- *Yes, because they have long necks and can reach food that the dome-backed tortoises cannot reach.*
- *Yes, if there is enough food for both species of tortoise.*
- *Yes, as long as they have different niches.*
- *No, because they would compete for limited food resources and only one species could be successful. The other would be eliminated over time.*
- *No, if the saddle-backed tortoise is not adapted to live in the new environment.*
- *No—the saddle-backed tortoise is well-adapted to a desert island ecosystem but is probably not well-adapted to a damp tropical island ecosystem.*

78. One credit is allowed for correctly stating *one* way the results obtained from these two conditions differed when you did this activity. Acceptable responses include but are not limited to: [1]

- *More seeds were collected with/without competition.*
- *It was easier to get food without competition.*
- *The greater the competition, the fewer the survivors.*

79. One credit is allowed for identifying *one* body system, other than the circulatory system, that becomes more active as a result of exercise, and for correctly stating *one* change the system you identified will undergo as a result of becoming more active. Acceptable responses include but are not limited to: [1]

- *Body system: respiratory—the person will breathe faster/inhale more oxygen gas*
- *Body system: excretory—the person will sweat more/exhale more carbon dioxide*
- *Body system: digestive—the rate of digestion will decrease*
- *Body system: locomotive—the joints and muscles will be very active*
- *Body system: nervous—reflex feedback loop activity will increase*
- *Body system: endocrine—insulin/cortisol/glucagon/testosterone secretion will increase*

80. One credit is allowed for correctly stating *one* appropriate hypothesis for this experiment. Acceptable responses include but are not limited to: [1]

- *If athletes preform the same exercises as nonathletes, the athletic students will have a lower heart rate during exercise.*
- *Athletic students have a lower heart rate during similar exercise than do nonathletic students.*
- *Nonathletes have a higher heart rate when exercising than do athletes.*

81. **2** *After exercise, the heart rates of the athletic students returned to resting pulse in four minutes* is the statement that is best supported by the data in the chart. The chart data indicate that the average resting pulse rate of athletes tested was 68 bpm (beats per minute). These data also indicate that the average heart rate for athletes four minutes after exercise had returned to 68 bpm.

WRONG CHOICES EXPLAINED

(1) *After exercise, the nonathletic students had a lower heart rate than the athletic students* is *not* the statement that is best supported by the data in the chart. The chart data indicate that the average after-exercise heart rates of nonathletic students were between 15 bpm and 27 bpm higher than that of athletic students

(3) *During exercise, both groups of students had the same increase above their resting pulse* is *not* the statement that is best supported by the data in the chart. The chart data indicate that the increase in average heart rate for athletes was +42 bpm (110 bpm – 68 bpm = 42 bpm). The increase in average heart rate for nonathletes was +58 bpm (130 bpm – 72 bpm = 58 bpm).

(4) *During exercise, the athletic students had a higher heart rate than the nonathletic students* is *not* the statement that is best supported by the data in the chart. The chart data indicate that at all points of measure, the average heart rates of athletic students was lower, not higher, than that of nonathletic students.

82. **2** To improve the validity of the conclusion reached in this experiment, the students should repeat the experiment *with a larger number of athletes and nonathletes*. Good scientific investigation is best done using large numbers of experimental subjects. This practice helps to eliminate experimental bias caused by outliers in the experimental groups being tested that would skew data and misrepresent experimental results.

WRONG CHOICES EXPLAINED

(1) It is *not* true that to improve the validity of the conclusion reached in this experiment, the students should repeat the experiment *disregarding any data that don't fit the hypothesis*. A hypothesis is an educated guess made

by the researcher, prior to the experiment, about the probable results of that experiment. The data gained from the experiment are analyzed and lead the researcher closer to scientific truths. Disregarding data that do not fit the hypothesis invalidates the experiment due to introduction of experimental bias. Disregarding data simply because they do not match the researcher's bias is never acceptable.

(3) It is *not* true that to improve the validity of the conclusion reached in this experiment, the students should repeat the experiment *comparing the heart rates and breathing rates of males and females*. Making such a change in focus would represent a new area of inquiry with a new independent variable (male/female), not an improvement of the experiment summarized in the question with the current independent variable (athlete/nonathlete).

(4) It is *not* true that to improve the validity of the conclusion reached in this experiment, the students should repeat the experiment *with the athletes doing different exercises than the nonathletes*. A basic tenet of scientific inquiry is that the conditions of an experiment should in all ways be identical except for the single independent variable being tested (the factor controlled by the researcher). In this experiment, the single independent variable is athlete/nonathlete. Adding different exercises to the experiment would introduce a second independent variable (exercise *X*/exercise *Y*) and would invalidate data resulting from the experiment.

83. One credit is allowed for correctly stating *one* conclusion the student would make about the unknown liquid based on the results and for supporting the answer. Acceptable responses include but are not limited to: [1]

- *The unknown liquid has starch in it but not glucose. The glucose indicator showed no change, but the starch indicator did.*
- *There is no glucose in the unknown liquid—the glucose indicator showed no change.*
- *There is starch in the unknown liquid—the indicator turned blue-black.*

84. One credit is allowed for correctly explaining why using new clothespins for some students and not others was an error in the experimental procedure. Acceptable responses include but are not limited to: [1]

- *Not using the same type of clothespins is introducing a new variable.*
- *New clothespins might have stronger/weaker springs.*
- *New clothespins would produce different results.*

85. One credit is allowed for correctly stating what factor causes the DNA fragments to move in this technique. Acceptable responses include but are not limited to: [1]

- *Electricity*
- *Electrical charge*
- *Electrical potential*
- *Attraction between opposite electrical charges*
- *Polarity*
- *Electron flow*
- *Potential difference*
- *Potential drop*
- *Voltage*

STANDARDS/KEY IDEAS	JUNE 2017 QUESTION NUMBERS	NUMBER OF CORRECT RESPONSES
STANDARD 1		
Key Idea 1: The central purpose of scientific inquiry is to develop explanations of natural phenomena in a continuing and creative process.	47, 54, 55, 72	
Key Idea 2: Beyond the use of reasoning and consensus, scientific inquiry involves the testing of proposed explanations involving the use of conventional techniques and procedures and usually requiring considerable ingenuity.		
Key Idea 3: The observations made while testing proposed explanations, when analyzed using conventional and invented methods, provide new insights into natural phenomena.	61, 63	
Laboratory Checklist	40, 44, 45, 46, 49	
STANDARD 4		
Key Idea 1: Living things are both similar to and different from each other and from nonliving things.	11, 18, 24, 27, 28, 29, 30*, 37, 39, 43, 53	
Key Idea 2: Organisms inherit genetic information in a variety of ways that result in continuity of structure and function between parents and offspring.	20, 34, 36, 38, 41, 42, 64, 66	
Key Idea 3: Individual organisms and species change over time.	8, 9, 22, 23, 25, 26, 31, 48	
Key Idea 4: The continuity of life is sustained through reproduction and development.	19, 52, 56, 57, 60	
Key Idea 5: Organisms maintain a dynamic equilibrium that sustains life.	13, 14, 15, 16, 17, 33, 50, 51, 58, 59, 62	
Key Idea 6: Plants and animals depend on each other and their physical environment.	2, 10, 12, 21, 67, 68, 69, 70, 71	
Key Idea 7: Human decisions and activities have a profound impact on the physical and living environment.	1, 3, 4, 5, 6, 7, 32, 35, 65	

*Question 30 eliminated

STANDARDS/KEY IDEAS	JUNE 2017 QUESTION NUMBERS	NUMBER OF CORRECT RESPONSES
REQUIRED LABORATORIES		
Lab 1: "Relationships and Biodiversity"	85	
Lab 2: "Making Connections"	74, 79, 80, 81, 82, 84	
Lab 3: "The Beaks of Finches"	73, 75, 76, 77, 78	
Lab 5: "Diffusion Through a Membrane"	83	

Examination August 2017

Living Environment

PART A

Answer all questions in this part. [30]

***Directions* (1–30): For *each* statement or question, record in the space provided the *number* of the word or expression that, of those given, best completes the statement or answers the question.**

1 A fruit fly is classified as a consumer rather than as a producer because it is unable to

1 reproduce asexually
2 synthesize its own food
3 release energy stored in organic molecules
4 remove wastes from its body 1_____

2 Which change is an example of maintaining dynamic equilibrium?

1 A plant wilts when more water is lost from the leaves than is lost by the roots.
2 A plant turns yellow when light levels are very low.
3 Insulin is released when glucose levels in the blood are high.
4 A person sweats when the environmental temperature is low. 2_____

3 Organisms contain compounds such as proteins, starches, and fats. The chemical bonds in these compounds can be a source of

1 amino acids
2 simple sugars
3 energy
4 enzymes

3____

4 Phosphorus is necessary for the growth of healthy plants. Scientists are developing plants that can grow in phosphorus-poor soil. Some of these new varieties, produced in a lab, make extra copies of a protein that helps them obtain more phosphorus from the soil. The process being used to develop these new varieties is most likely

1 paper chromatography
2 natural selection
3 direct harvesting
4 genetic engineering

4____

5 Which life function is *not* necessary for an individual organism to stay alive?

1 nutrition
2 reproduction
3 regulation
4 excretion

5____

6 Lobsters prey on sea hares, which are marine animals. The lobsters find their prey through a sense of smell. The sea hares defend themselves by squirting ink at the lobster, as shown in the photo below. The ink sticks to the lobster, interfering with its sense of smell.

The most likely reason the sea hare can escape is because the sea hare ink

1 pushes the sea hare away rapidly as the ink is expelled
2 blocks a receptor on certain cells in the lobster
3 causes the lobster to change its prey
4 prevents movement of the lobster 6_____

7 Which statement is an accurate description of genes?

1 Proteins are made of genes and code for DNA.
2 Genes are made of proteins that code for nitrogen bases.
3 DNA is made of carbohydrates that code for genes.
4 Genes are made of DNA and code for proteins. 7_____

8 The bobolink is a small blackbird that nests in fields of tall grass. It breeds in the summer across much of southern Canada and the northern United States. It migrates long distances, wintering in southern South America. The numbers of these birds are declining due to disruption of the areas where they live.

In order to save these birds from extinction, the best course of action would be to

1 prevent the birds from migrating to South America
2 encourage farmers to let their hay fields undergo succession
3 work to protect bobolink habitats in South and North America
4 capture all the bobolinks and keep them safe in zoos

8 ____

9 A child with cystic fibrosis has an altered protein in his cells that stops chloride ions from leaving the cells. This protein most likely affects the functioning of

1 cell membranes
2 nuclei
3 mitochondria
4 ribosomes

9_____

10 Which row in the chart below shows a direct relationship that can exist between two living organisms?

Row	Relationship
(1)	producer – carnivore
(2)	predator – prey
(3)	parasite – prey
(4)	carnivore – host

10_____

11 Scientists have studied the return of plant life on Mount St. Helens ever since the volcano erupted in 1980. Wildflowers began colonizing the area, followed by shrubs and small trees. Scientists predict that it will likely take hundreds of years before the area returns to a forest dominated by fir and hemlock trees. These changes are an example of

1 humans degrading an ecosystem by removing wildflowers
2 the loss of genetic variation in a plant species
3 the growth of a forest through ecological succession
4 the biological evolution of wildflowers, shrubs, and trees

11_____

12 The most likely result of completely removing carbon dioxide from the environment of a plant is that sugar production will

1 continue at the same rate
2 increase and oxygen production will also increase
3 increase and oxygen production will stay the same
4 decrease and eventually stop

12_____

13 Before a new shopping center can be built on previously undeveloped land, the builders must submit a proposal to the local government for approval. Which statement identifies an environmental concern associated with the development of the shopping center?

1 Building the center would decrease resources needed by local organisms.
2 The new shopping center would increase competition with already existing businesses.
3 Building the center would decrease the amount of pollution in the area.
4 The new shopping center would increase the biodiversity of the area.

13_____

14 Homeowners have been encouraged to learn how to identify invasive plants and to remove them if they find them. The most likely reason for removing invasive plants is to

1 allow only one type of native plant to grow
2 preserve biodiversity
3 eliminate unfamiliar food sources
4 increase the rate of ecological succession

14_____

15 Which row in the chart below correctly pairs a human activity with its impact on the environment?

Row	Human Activity	Impact
(1)	decrease in the use of pesticides	erosion of rock in the soil
(2)	increase in housing developments	improvement in air quality
(3)	increase in human population	reduction in water usage
(4)	decrease in recycling	reduction in amount of available resources

15_____

16 The diagram below represents some steps in a procedure used in the field of biotechnology.

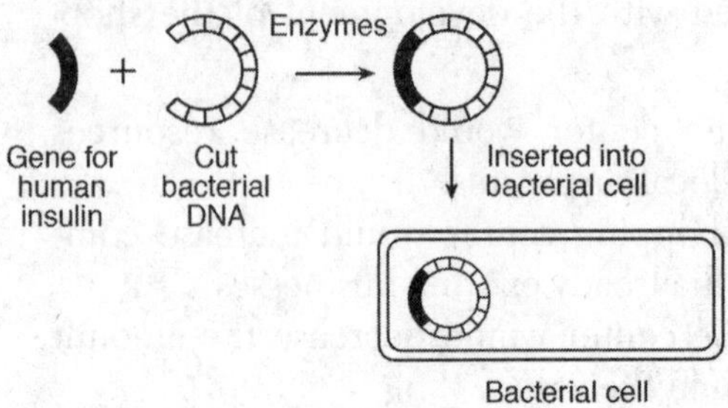

This bacterial cell can now be used to produce

1 the bacterial gene for insulin that can be inserted into humans
2 human genes for enzymes that can be inserted into humans
3 insulin that can be used by humans
4 enzymes necessary to treat human diseases 16_____

17 The graph below represents the number of brown and green beetles collected in a particular ecosystem.

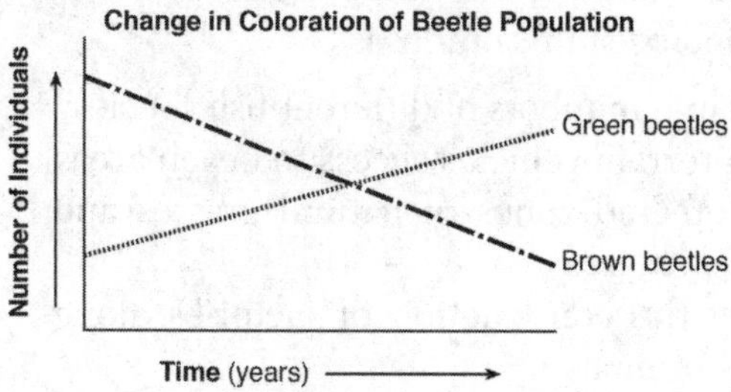

The change observed in the number of green and brown beetles in the population is most likely due to

1 natural selection
2 selective breeding
3 gene manipulation
4 a common ancestor 17_____

18 A reproductive system is represented in the diagram below.

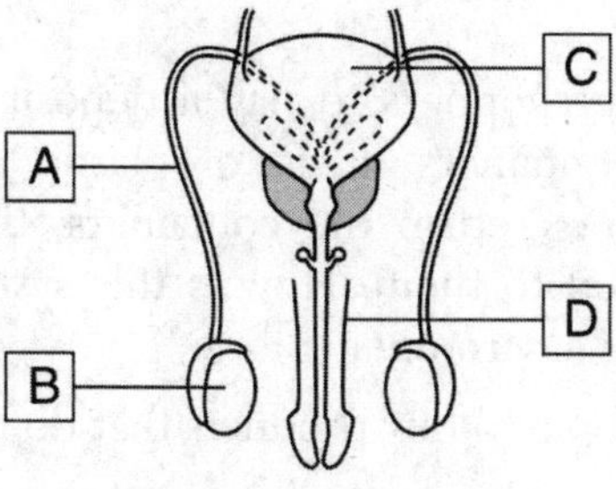

Which structure is correctly paired with its reproductive function?

1 *A* – pathway of gametes
2 *B* – synthesis of progesterone
3 *C* – production of sperm
4 *D* – regulation of homeostasis 18_____

19 For centuries, humans have used resources from coastal areas and open ocean waters. An example of an activity that would promote the conservation of coastal areas and ocean resources is

1 harvesting large numbers of different fish species
2 allowing all-terrain vehicles access to beach areas
3 creating protected zones of natural grasses and shrubs in beach areas
4 encouraging the construction of factories along the ocean shoreline

19_____

20 Which activity would eventually result in a stable ecosystem?

1 deforestation in an area to increase space for the species living there
2 mowing a large field so it can be used for recreation
3 allowing native plants to grow undisturbed in an abandoned field
4 spraying pesticides on a field at the end of each growing season

20_____

21 Some states require shoppers to pay a deposit on certain beverage containers made of plastic and glass. When shoppers return the containers, their deposits are returned to them. How is this system intended to help the environment?

1 It encourages people to buy products that do not have a deposit.
2 It reduces the amount of money shoppers actually spend.
3 It reduces the amount of plastics and glass put into landfills.
4 It forces manufacturers to reduce air pollution when they are making the containers.

21_____

22 The diagram below represents a food web.

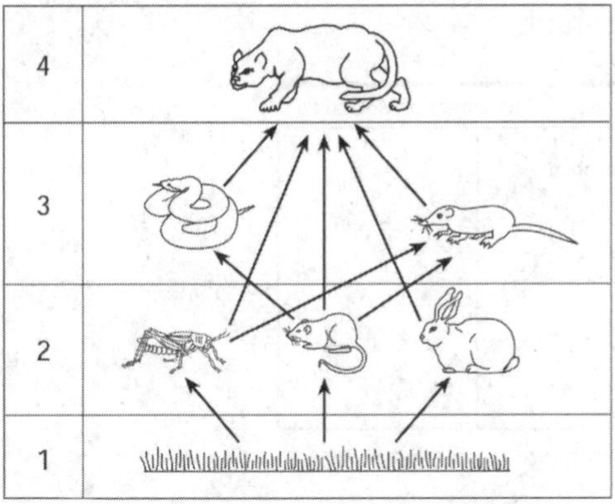

Which level contains organisms that carry out autotrophic nutrition?

1 1
2 2
3 3
4 4

22_____

23 Mad cow disease is a fatal disease that destroys brain tissue. Researchers have found that a prion protein, which is an abnormally constructed molecule, is responsible. Which statement best describes the characteristics a protein must have to function correctly?

1 A protein is a long chain of amino acids folded into a specific shape.
2 A protein is a long chain of simple sugars folded into a specific shape.
3 A protein is made of amino acids synthesized into a short, circular chain.
4 A protein is made of simple sugars synthesized into a short, circular chain.

23_____

24 The diagram below represents the results of the net movement of a specific kind of molecule across a living cell membrane.

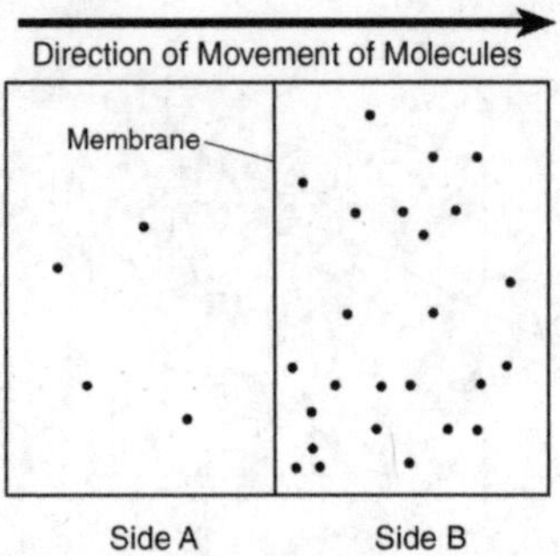

The movement of molecules from side *A* to side *B* is an example of the process of

1 active transport
2 chromatography
3 cellular respiration
4 diffusion 24____

25 Several companies now offer DNA "banking services," where DNA is extracted from a pet and is stored so that a "replacement pet" might be produced using cloning techniques when the original pet dies. Which statement best explains why the replacement pets that are produced in this way might *not* look or act like the original?

1 The new animal must get the DNA from two different parents, not just one cell.
2 Mutations could occur that change the cloned animal into a completely different species.
3 Recombination of the cells as they are cloned will make the resulting pet act differently.
4 The environment could influence how genes are expressed, changing how the animal looks and acts. 25____

26 It is recommended that people avoid excessive use of tanning beds. Exposure to the radiation emitted by tanning beds can cause skin cancer. This cancer is the direct result of a

1 change in a starch molecule
2 mutation in the genetic material
3 mutation in a protein
4 change in a fat molecule

26____

27 The diagram below represents a developing fetus in a human.

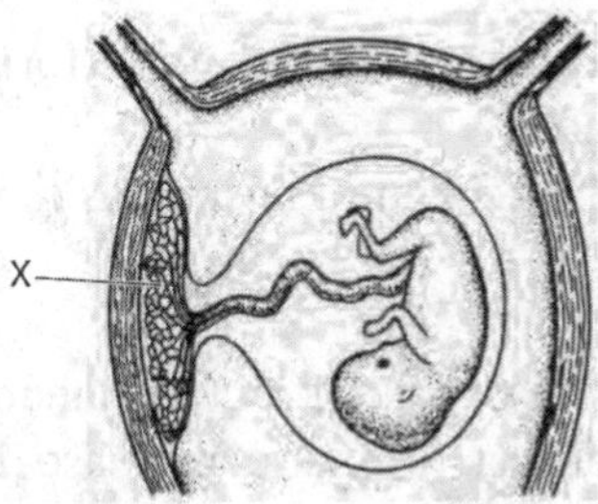

What would most likely happen if structure *X* were damaged in the early stages of pregnancy?

1 The genes from the mother would not be turned on in the fetus.
2 The nutrients necessary for development would not be able to reach the fetus.
3 The fertilized egg would not be able to travel from the ovary to the uterus.
4 Development would take longer since the fetus would have to synthesize nutrients.

27____

28 The reproductive structure in a female mammal that produces sex cells is the

1 ovary
2 testes
3 uterus
4 placenta

28____

29 Fungi are decomposers that play an important role in the maintenance of an ecosystem. The role of fungi is important because they

1 synthesize energy-rich compounds that are directly used by producers
2 break down materials that can then be used by other organisms
3 limit the number of plants that can perform photosynthesis in an area
4 are competitors of other consumers such as herbivores

29____

30 In 2011 and 2012, scientists working on the Banana River in Florida recorded a dramatic increase in the number of manatee deaths. Over the past 50 years, this area has also seen the human population increase by more than 500,000 people. It is believed that pollution from numerous sewage tanks leaked into the water, eliminating the manatees' food source, replacing it with an alga that is toxic to the manatee. This is an example of

1 a natural cycle in an ecosystem
2 the effect of increased biodiversity on an ecosystem
3 direct harvesting in an ecosystem
4 human actions altering ecosystems with serious consequences

30____

PART B–1

Answer all questions in this part. [13]

***Directions* (31–43): For *each* statement or question, record in the space provided the *number* of the word or expression that, of those given, best completes the statement or answers the question.**

31 A student performed an experiment to see if water temperature affects the level of activity in aquatic snails. The student set up four tanks with five snails in each tank. All four of the setups were identical in every way, except for the temperature of the water. In order to make the conclusions more valid, the student could

1 alter the pH of the water
2 change the size of the tank
3 carry out the experiment for a shorter period of time
4 use a larger number of snails

31____

32 The following events occur during sexual reproduction:

A. mitosis
B. meiosis
C. fertilization
D. birth

Which sequence represents the correct order of these events during sexual reproduction?

1 $A \rightarrow C \rightarrow B \rightarrow D$
2 $B \rightarrow C \rightarrow A \rightarrow D$
3 $C \rightarrow B \rightarrow A \rightarrow D$
4 $B \rightarrow A \rightarrow C \rightarrow D$

32____

33 A broad body of evidence, subject to revisions, supported by different kinds of scientific investigations and often involving the contributions of scientists from different disciplines is necessary to develop

1 an inference
2 a fact
3 a theory
4 a prediction

33____

34 The diagrams below represent portions of two genes that code for leaf structure in the same species of clover. Gene 1 was taken from the cells of a clover plant with 3 leaves and gene 2 was taken from the cells of a clover plant with 4 leaves.

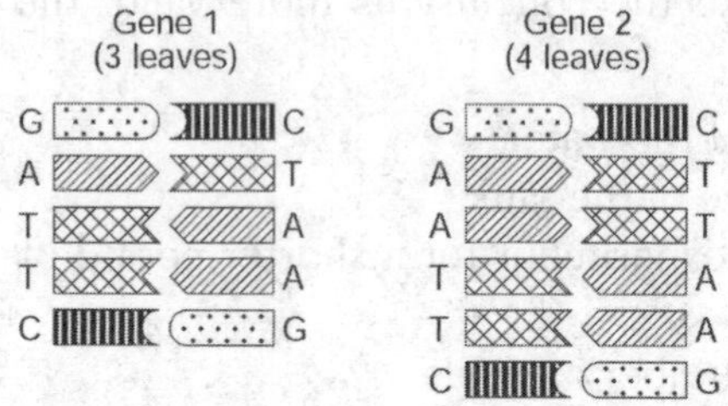

The clover plant having gene 2 (4 leaves) was most likely the result of

1 an insertion
2 a deletion
3 a substitution
4 normal replication

34____

35 Increased concern over the number of heat-related illnesses among football players has led to a possible change in uniform design. Shoulder pads were designed that constantly blew cool, dry air underneath the shoulder pads. Tests showed that the use of the device during rest and recovery periods resulted in a reduction of body temperature and heart rate. This new device would help the athlete to

1 control the rate of muscle activity
2 increase muscle strength
3 maintain homeostasis
4 eliminate the release of heat from the body 35_____

Base your answers to questions 36 and 37 on the information and data table below and on your knowledge of biology.

A student wanted to investigate the effect of light on the rate of ripening of tomatoes. She set up four pots of the same size with identical amounts of soil, water, and type of tomato plants. Each plant was exposed to a different intensity of light as shown in the table below.

Plant	Light Intensity (lumens)	
1	0	
2	1000	
3	5000	
4	10,000	

36 To report the final results, which label would be most appropriate for the third column of the data table?

1 Height of Tomato Plants (cm)
2 Average Ripening Time (days)
3 Average Weight of Tomatoes per Plant (grams)
4 Acidity of Tomatoes (pH) 36____

37 The independent variable in this experiment is the

1 type of tomato plant
2 amount of soil provided
3 color of tomatoes
4 light intensity 37____

Base your answers to questions 38 and 39 on the diagram below and on your knowledge of biology. The diagram illustrates activities taking place in the body of a human.

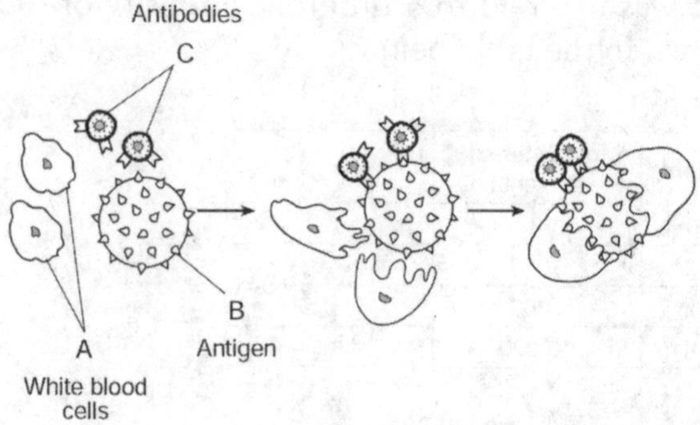

38 Vaccinations usually stimulate the body to produce more of

1 structure *A*, only
2 structure *B*, only
3 structures *A* and *C*, only
4 structures *A*, *B*, and *C* 38_____

39 Which structure normally stimulates an allergic response?

1 *A*, only
2 *B*, only
3 *C*, only
4 *A*, *B*, and *C* 39_____

---

40 Which population in the chart below has the best chance for survival in a rapidly changing environment?

Population	Type of Reproduction	Average Life Span of Individuals	Total Number of Offspring Produced
(1)	sexual	13 days	100
(2)	asexual	13 days	100
(3)	sexual	12 weeks	25
(4)	asexual	12 weeks	25

40_____

41 The table below represents a segment of a DNA molecule found in a stomach cell, both before and after undergoing replication.

DNA Segment Before and After Replication

Before replication	TGT	ATG	AAA	CAC	AAT	TAT
After replication	TGT	ATT	AAA	CAC	AAT	TTT

Which statement best describes a change that would most likely be observed in the cells formed as a result of this mitotic division?

1 An enzyme the cell produces might no longer function.

2 The cells would begin to form gametes to be released.

3 Many new hormones would be synthesized by the cells.

4 Chloroplasts would be produced by the ribosomes. 41 _____

Base your answers to questions 42 and 43 on the information and diagram below and on your knowledge of biology.

The setup below shows four test tubes. Tube 1 contains water only. Tube 2 contains a live snail. Tube 3 contains a live green water plant. Tube 4 contains both a live green water plant and a live snail.

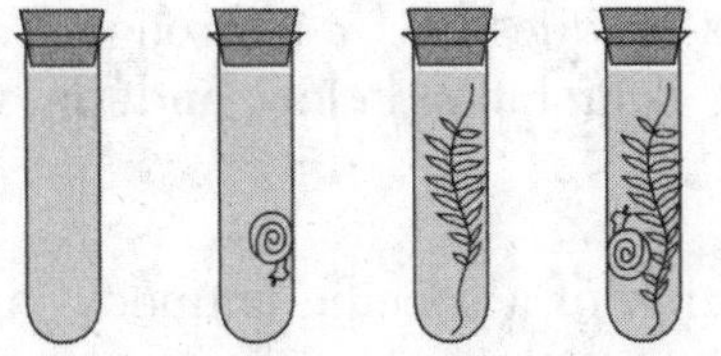

42 In this setup, which tubes contain at least one organism carrying on cellular respiration?

1 tubes 1 and 2, only
2 tubes 2 and 4, only
3 tubes 3 and 4, only
4 tubes 2, 3, and 4, only

42_____

43 Which compound that directly provides energy in living cells is being produced in every tube where cellular respiration is occurring?

1 oxygen
2 glucose
3 DNA
4 ATP

43_____

PART B–2

Answer all questions in this part. [12]

***Directions* (44–55): For those questions that are multiple choice, record in the space provided the *number* of the choice that, of those given, best completes each statement or answers each question. For all other questions in this part, follow the directions given and record your answers in the spaces provided.**

Base your answers to questions 44 through 47 on the information and data table below and on your knowledge of biology.

The concentration of a specific antibody in the blood of an individual was measured at various times over a period of 50 days. The results obtained are shown in the data table below.

Antibody Concentration in an Individual

Day	**Antibody Concentration in Arbitrary Units** (arb. units)
5	0
10	110
16	120
25	10
35	200
45	390
50	200

***Directions* (44–45): Using the information in the data table, construct a line graph on the grid, following the directions below.**

44 Mark an appropriate scale, without any breaks in the data, on each labeled axis. [1]

45 Plot the data on the grid. Connect the points and surround each point with a small circle. [1]

Example: 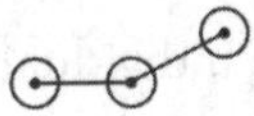

Antibody Concentration in an Individual

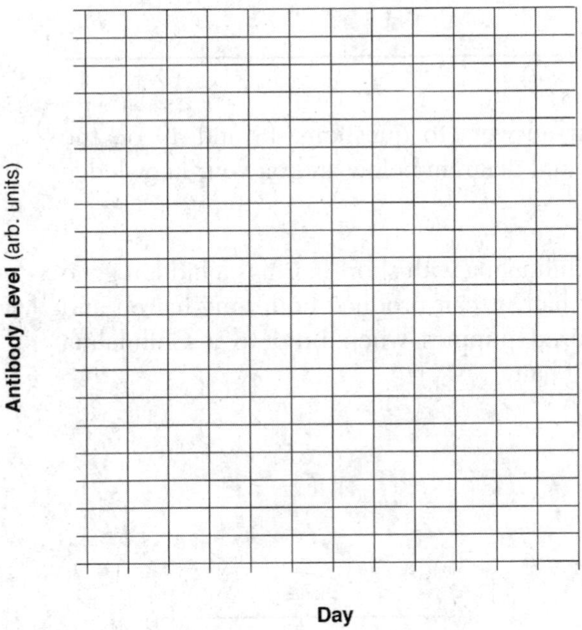

46 State *one* reason for the change in antibody production during the first 10 days. [1]

____

____

Note: The answer to question 47 should be recorded in the space provided.

47 The antibody level (in arb. units) of the individual on day 30 is closest to

1 30
2 70
3 110
4 160

47____

Base your answers to questions 48 and 49 on the information and diagram below and on your knowledge of biology.

If a Chihuahua with short hair has a hidden gene for long hair, it can produce both long-haired and short-haired puppies when bred to a Chihuahua with long hair.

48 A family decides that they want to produce Chihuahuas with long hair. Identify a procedure that could be used to make sure that the puppies all have long hair. [1]

____

Note: The answer to question 49 should be recorded in the space provided.

49 A Chihuahua is born having a trait that is different from either of its parents. A possible explanation for the difference is that the Chihuahua puppy

1 was produced as a result of the recombination of genes during sexual reproduction
2 was produced as a result of the process of asexual reproduction
3 inherited a gene from one of its grandparents and not its parents
4 had a mutation that occurred after it was born 49______

Base your answers to questions 50 and 51 on the diagram below and on your knowledge of biology. The diagram represents a technique used by scientists today to maintain the genetic makeup of an organism.

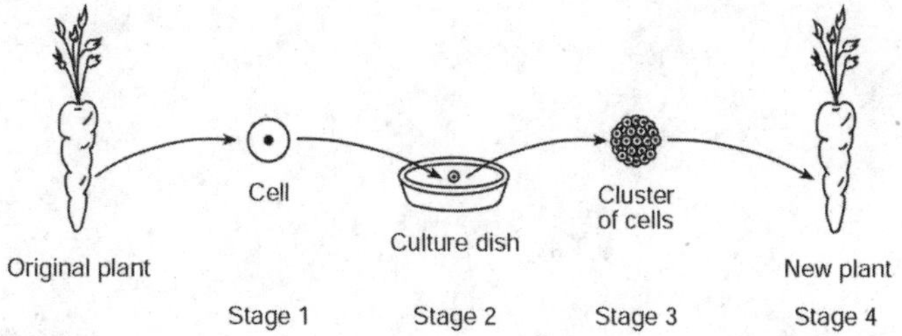

Note: The answer to question 50 should be recorded in the space provided.

50 Which graph below best represents the DNA content found in each cell in each of the stages in the diagram above?

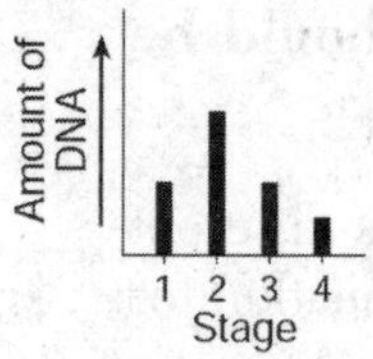

1

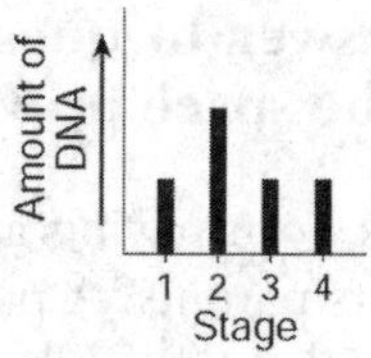

3

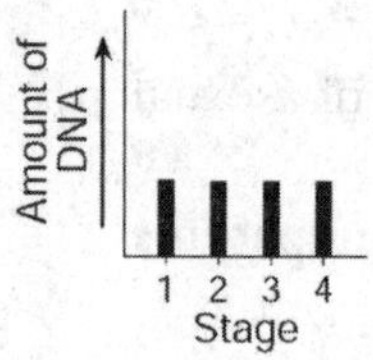

2

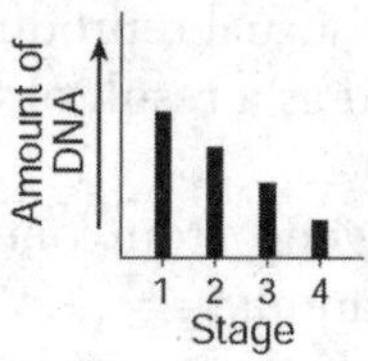

4

50______

51 Describe *one* specific reason why scientists would want to maintain the genetic makeup of a particular plant. [1]

____

____

Base your answers to questions 52 and 53 on the information below and on your knowledge of biology.

Breast Cancer Research

Most deaths that are a result of breast cancer occur because the cancer cells metastasize (spread) from the breast to other organs. As they metastasize, cancer cells travel through the bloodstream.

MicroRNA molecules are involved in both the movement and control of metastasized cells. One microRNA, known as miR-7, shuts down a protein that helps cancer cells travel through the blood.

Understanding how miR-7 interacts with cancer cells may lead to new treatments for certain types of cancer. Since certain levels of miR-7 expression can also stimulate the development of cancer cells, the use of miR-7 to treat cancer will have to be studied in more detail. Researchers are hoping that eventually levels of miR-7 will be used to diagnose, treat, and prevent the spread of cancer in an individual.

52 State *one negative* effect of using miR-7 as the only treatment for breast cancer. [1]

____

____

53 State *one* way cancer cells are different from normal body cells. [1]

____

____

Base your answers to questions 54 and 55 on the information and diagram below and on your knowledge of biology.

Each body cell contains the same genetic information, but can differ in appearance and size. The diagram below shows three different types of cells found in the human body.

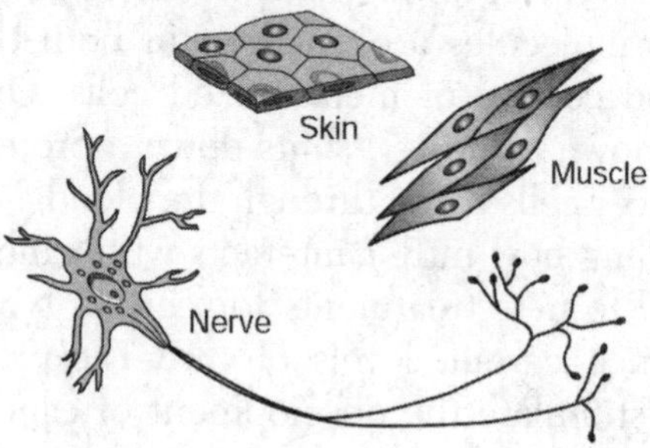

54 Identify *one* similarity, other than the genetic information, that these body cells have. [1]

____

55 Explain why differences in these human body cells are a biological advantage. [1]

____

____

PART C

Answer all questions in this part. [17]

***Directions* (56–72): Record your answers in the spaces provided.**

Base your answers to questions 56 through 58 on the information below and on your knowledge of biology.

Owl vs. Owl

Barred owl

Spotted owl

Federal wildlife officials plan to dispatch armed bird specialists into forests of the Pacific Northwest starting this fall to shoot one species of owl to protect another that is threatened with extinction. …

…"If we don't manage barred owls, the probability of recovering the spotted owls goes down significantly," said Paul Henson, Oregon state supervisor for Fish and Wildlife. The agency's preferred course of action calls for killing 3,603 barred owls in four study areas in Oregon, Washington and northern California over the next four years. …

…Mr. Henson said unless barred owls are brought under control, the spotted owl in coming decades might disappear from Washington's northern Cascade Range and Oregon's Coast Range, where the barred owl incursion [takeover] has been greatest.

The northern spotted owl was listed as a threatened species in 1990. Barred owls are bigger, more aggressive and less picky about food. Barred owls now cover the spotted owl's range, in some places outnumbering them as much as 5-to-1.

Source: Associated Press, 7/26/13

56 Describe how the barred owl population is having a *negative* effect on the spotted owl population. [1]

____

____

57 Explain why it is important to protect the spotted owl from extinction. [1]

____

____

58 Certain groups oppose the plan to kill barred owls, in part because they feel it will not solve the problem. They recommend that the focus should be on protecting the habitat of the spotted owl. Describe the role that the habitat plays in the survival of an animal species such as the spotted owl. [1]

____

____

Base your answer to question 59 on the information below and on your knowledge of biology.

The 1990 Federal Clean Air Act requires New York State to conduct an emissions test on most gasoline-powered automobiles in order to help reduce harmful emissions. Vehicles that fail this test must be repaired and pass inspection before they can be driven on the road. Some people did not support this legislation.

59 State *one* advantage and *one* disadvantage of automobile emission testing. [1]

Advantage: ______________________________

____

Disadvantage: ______________________________

____

Base your answers to questions 60 through 63 on the information below and on your knowledge of biology.

Enzyme Investigation

An enzyme was isolated from digestive juices taken from the small intestine. An experiment was set up to test the ability of the enzyme to break down protein. Two test tubes, labeled *A* and *B*, were placed in a hot water bath at 37°C, human body temperature.

Test tube *A* contained only protein and test tube *B* contained protein and the enzyme. The chart below shows the set-up.

Test Tube	Contents
A	protein
B	protein, enzyme

After two hours, the contents of both test tubes were analyzed. Test tube *A* showed only the presence of protein. Test tube *B* showed the presence of the end products of protein digestion, indicating the enzyme had successfully broken down the protein.

60 Identify the end products of protein digestion that made up the contents of test tube *B* after the two hours. [1]

____

61 Explain the importance of temperature in the functioning of enzymes. [1]

____

____

62 State what the result would be if the same enzyme that was added to test tube *B* was added to a test tube containing starch. Support your answer. [1]

____

____

63 In the digestive system many large molecules, such as proteins, are broken down into much smaller molecules. State what happens to these smaller molecules following digestion. [1]

____

____

Base your answers to question 64–66 on the information below and on your knowledge of biology.

Secondhand Smoke and Estrogen

A fertility researcher conducted a study of pregnant women. The researcher's hypothesis was that the estrogen levels of pregnant women who were exposed to daily secondhand cigarette smoke would be higher than estrogen levels of pregnant women not exposed to daily secondhand smoke.

The researcher measured the estrogen levels of eight pregnant women each week throughout their pregnancy. Four of the women lived in houses with heavy smokers, the other four did not. The women's ages varied from 19 to 42 years old. Six of the women were pregnant with girls, one was pregnant with a boy, and one was pregnant with twin boys. The research was submitted for peer review.

64–66 Analyze this experiment. In your answer, be sure to:

- identify *one* error in the researcher's experimental design [1]
- identify *one* way, other than affecting estrogen levels, that secondhand smoke could affect a developing embryo [1]
- explain why the process of peer review is an important step in this research [1]

____

____

____

____

____

____

Base your answers to questions 67 through 69 on the information and passage below and on your knowledge of biology.

Snowy Owls Move to the South

Snowy owls are large white birds that normally inhabit the cold northern regions of Canada. Recently, scientists and birdwatchers have sighted the snowy owls much farther south than usual.

When snowy owls are in northern areas, they feed on lemmings (small rodents). When lemmings are not available, as in the areas further south, the owls will seek out mice or rabbits as their food source.

Several snowy owls migrated into an area represented by the food web below.

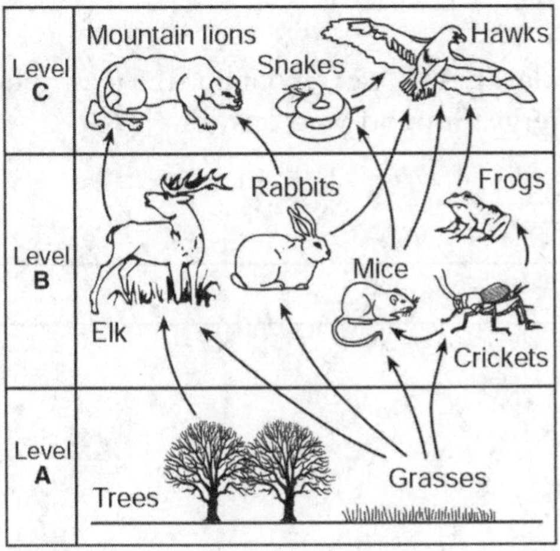

67 Identify *one* population of organisms shown in the food web, other than rabbits or mice, that would likely be affected by the introduction of the snowy owls and explain why their population would be affected. [1]

Population affected:____________________

____

____

68 Identify *one* condition that might cause snowy owls to leave their usual habitat and move to another area. [1]

____

____

69 State which level, *A*, *B*, or *C*, contains the *least* total available energy. Support your answer. [1]

Level:____________

____

____

Base your answers to questions 70 through 72 on the information below and on your knowledge of biology.

Pocket Mice

Pocket mice are small rodents that feed mainly at night and are preyed upon by owls, hawks, and snakes. Scientists studied pocket mice living on dark volcanic rock in both New Mexico and fifty miles away in Arizona. They recorded their data in the chart below.

Number of Mice on Dark Volcanic Rock

Year	New Mexico		Arizona	
	Light Fur	Dark Fur	Light Fur	Dark Fur
2000	120	122	16	125
2001	140	136	8	140
2002	134	130	6	135
2003	115	120	12	115
2004	122	126	8	129

70 State *one* possible hypothesis that would explain the differences in the observed data between the two locations. [1]

____

____

71 Dark fur color in pocket mice is the result of a mutation. Scientists analyzed the sequence of bases in the gene known to play a role in fur color and discovered that the mutation was identical in both the New Mexico and Arizona mouse populations. Explain how it is possible for these two different populations to have identical gene sequences for dark fur color. [1]

____

____

72 Explain what is meant by the statement: "While mutations are random, natural selection is not." [1]

____

____

PART D

Answer all questions in this part. [13]

Directions **(73–85): For those questions that are multiple choice, record in the space provided the *number* of the choice that, of those given, best completes the statement or answers the question. For all other questions in this part, follow the directions given and record your answers in the spaces provided.**

Note: The answer to question 73 should be recorded in the space provided.

73 The diagram below represents evolutionary pathways of seven groups of organisms alive today.

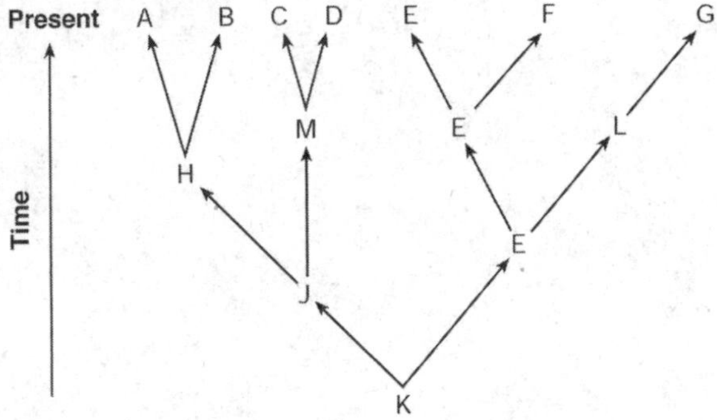

Which two living species would be expected to have the most similar proteins?

1 *A* and *C*
2 *B* and *C*
3 *E* and *F*
3 *H* and *M*

73 ____

Note: The answer to question 74 should be recorded in the space provided.

74 Scientists recently discovered that three different types of squid, a marine animal, previously thought to be three different species, were actually all members of one species. Their earlier ideas were based on using squid carcasses (dead bodies). The new, more accepted classification is most probably based on an analysis of

1 a greater number of squid carcasses
2 the feeding habits of the three different species
3 a number of newly found squid fossils
4 the DNA present in the cells of squid

74 ____

Note: The answer to question 75 should be recorded in the space provided.

75 The diagram below represents a laboratory experiment involving sucrose and water molecules in a cellophane bag which functions in the same way as dialysis tubing.

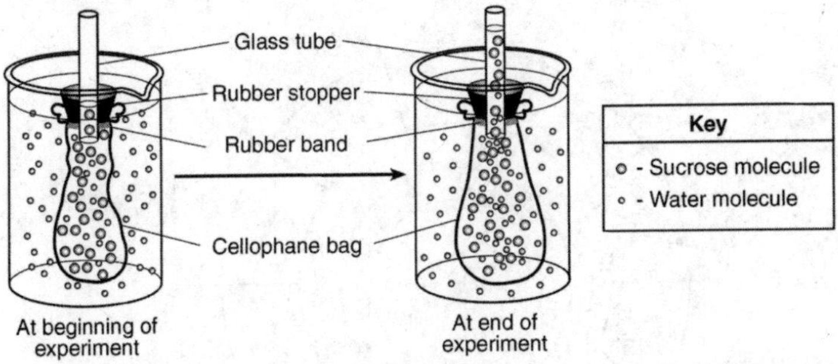

Which statement correctly explains the rise of liquid in the tube at the end of the experiment?

1 The concentration of sucrose molecules increased as water molecules entered the bag. This concentration increase pushed the liquid up the tube.
2 Water entered the bag due to the lower concentration of water inside. The extra water pushed the liquid up the tube as the bag filled.
3 Sucrose indicator entered the bag and reacted with the sucrose molecules. The reaction made the bag increase in size and pushed the liquid up the tube.
4 Sucrose molecules moved out of the bag and up the tube while water moved out, causing the rise of liquid in the tube. 75____

Base your answers to questions 76 through 78 on the diagram below that shows variations in the beaks of finches in the Galapagos Islands and on your knowledge of biology.

Variations in Beaks of Galapagos Islands Finches

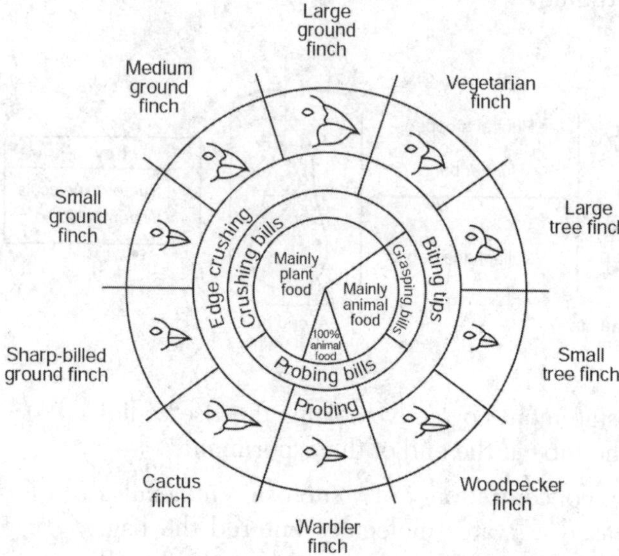

from: *Galapagos: A Natural History Guide*

Note: The answer to question 76 should be recorded in the space provided.

76 Which row correctly pairs a finch species with its primary nutritional role and bill type?

Row	Finch	Bill Type	Nutritional Role
(1)	cactus finch	probing bill	carnivore
(2)	medium ground finch	grasping bill	herbivore
(3)	large tree finch	crushing bill	herbivore
(4)	warbler finch	probing bill	carnivore

76 ____

77 In certain years, the Galapagos plants produce many tube-shaped flowers rich in nectar. Identify the finch that is best adapted to feed on the nectar within those flowers. Support your answer. [1]

---

78 The number of small tree finches is increasing on an island inhabited by a large population of small ground finches. State *one* reason why the population of small ground finches has *not* been affected by the increasing number of small tree finches. [1]

---

79 Explain why glucose molecules can cross a cell membrane and starch molecules can *not*. [1]

Base your answers to questions 80 through 82 on the information below and on your knowledge of biology.

Progressive Resistance Exercise

Progressive resistance exercise (PRE) is a method of increasing the ability of muscles to generate force. The principles of PRE for increasing force production in muscles have remained unchanged for almost 60 years. These principles are (1) to perform a small number of repetitions until fatigued, (2) to allow sufficient rest between exercises for recovery, and (3) to increase the resistance as the ability to generate force increases. Traditionally, PRE has been used by young, healthy adults to improve athletic performance.

A student decided to incorporate PRE into his exercise program. He did not know how to determine when he had allowed sufficient rest between exercises for recovery. He hypothesized that waiting for his pulse to return to normal would probably be a good indication.

80 Explain why allowing his pulse rate to return to normal might be a good indication that he had waited long enough for recovery. [1]

____

____

Note: The answer to question 81 should be recorded in the space provided.

81 Students wanted to try PRE to increase their ability to rapidly squeeze a clothespin. They thought if they could do this, they could challenge another class to a clothespin squeezing competition and win. Which steps should the students take to follow the principles of PRE?

1 Measure their pulse rate after squeezing the clothespin until fatigued. Then increase the resistance of the clothespin for the next trial.
2 Squeeze a clothespin until fatigued, rest, and repeat. Over time, they should gradually increase the resistance of the clothespins they are squeezing.
3 Measure their pulse rate, squeeze a clothespin for one minute, rest, and measure their pulse rate.
4 Squeeze a clothespin for as long as they can, measure their pulse rate, rest, eat some candy. Increase the resistance of the clothespin for the next trial.

81 ____

Note: The answer to question 82 should be recorded in the space provided.

82 Students following the principles of PRE monitored their ability to lift weights. Which observation would indicate that their exercise program was successful?

1 They could eventually lift heavier weights than when they started.
2 Their pulse rate increased more rapidly as they kept lifting weights.
3 The number of weights their group could lift during competition decreased.
4 Males and females could lift the same weight an equal number of times during competition. 82____

---

83 Using the axes on the graph below, sketch a line graph showing the changes in heart rate of a person who is walking slowly, then begins running, and then sits down to rest for a few minutes. [1]

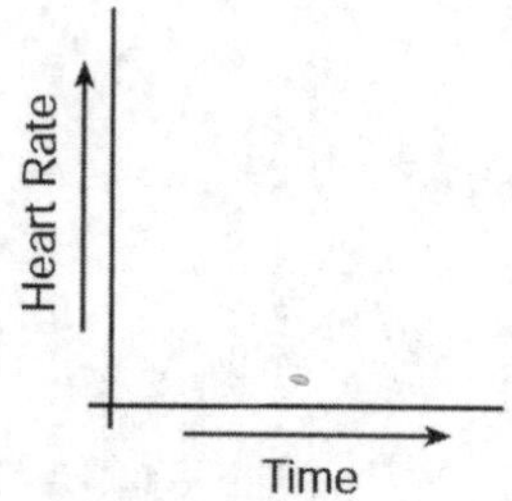

84 Identify *one* waste product that is released during exercise. Explain how this waste product leaves the body. [1]

Waste product: ________________________________

____

____

85 State *one* way scientists could use the banding patterns produced by gel electrophoresis. [1]

____

____

Answers August 2017

Living Environment

Answer Key

PART A

1. 2
2. 3
3. 3
4. 4
5. 2
6. 2
7. 4
8. 3
9. 1
10. 2
11. 3
12. 4
13. 1
14. 2
15. 4
16. 3
17. 1
18. 1
19. 3
20. 3
21. 3
22. 1
23. 1
24. 1
25. 4
26. 2
27. 2
28. 1
29. 2
30. 4

PART B–1

31. 4
32. 2
33. 3
34. 1
35. 3
36. 2
37. 4
38. 3
39. 2
40. 1
41. 1
42. 4
43. 4

PART B–2

44. *See* Answers Explained.
45. *See* Answers Explained.
46. *See* Answers Explained.
47. 3
48. *See* Answers Explained.
49. 1
50. 2
51. *See* Answers Explained.
52. *See* Answers Explained.
53. *See* Answers Explained.
54. *See* Answers Explained.
55. *See* Answers Explained.

PART C. *See* **Answers Explained**.

PART D

73. 3
74. 4
75. 2
76. 4
77. *See* Answers Explained.
78. *See* Answers Explained.
79. *See* Answers Explained.
80. *See* Answers Explained.
81. 2
82. 1
83. *See* Answers Explained.
84. *See* Answers Explained.
85. *See* Answers Explained.

Answers Explained

PART A

1. **2** A fruit fly is classified as a consumer rather than as a producer because it is unable to *synthesize its own food*. Fruit flies, like all animal species, gain their nutrition by consuming the organic tissues of other plants and/or animals. They are incapable of manufacturing organic food molecules to use as food as green plants do.

WRONG CHOICES EXPLAINED

(1) It is *not* true that a fruit fly is classified as a consumer rather than as a producer because it is unable to *reproduce asexually*. Fruit flies, like all animal species, reproduce sexually, not asexually. However, this is not the reason they are classified as consumers.

(3), (4) It is *not* true that a fruit fly is classified as a consumer rather than as a producer because it is unable to *release energy stored in organic molecules* or because it is unable to *remove wastes from its body*. Fruit flies, like all animal species, release energy from organic molecules via the process of respiration and remove wastes from their bodies via the process of excretion.

2. **3** *Insulin is released when glucose levels in the blood are high* is the change that is an example of maintaining dynamic equilibrium. This feedback mechanism ensures that excess glucose in the blood is removed and stored in the liver as glycogen. By this means, body chemistry is restored to a stable condition that supports life.

WRONG CHOICES EXPLAINED

(1) *A plant wilts when more water is lost from the leaves than is lost by the roots* is *not* the change that is an example of maintaining dynamic equilibrium. A plant wilts when its tissues contain less water than is needed to keep its cells rigid. During extremely dry, hot periods, a plant may lose water from both its leaves and its roots. By wilting, the plant closes stomates and lenticels as a means of conserving water until water concentration is restored to a stable condition necessary to sustain life. Therefore, a plant wilting may be considered to be an example of maintenance of homeostasis.

[NOTE: This is a possible correct answer to the question as written. However, it is not the best answer.]

(2) *A plant turns yellow when light levels are very low* is *not* the change that is an example of maintaining dynamic equilibrium. A plant kept in very low light conditions may temporarily lose its ability to synthesize chlorophyll, a green plant pigment that enables the photosynthetic reactions. As a result,

its leaves may yellow until proper light conditions are restored. It is unclear whether this process represents a means of maintaining homeostasis for the plant.

(4) *A person sweats when the environmental temperature is low* is *not* the change that is an example of maintaining dynamic equilibrium. A person sweats when the body's internal temperature rises above the normal 98.6°F (37°C). As sweat evaporates from the skin surface, the cooling action of this evaporation helps to cool the blood until body temperature is restored to a stable condition necessary to sustain life. The environmental temperature has little effect on this process other than to accelerate or retard it. Therefore, a person sweating may be considered to be an example of maintenance of homeostasis.

[NOTE: This is a possible correct answer to the question as written. However, it is not the best answer.]

3. **3** The chemical bonds in these compounds can be a source of *energy*. Chemical bonds are formed when atoms link together by completing the atoms' electron shells. The chemical bonds of organic molecules can be broken by the body's respiratory process and their energy transferred to molecules of ATP. ATP transports this energy to the body's tissues for use in biochemical processes that require it.

WRONG CHOICES EXPLAINED

(1), (2), (4) The chemical bonds in these compounds can *not* be a source of *amino acids*, *simple sugars*, or *enzymes*. These substances may be formed in the body as a result of the life processes of nutrition (amino acids, simple sugars) or synthesis (enzymes). Chemical bonds cannot be sources of these substances under any circumstances.

4. **4** The process being used to develop these new varieties is most likely *genetic engineering*. Genetic engineering is a laboratory technique in which a gene for a desired trait is snipped from the DNA of a donor cell and inserted into the genome of a recipient cell. If such a phosphorus-extracting gene could be located in a donor organism's genome, it could be snipped and inserted into the genomes of fertilized plant ova of the recipient plant species. The resulting recipient plants would then contain this gene in all of their cells and theoretically would be capable of passing this gene on to their offspring.

WRONG CHOICES EXPLAINED

(1) The process being used to develop these new varieties is *not* most likely *paper chromatography*. Paper chromatography is a laboratory technique that is used to help researchers separate mixtures of organic chemicals such as plant pigments. This technique is not used to manipulate genetic traits.

(2) The process being used to develop these new varieties is *not* most likely *natural selection*. Natural selection, sometimes referred to as survival of the fittest, postulates that certain members of a species are better adapted to their environment than others. Natural selection is assumed to be a natural process, not a laboratory technique.

(3) The process being used to develop these new varieties is *not* most likely *direct harvesting*. Direct harvesting is a human practice in which plants or animals are removed (harvested) from a natural environment for their economic value. This practice is not used to manipulate genetic traits.

5. **2** *Reproduction* is the life function that is *not* necessary for an individual organism to stay alive. Although reproduction is essential for survival of the species, it is not necessary for an individual's survival on a day-to-day basis.

WRONG CHOICES EXPLAINED

(1), (3), (4) It is *not* true that *nutrition*, *regulation*, or *excretion* are the life functions that are *not* necessary for an individual organism to stay alive. These life functions enable the body to process food, control and coordinate bodily functions, and remove metabolic wastes, respectively. Each of these life functions is essential for the body's ability to maintain homeostasis and therefore to sustain life on a day-to-day basis.

6. **2** The most likely reason the sea hare can escape is because the sea hare ink *blocks a receptor on certain cells in the lobster*. The paragraph states that the "lobsters find their prey through a sense of smell." The biological mechanism by which this occurs involves odor-producing chemicals that are constantly released by the sea hare into its water environment. These chemicals diffuse in the water and are sensed by the lobster by means of odor sensors located on its antennae. These odor sensors are made up of cells whose membranes contain receptor molecules capable of detecting the chemical in the water. Once these receptors are blocked by the sea hare's ink, the lobster can no longer detect the sea hare through its sense of smell.

WRONG CHOICES EXPLAINED

(1), (3), (4) It is *not* true that the most likely reason the sea hare can escape is because the sea hare ink *pushes the sea hare away rapidly as the ink is expelled*, *causes the lobster to change its prey*, or *prevents movement of the lobster*. None of these responses contain information that is indicated in the paragraph as being the means by which the sea hare avoids predation by the lobster.

7. **4** *Genes are made of DNA and code for proteins* is the statement that is an accurate description of genes. Genes located on chromosomes are composed of deoxyribonucleic acid (DNA) molecules. DNA molecules contain

unique sequences of nitrogen bases (A, T, G, and C) that form a code that regulates the synthesis of specific protein molecules essential to the survival of the organism.

WRONG CHOICES EXPLAINED

(1), (2), (3) *Proteins are made of genes and code for DNA*, *genes are made of proteins that code for nitrogen bases*, and *DNA is made of carbohydrates that code for genes* are *not* the statements that are accurate descriptions of genes. Each of these statements contains one or more incorrect facts about the nature of genes, proteins, or DNA.

8. **3** In order to save these birds from extinction, the best course of action would be to *work to protect bobolink habitats in South and North America*. First, it is necessary to determine the bobolinks' preferred geographical zones and environmental conditions. Next, efforts should be made to ensure that these areas be preserved undeveloped in perpetuity (forever). These steps would help to ensure that the birds have undisturbed nesting and feeding habitats for centuries to come and would allow them to recover as a species.

WRONG CHOICES EXPLAINED

(1), (4) It is *not* true that in order to save these birds from extinction, the best course of action would be to *prevent the birds from migrating to South America* or to *capture all the bobolinks and keep them safe in zoos*. Even if these actions were possible, neither of them would assist the bobolink species to recover from the brink of extinction. Rather, these actions would significantly disrupt the life cycle of bobolinks and would help to ensure their eventual extinction.

(2) It is *not* true that in order to save these birds from extinction, the best course of action would be to *encourage farmers to let their hay fields undergo succession*. The paragraph states that bobolinks nest in fields of tall grasses such as hay fields. If hay fields were allowed to undergo succession, many of them would revert to forest lands with environmental conditions different from those of grassy fields. This action would significantly limit the nesting range of bobolinks and might contribute to their eventual extinction.

9. **1** This protein most likely affects the functioning of *cell membranes*. The cell membrane functions to regulate the entry into and exit out of the cell of materials essential for life or dangerous to the health of the cell. A protein known as CFTR located on the cell membrane regulates the flow of chloride ion into and out of the cell. Genetic alteration of CFTR can impede the normal flow of chloride ion out of the cell. It is implied by the stem of the question that too high a concentration of chloride ion in the cell is a probable contributing factor in the development of cystic fibrosis in humans. Therefore, it is likely that this abnormal protein affects the functioning of cell membranes.

WRONG CHOICES EXPLAINED

(2) This protein does *not* most likely affect the functioning of *nuclei*. Nuclei contain genetic material in the form of chromosomes. Nuclei and their contents would not be directly affected by the condition caused by the alteration of this protein.

(3) This protein does *not* most likely affect the functioning of *mitochondria*. Mitochondria contain the enzymes that catalyze the respiratory reactions. Mitochondria and their contents would not be directly affected by the condition caused by the alteration of this protein.

(4) This protein does *not* most likely affect the functioning of *ribosomes*. Ribosomes facilitate the protein synthesis reactions. Ribosomes and their contents would not be directly affected by the condition caused by the alteration of this protein.

10. **2** Row *2* in the chart shows a direct relationship that can exist between two living organisms. The term "predator" refers to animal species that hunt and consume the bodies of other animals for their food. The term "prey" refers to animal species that are consumed by one or more predator species. A nutritional relationship typically exists between predators and prey in a natural ecosystem. This relationship exists between the nutritional roles of the living organisms, not between the living organisms themselves.

WRONG CHOICES EXPLAINED

(1) Row *1* in the chart does *not* show a direct relationship that can exist between two living organisms. The term "producer" refers to plant species that are capable of manufacturing their own food. The term "carnivore" refers to animal species that consume the bodies of other animal species for their food. No direct nutritional relationship exists between producers and carnivores in a natural ecosystem.

(3) Row *3* in the chart does *not* show a direct relationship that can exist between two living organisms. The term "parasite" refers to plant and animal species that live in or on the bodies of other organisms and use the tissues of those organisms for food. The term "prey" refers to animal species that are consumed by one or more predator species. No direct nutritional relationship exists between parasites and prey in a natural ecosystem

(4) Row *4* in the chart does *not* show a direct relationship that can exist between two living organisms. The term "carnivore" refers to animal species that consume the bodies of other animal species for their food. The term "host" refers to plant and animal species that serve as sources of food for one or more parasite species. No direct nutritional relationship exists between carnivores and hosts in a natural ecosystem.

11. **3** These changes are an example of *the growth of a forest through ecological succession*. Ecological succession refers to the series of changes that occur to the dominant plant community of an area that has been sig-

nificantly affected by an environmental change. When Mount St. Helens exploded, it destroyed the forest ecosystem on the mountain and on hundreds of square miles surrounding the mountain. This ecosystem is now slowly recovering through the process of ecological succession.

WRONG CHOICES EXPLAINED

(1), (2), (4) These changes are *not* an example of *humans degrading an ecosystem by removing wildflowers*, *the loss of genetic variation in a plant species*, or *the biological evolution of wildflowers, shrubs, and trees*. There is no mention in the paragraph about direct harvesting, genetic variation, or biological evolution.

12. **4** The most likely result of completely removing carbon dioxide from the environment of a plant is that sugar production will *decrease and eventually stop*. Carbon dioxide is an inorganic compound required for the process of photosynthesis to occur in green plants. The products of photosynthesis are glucose, which is a simple sugar, and oxygen gas. In the event that all carbon dioxide is removed from the closed environment of a plant, the photosynthetic process and the production of sugar and oxygen will cease in that plant.

WRONG CHOICES EXPLAINED

(1), (2), (3) The most likely result of completely removing carbon dioxide from the environment of a plant is *not* that sugar production will *continue at the same rate*, *increase and oxygen production will also increase*, or *increase and oxygen production will stay the same*. The rates of sugar and oxygen production will both decrease and eventually stop.

13. **1** *Building the center would decrease resources needed by local organisms* is the statement that identifies an environmental concern associated with the development of the shopping center. The statement raises a question about the loss of habitat, food, nesting space, and other resources that will occur when trees and other vegetation are removed and the site developed with structures, parking lots, and other construction items.

WRONG CHOICES EXPLAINED

(2) *The new shopping center would increase competition with already existing businesses* is *not* the statement that identifies an environmental concern associated with the development of the shopping center. This is an economic, not an environmental, concern.

(3) *Building the center would decrease the amount of pollution in the area* is *not* the statement that identifies an environmental concern associated with the development of the shopping center. The traffic and activities of the center would increase, not decrease, pollution.

(4) *The new shopping center would increase the biodiversity of the area* is *not* the statement that identifies an environmental concern associated with the development of the shopping center. The traffic and activities of the center would decrease, not increase, biodiversity.

14. **2** The most likely reason for removing invasive plants is to *preserve biodiversity*. Invasive plants often crowd out and eliminate native species from their natural habitats. They effectively reduce biodiversity in these habitats, an action that destabilizes the ecosystem.

WRONG CHOICES EXPLAINED

(1) The most likely reason for removing invasive plants is *not* to *allow only one type of native plant to grow*. Removing invasive species allows many native plant species to grow, which increases biodiversity.

(3) The most likely reason for removing invasive plants is *not* to *eliminate unfamiliar food sources*. Invasive plants may produce berries, seeds, and other parts that provide food for native animals. However, removing these plants is not done for the purpose of eliminating these food sources.

(4) The most likely reason for removing invasive plants is *not* to *increase the rate of ecological succession*. Ecological succession refers to the series of changes that occur to the dominant plant community of an area that has been significantly affected by an environmental change. However, removing these plants is not done for the purpose of increasing the rate of succession.

15. **4** Row *4* correctly pairs a human activity with its impact on the environment. Recycling discarded items made of plastic, metal, and paper reduces the amount of new material that needs to be mined or harvested. When recycling is reduced, more of these resources must be gathered as raw materials for manufacturing, thus reducing the amount of available resources.

WRONG CHOICES EXPLAINED

(1) Row *1* does *not* correctly pair a human activity with its impact on the environment. Reducing the use of pesticides would have no known effect on the rate of erosion of rock in the soil.

(2) Row *2* does *not* correctly pair a human activity with its impact on the environment. Increasing housing developments will degrade, not improve, air quality.

(3) Row *3* does *not* correctly pair a human activity with its impact on the environment. Increasing the human population will increase, not reduce, water usage worldwide.

16. **3** This bacterial cell can now be used to produce *insulin that can be used by humans*. The process illustrated in the diagram is genetic engineering, a laboratory technique in which a gene for a desired trait is snipped from the DNA of a donor cell and inserted into the genome of a recipient cell. The

answer to this question is clearly provided in the diagram, which identifies a "gene for human insulin" being inserted into the genome of a bacterial cell.

WRONG CHOICES EXPLAINED

(1) It is *not* true that this bacterial cell can now be used to produce *the bacterial gene for insulin that can be inserted into humans*. Bacteria do not naturally display the genetic ability to produce insulin.

(2), (4) It is *not* true that this bacterial cell can now be used to produce *human genes for enzymes that can be inserted into humans* or the *enzymes necessary to treat human diseases*. The gene illustrated as being spliced into the bacterial cell is the gene for insulin. Insulin is a hormone, not an enzyme. The enzymes noted in the diagram are restriction enzymes needed to catalyze the splicing reaction.

17. **1** The change observed in the number of green and brown beetles in the population is most likely due to *natural selection*. In the diagram, a species population of beetles of varying colors is illustrated. In the original population sample, about 80% of the beetles have brown bodies, while about 20% have green bodies. Over time, the proportion of brown beetles declines and the proportion of green beetles increases until these proportions are virtually reversed. We must assume that some selection pressure in this ecosystem (not identified) is eliminating the brown beetles and allowing the green beetles to become the more successful variation in the species population. The green beetles survive, reproduce, and pass on their genes for green coloration to succeeding generations. This represents natural selection and biological evolution at work in the present day.

WRONG CHOICES EXPLAINED

(2) The change observed in the number of green and brown beetles in the population is most likely *not* due to *selective breeding*. Selective breeding is a technique used by animal and plant breeders in which organisms with desirable traits are cross-bred with the hope of producing offspring that also display those traits. Selective breeding does not explain the phenomena illustrated in the diagram.

(3) The change observed in the number of green and brown beetles in the population is most likely *not* due to *gene manipulation*. Gene manipulation is a laboratory technique in which a gene for a desired trait is snipped from the DNA of a donor cell and inserted into the genome of a recipient cell. Gene manipulation does not explain the phenomena illustrated in the diagram.

(4) The change observed in the number of green and brown beetles in the population is most likely *not* due to *a common ancestor*. A common ancestor is a species of organism that existed in the past and that provided the fundamental genetic makeup of two or more existing species. Common ancestry does not explain the phenomena illustrated in the diagram.

18. **1** Structure *A* is correctly paired with its reproductive function. Structure *A* is the vas deferens, which functions to transport sperm cells (male gametes) from the testis to the urethra during sexual intercourse.

WRONG CHOICES EXPLAINED

(2) Structure *B* is *not* correctly paired with its reproductive function. Structure *B* is the testis, which functions to produce mature sperm cells necessary for the reproductive process. Small amounts of progesterone are produced in men in the adrenal glands.

(3) Structure *C* is *not* correctly paired with its reproductive function. Structure *C* is the urinary bladder, which functions to store urine prior to elimination. The urinary bladder is an excretory, not a reproductive, structure.

(4) Structure *D* is *not* correctly paired with its reproductive function. Structure *D* is the penis, which functions to insert mature sperm cells into the female reproductive tract during sexual intercourse.

19. **3** An example of an activity that would promote the conservation of coastal areas and ocean resources is *creating protected zones of natural grasses and shrubs in beach areas*. This is an example of a positive human activity designed to reduce beach erosion and storm surge that can destroy natural habitat and threaten human communities that depend on that habitat for economic growth.

WRONG CHOICES EXPLAINED

(1), (2), (4) It is *not* true that an example of an activity that would promote the conservation of coastal areas and ocean resources is *harvesting large numbers of different fish species*, *allowing all-terrain vehicles access to beach areas*, or *encouraging the construction of factories along the ocean shoreline*. These are examples of negative human activities that would serve to accelerate the destruction of habitat, allow the erosion of beaches, threaten human communities, increase environmental pollution, and reduce biodiversity in these areas.

20. **3** *Allowing native plants to grow undisturbed in an abandoned field* is the activity that would eventually result in a stable ecosystem. In this field, cultivated crops would be replaced by native grasses, shrubs, and small trees. Eventually the field would revert to the dominant plant climax community typical of the area in which it is located. In New York State, this might be a mixed oak-maple-beech forest community. This climax community would support thousands of other plant and animal species and would be self-perpetuating unless significantly disturbed by flood, fire, wind, or other catastrophic occurrence.

WRONG CHOICES EXPLAINED

(1), (2), (4) *Deforestation in an area to increase space for the species living there*, *mowing a large field so it can be used for recreation*, or *spraying pesticides on a field at the end of each growing season* are *not* the activities that would eventually result in a stable ecosystem. These are examples of negative human activities that would serve to accelerate the destruction of habitat, increase environmental pollution, and reduce biodiversity in these areas. Each of these activities would result in a less stable ecosystem.

21. **3** *It reduces the amount of plastics and glass put into landfills* is the statement that correctly describes how this system is intended to help the environment. In addition to reducing pressure on limited landfill space, the materials in these containers can be recycled into new products. This reduces pressure on our limited natural resources, such as petroleum (for plastics) and aluminum (for cans), as well as on the energy sources that would otherwise be needed to convert these raw materials into new products.

WRONG CHOICES EXPLAINED

(1) *It encourages people to buy products that do not have a deposit* is *not* the statement that correctly describes how this system is intended to help the environment. This may or may not prove to be true in practice as it represents a personal choice by the consumer. Participation in recycling programs, with or without deposit laws, is a human activity that has a positive effect on the environment.

(2) *It reduces the amount of money shoppers actually spend* is *not* the statement that correctly describes how this system is intended to help the environment. It is probable that any costs related to this program are passed on to consumers by manufacturers and retailers. It is likely that consumer costs are increased, not reduced, by these laws. That contribution is small compared to the value of environmental improvements that result from them.

(4) *It forces manufacturers to reduce air pollution when they are making the containers* is *not* the statement that correctly describes how this system is intended to help the environment. These laws do not force manufacturers to do anything more than ensure that the containers are labeled in such a way that allows consumers to return them and recover their deposit.

22. **1** Level *1* contains organisms that carry out autotrophic nutrition. Autotrophic (self-feeding) nutrition is a process carried out primarily by green plants, also known as producers. This process uses inorganic raw materials (CO_2 and H_2O) to synthesize glucose ($C_6H_{12}O_6$) using the energy of sunlight. Glucose is used as food energy by the plants and the excess is available to be consumed by herbivores (plant eaters) in the food web.

WRONG CHOICES EXPLAINED

(2), (3), (4) Levels 2, 3, and 4 do *not* contain organisms that carry out autotrophic nutrition. These levels contain heterotrophic (other-feeding) organisms that derive their food energy by consuming the bodies of other organisms. The arrows in the diagram indicate the nutritional relationships and the flow of energy in this food web.

23. **1** *A protein is a long chain of amino acids folded into a specific shape* is the statement that best describes the characteristics a protein must have to function correctly. The specific shapes of normal proteins allow them to perform specific functions in the cell that promote the survival of the cell and the organism of which they are a part.

WRONG CHOICES EXPLAINED

(2), (4) *A protein is a long chain of simple sugars folded into a specific shape* and *a protein is made of simple sugars synthesized into a short, circular chain* are *not* the statements that best describe the characteristics a protein must have to function correctly. Proteins are synthesized as long chains of amino acids connected in a specific sequence, not simple sugars.

(3) *A protein is made of amino acids synthesized into a short, circular chain* is *not* the statement that best describes the characteristics a protein must have to function correctly. Proteins are synthesized as long chains of amino acids connected in a specific sequence. These chains then fold into particular shapes dictated by the attractive and repulsive forces within the amino acid chain.

24. **1** The movement of molecules from side *A* to side *B* is an example of the process of *active transport*. Active transport is a process in which a substance moves against the concentration gradient from an area of lower relative concentration to an area of higher relative concentration of that substance. This process requires the expenditure of cellular energy. In the diagram, molecules of a substance are in low relative concentration on side *A* and are shown moving across a membrane to side *B*, which has a higher relative concentration of the molecules. Energy must be involved in this movement against the concentration gradient. (The source of this energy is not indicated.)

WRONG CHOICES EXPLAINED

(2) The movement of molecules from side *A* to side *B* is *not* an example of the process of *chromatography*. Chromatography is a laboratory technique that is used to help researchers study mixtures of organic chemicals such as plant pigments. This technique relies on the processes of capillarity and diffusion to enable its operation.

(3) The movement of molecules from side *A* to side *B* is *not* an example of the process of *cellular respiration*. Cellular respiration is a process by

which living cells derive energy from the chemical bonds of glucose and store it temporarily in molecules of ATP. ATP is a source of energy that enables active transport to occur in cells.

(4) The movement of molecules from side *A* to side *B* is *not* an example of the process of *diffusion*. Diffusion is a process in which a substance moves with the concentration gradient from an area of higher relative concentration to an area of lower relative concentration of that substance. This process does not require the expenditure of cellular energy. In the diagram, the movement of molecules is shown to be against, not with, the concentration gradient.

25. **4** *The environment could influence how genes are expressed, changing how the animal looks and acts* is the statement that best explains why the replacement pets that are produced in this way might *not* look or act like the original. Geneticists have studied the environmental effects of gene expression in plants and animals for many years. Their studies have shown clear examples of these effects and have established the phenomenon as a scientific likelihood in cases involving identical twins and clones.

WRONG CHOICES EXPLAINED

(1) *The new animal must get the DNA from two different parents, not just one cell* is *not* the statement that best explains why the replacement pets that are produced in this way might *not* look or act like the original. The cloning technique involves the removal of a diploid nucleus from a donor organism and its insertion into a denucleated fertilized egg cell of the same or related species. By definition, this diploid nucleus would contain the DNA of the two parents of the donor.

(2), (3) *Mutations could occur that change the cloned animal into a completely different species* and *recombination of the cells as they are cloned will make the resulting pet act differently* are *not* the statements that best explain why the replacement pets that are produced in this way might *not* look or act like the original. Mutation and recombination are not processes at work in the cloning technique. These are nonsense distracters.

26. **2** This cancer is the direct result of a *mutation in the genetic material*. Mutations are random events in which genetic material is altered though the effects of mutagenic agents including radiation. In order for a cell to become cancerous, mutations must occur to the genes that control mitotic cell division. Once these genes are compromised in a cell, the mutated cell's division becomes rapid and abnormal, resulting in a mass of cancerous cells that crowd out and kill normal tissues.

WRONG CHOICES EXPLAINED

(1), (3), (4) This cancer is *not* the direct result of a *change in a starch molecule*, a *mutation in a protein*, or a *change in a fat molecule*. Starches, proteins, and fats are organic compounds found naturally in humans and other organisms. Alteration of these compounds would have no effect on the occurrence of cancer in these organisms.

27. **2** *The nutrients necessary for development would not be able to reach the fetus* is the event that would be most likely to happen if structure *X* were damaged in the early stages of pregnancy. Structure *X* represents the placenta that forms in the wall of the uterus during pregnancy. The placenta allows for the diffusion of nutrients, oxygen, and other essential materials from the mother's body into the tissues of the fetus during its development. If the placenta were damaged early in pregnancy, this diffusion would be slowed or stopped, possibly resulting in the death of the fetus.

WRONG CHOICES EXPLAINED

(1) *The genes from the mother would not be turned on in the fetus* is *not* the event that would be most likely to happen if structure *X* were damaged in the early stages of pregnancy. Genes in the fetus are turned on or off as its tissues differentiate during development. This process occurs regardless of the condition of the placenta as long as the fetus survives.

(3) *The fertilized egg would not be able to travel from the ovary to the uterus* is *not* the event that would be most likely to happen if structure *X* were damaged in the early stages of pregnancy. In this diagram, the fertilized egg has already moved from the ovary to the uterus.

(4) *Development would take longer since the fetus would have to synthesize nutrients* is *not* the event that would be most likely to happen if structure *X* were damaged in the early stages of pregnancy. The human fetus is not capable of synthesizing nutrients under any conditions. Nutrients can reach the developing fetus only through the placental connection.

28. **1** The reproductive structure in a female mammal that produces sex cells is the *ovary*. The mammalian ovary contains thousands of immature primary sex cells (eggs) that periodically undergo meiotic cell division and maturation prior to being released into the oviduct for fertilization.

WRONG CHOICES EXPLAINED

(2) The reproductive structure in a female mammal that produces sex cells is *not* the *testes*. The testes are reproductive organs that produce sex cells (sperm) in male, not female, mammals.

(3) The reproductive structure in a female mammal that produces sex cells is *not* the *uterus*. The uterus is an organ in female mammals that supports fetal development by providing an environment favorable for that development.

(4) The reproductive structure in a female mammal that produces sex cells is *not* the *placenta*. The placenta supports fetal development in female mammals by providing nutrients, oxygen, and other materials essential for that development.

29. **2** The role of fungi is important because they *break down materials that can then be used by other organisms*. Fungi (such as mushrooms, molds, mildews, and bracket fungi) grow in the moist tissues of dead organisms (such as dead leaves and trees) and secrete digestive enzymes that break down those tissues into simpler molecules. The fungi absorb some of these simple molecules and use them as nutrients for their life functions. In the process, the complex tissues of those dead organisms are broken down into small particles that enrich the soil and can be further broken down by other soil-living organisms (such as earthworms, insects, and bacteria). Eventually, nitrogen-rich compounds are produced in the soil that act as natural fertilizer for green plants.

WRONG CHOICES EXPLAINED

(1) It is *not* true that the role of fungi is important because they *synthesize energy-rich compounds that are directly used by producers*. Fungi are not capable of synthesizing energy-rich compounds. Producers such as green plants manufacture energy-rich compounds through the photosynthetic reactions, not from the by-products of decomposition.

(3) It is *not* true that the role of fungi is important because they *limit the number of plants that can perform photosynthesis in an area*. In a balanced ecosystem, the presence of fungi enhances the growth and numbers of green plants living there.

(4) It is *not* true that the role of fungi is important because they *are competitors of other consumers such as herbivores*. Herbivores (plant eaters) consume green plants growing in their environment, not dead and decomposing plant matter such as dead leaves and trees. There is no direct competition between fungi and other decomposers in a natural environment.

30. **4** This is an example of *human actions altering ecosystems with serious consequences*. Any time a naturally occurring species is negatively affected by human action, the consequences for that species and for the ecosystem can be great. In this case, it is believed that human sewage runoff is at least partially responsible for the manatee deaths.

WRONG CHOICES EXPLAINED

(1) This is *not* an example of *a natural cycle in an ecosystem*. Although there are several natural cycles at work in this example, the manatee deaths are believed to be caused by the overwhelming amount of human sewage runoff and decomposition in this area. This is an example of an unnatural, not a natural, condition.

(2) This is *not* an example of *the effect of increased biodiversity on an ecosystem*. In this example, the manatee deaths represent a decrease, not an increase, of biodiversity of this area.

(3) This is *not* an example of *direct harvesting in an ecosystem*. Direct harvesting is a human practice in which plants or animals are removed

(harvested) from a natural environment for their economic value. In this example, the manatees are not being harvested intentionally but, instead, are being killed by human negligence.

PART B–1

31. **4** In order to make the conclusions more valid, the student could *use a larger number of snails*. Good scientific investigation is best done using large numbers of experimental subjects. This practice helps to eliminate experimental bias caused by outliers in the experimental groups being tested that would skew data and misrepresent experimental results.

WRONG CHOICES EXPLAINED

(1), (2), (3) It is *not* true that in order to make the conclusions more valid, the student could *alter the pH of the water*, could *change the size of the tank*, or could *carry out the experiment for a shorter period of time*. Redesigning the experiment by changing these variables could drastically change experimental results. All variables should be kept the same in repeated experimental trials in order to avoid this situation.

32. **2** Sequence $B \rightarrow C \rightarrow A \rightarrow D$ represents the correct order of these events during sexual reproduction. The sequence in mammals begins with meiosis (*B*), a type of cell division that results in the production of haploid (*n*) gametes (egg and sperm cells). Next in the sequence is fertilization (*C*), during which haploid gametes fuse to produce a diploid (2*n*) zygote. This is followed by rapid mitotic cell division (*A*) and differentiation required to form the embryo. Finally, after a period of growth and development of the embryo, birth (*D*) of the offspring occurs.

WRONG CHOICES EXPLAINED

(1), (3), (4) Sequences $A \rightarrow C \rightarrow B \rightarrow D$, $C \rightarrow B \rightarrow A \rightarrow D$, and $B \rightarrow A \rightarrow C \rightarrow D$ do *not* represent the correct order of these events during sexual reproduction. Each of these responses contains one or more events out of correct order.

33. **3** A broad body of evidence, subject to revisions, supported by different kinds of scientific investigations and often involving the contributions of scientists from different disciplines is necessary to develop *a theory*. This is the essential definition of a theory that is developed by competent scientific methods. Theories depend on the results of many different experiments and observations testing many related hypotheses.

WRONG CHOICES EXPLAINED

(1) It is *not* true that a broad body of evidence, subject to revisions, supported by different kinds of scientific investigations and often involving the contributions of scientists from different disciplines is necessary to develop *an inference*. An inference is defined as a conclusion reached on the basis of reasoning. Inferences relate to the results of a single experiment or a set of experiments testing the same hypothesis.

(2) It is *not* true that a broad body of evidence, subject to revisions, supported by different kinds of scientific investigations and often involving the contributions of scientists from different disciplines is necessary to develop *a fact*. A fact is defined as an objective and verifiable observation. Facts are interpreted by hypotheses that must be tested, inferences that are drawn, and theories that are developed.

(4) It is *not* true that a broad body of evidence, subject to revisions, supported by different kinds of scientific investigations and often involving the contributions of scientists from different disciplines is necessary to develop *a prediction*. A prediction is defined as a forecast of the results that will be obtained in a single experiment. Predictions are also known as hypotheses.

34. **1** The clover plant having gene 2 (4 leaves) was most likely the result of *an insertion*. An examination of the diagrams reveals that gene 2 differs from gene 1 by the addition of one nucleotide pair (A-T). This addition is commonly known as insertion. The extra nucleotide pair changes the codon of this gene such that it will code for an altered protein. This altered protein will result in the formation of four leaves rather than three leaves in the clover plant.

WRONG CHOICES EXPLAINED

(2) The clover plant having gene 2 (4 leaves) was *not* most likely the result of *a deletion*. A deletion mutation would show one or more fewer, not more, nucleotide pairs in gene 2.

(3) The clover plant having gene 2 (4 leaves) was *not* most likely the result of *a substitution*. A substitution mutation would show one or more different, not more or fewer, nucleotide pairs in gene 2.

(4) The clover plant having gene 2 (4 leaves) was *not* most likely the result of *normal replication*. A normal replication would show no change in type or number of nucleotide pairs in gene 2.

35. **3** This new device would help the athlete to *maintain homeostasis*. Homeostasis (steady state) is a term used to refer to the dynamic equilibrium that occurs in living things when metabolic activities are balanced in support of life. By cooling the body, this device helps the athlete maintain normal body temperature, which helps the body maintain homeostasis.

WRONG CHOICES EXPLAINED

(1) This new device would *not* help the athlete to *control the rate of muscle activity*. The athlete controls muscle activity through his/her nervous system, not by use of a device that controls body temperature.

(2) This new device would *not* help the athlete to *increase muscle strength*. Muscle strength is increased through focused exercise, not by use of a device that controls body temperature.

(4) This new device would *not* help the athlete to *eliminate the release of heat from the body*. This situation would allow uncontrolled increase of body temperature and could result in heat stroke or death of the athlete if not corrected.

36. **2** To report the final results, the label that would be most appropriate for the third column of the data table is *Average Ripening Time (days)*. The description of the experiment clearly states that the student wants to gather data on "the effect of light on the rate of ripening of tomatoes." The rate of ripening of tomatoes is the dependent variable (the variable dependent on the independent variable) in this experiment.

WRONG CHOICES EXPLAINED

(1), (3), (4) It is *not* true that to report the final results, the label that would be most appropriate for the third column of the data table is *Height of Tomato Plants (cm)*, *Average Weight of Tomatoes per Plant (grams)*, or *Acidity of Tomatoes (pH)*. None of these variables is mentioned in the description of this experiment as being of interest to the student.

37. **4** The independent variable in this experiment is the *light intensity*. The description of the experiment clearly states that the student wants to gather data on "the effect of light on the rate of ripening of tomatoes." The second column in the table is titled "Light Intensity (lumens)" and shows four different levels of light intensity to be manipulated by the student. This variable is under the control of the researcher. So it is the independent variable in this experiment.

WRONG CHOICES EXPLAINED

(1), (2), (3) The independent variable in this experiment is *not* the *type of tomato plant*, the *amount of soil provided*, or the *color of tomatoes*. None of these variables is mentioned in the description of this experiment as being of interest to the student. These values should be maintained constant for all setups in this experiment.

38. **3** Vaccinations usually stimulate the body to produce more of *structures A and C, only*. Vaccines contain dead or weakened pathogens that are purposely injected into the body as a means of introducing the pathogen's antigens (*B*) into the bloodstream. This action stimulates the body's immune

response, resulting in the production of specific antibodies (*C*) and white blood cells (*A*) that will recognize and counteract the pathogen. This immune response creates a permanent natural immunity to the pathogen and the disease it carries.

WRONG CHOICES EXPLAINED

(1), (2), (4) Vaccinations do *not* usually stimulate the body to produce more of *structure A, only*; *structure B, only*; or *structures A, B, and C*. Each of these responses contains one or more incorrect items.

39. **2** Structure *B, only* normally stimulates an allergic response. Structure *B* is an antigen, which is a foreign protein that may be produced by allergins (such as ragweed pollen, animal dander, mold spores, and other natural substances). When introduced into the body, these antigens are recognized by the immune system as foreign invaders. The immune system responds to these antigens by stimulating the production of fluids (such as lymph, tears, and mucus) and raising body temperature in an attempt to eliminate the antigens from the body. In combination, these responses represent an allergic reaction.

WRONG CHOICES EXPLAINED

(1), (3), (4) It is *not* true that structure *A, only*; *C, only*; or structures *A, B, and C* normally stimulate an allergic response. The body's immune system does not normally react to its own products (substances *A* and *C*) as foreign invaders.

40. **1** Population *1* in the chart has the best chance for survival in a rapidly changing environment. Population *1* reproduces sexually, ensuring that the population maximizes variation caused by genetic recombination. The population's short life span and large number of offspring ensures that several large generations can be produced in a short time frame. In combination, these reproductive characteristics help to maximize the species' ability to adjust and evolve in the face of changing environmental conditions (selection pressures).

WRONG CHOICES EXPLAINED

(2), (4) It is *not* true that population *2* or population *4* in the chart has the best chance for survival in a rapidly changing environment. Both of these populations reproduce asexually, a reproductive process that minimizes genetic variation. In addition, population *4* has a relatively long life span and produces a relatively small number of offspring during its lifetime. So its evolutionary reaction to rapidly changing environmental conditions would be slow as compared to population *1*.

(3) It is *not* true that population *3* in the chart has the best chance for survival in a rapidly changing environment. Although this population produces

sexually, its relatively long life span and relatively small number of offspring would limit its evolutionary reaction to rapidly changing environmental conditions as compared to population *1*.

41. **1** *An enzyme the cell produces might no longer function* is the statement that best describes a change that would most likely be observed in the cells formed as a result of this mitotic division. According to the information presented in the table, the normally flawless replication of this DNA segment went awry, producing substitution errors in the second and sixth triplet codons. As a result, the enzyme coded by this segment will also contain errors in its amino acid sequence. The resulting flawed enzyme may or may not function properly depending on whether or not these errors occur in the enzyme's active site.

WRONG CHOICES EXPLAINED

(2), (4) *The cells would begin to form gametes to be released* and *chloroplasts would be produced by the ribosomes* are *not* the statements that best describe a change that would most likely be observed in the cells formed as a result of this mitotic division. A single pair of substitution errors in a short DNA sequence would not be capable of altering the cell's tissue type or its production of specialized cells or organelles. These are nonsense distracters.

(3) *Many new hormones would be synthesized by the cells* is *not* the statement that best describes a change that would most likely be observed in the cells formed as a result of this mitotic division. Although it is remotely possible that the new sequence of codons may result in a new and active hormone, it is very unlikely that it would result in many new hormones.

42. **4** In this setup, *tubes 2, 3, and 4, only* contain at least one organism carrying on cellular respiration. All organisms, whether plant, animal, or other, must carry on a form of cellular respiration to provide ATP energy to carry out its other life functions. Only tube *1* (which contains water without any organisms such as algae) is free of any living organisms that carry on cellular respiration.

WRONG CHOICES EXPLAINED

(1), (2), (3) It is *not* true that in this setup, *tubes 1 and 2, only*; *tubes 2 and 4, only*; or *tubes 3 and 4, only* contain at least one organism carrying on cellular respiration. Each of these responses contains one or more incorrect items.

43. **4** *ATP* is the compound that directly provides energy in living cells and is being produced in every tube where cellular respiration is occurring. The respiratory reactions systematically break down energy-rich molecules of glucose and transfer their chemical bond energy to molecules of ATP. These ATP molecules then transfer their energy to biochemical reactions in the

cell that drive the cell's metabolism. Without this essential energy transfer enabled by ATP, the cell would quickly die.

WRONG CHOICES EXPLAINED

(1) *Oxygen* is *not* the compound that directly provides energy in living cells and is being produced in every tube where cellular respiration is occurring. Oxygen gas is not a compound but an energy-poor element with the molecular formula O_2. Oxygen serves as a raw material in the (aerobic) cellular respiratory reactions. Oxygen is being produced in tubes *3* and *4* as a result of photosynthesis, not cellular respiration.

(2) *Glucose* is *not* the compound that directly provides energy in living cells and is being produced in every tube where cellular respiration is occurring. Glucose is an energy-rich compound that serves as a raw material in the aerobic respiratory reactions. Glucose provides energy indirectly, not directly, to living cells through the process of cellular respiration. Glucose is being produced in tubes *3* and *4* as a result of photosynthesis, not cellular respiration.

(3) *DNA* is *not* the compound that directly provides energy in living cells and is being produced in every tube where cellular respiration is occurring. DNA is a complex compound that serves no direct role in the cellular respiratory reactions or in energy transfer in the cell. DNA is being produced in tubes *2*, *3*, and *4* as a result of replication, not cellular respiration.

PART B–2

44. One credit is allowed for correctly marking an appropriate scale, without any breaks in the data, on each labeled axis. [1]

45. One credit is allowed for correctly plotting the data on the grid, connecting the points, and surrounding each point with a small circle. [1]

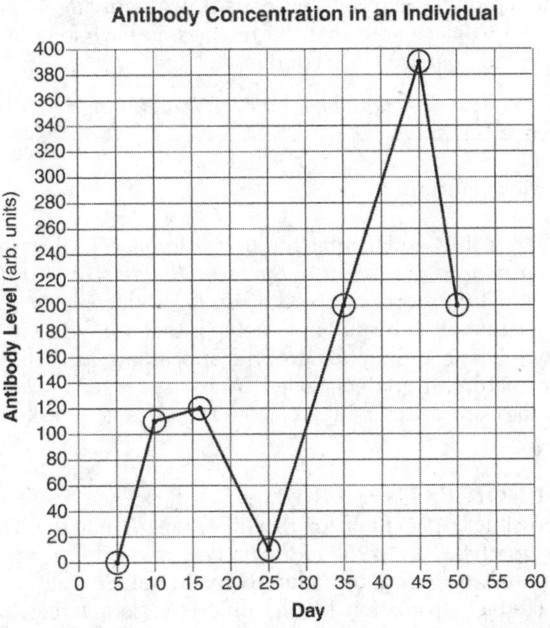

46. One credit is allowed for correctly stating *one* reason for the change in antibody production during the first 10 days. Acceptable responses include but are not limited to: [1]

- *The individual was exposed to a disease.*
- *The individual received a vaccination.*
- *The individual got sick.*
- *The person was exposed to specific antigens.*
- *The body is producing an immune response.*
- *More white blood cells were produced.*

47. **3** The antibody level (in arb. units) of the individual on day 30 is closest to *110*. An examination of the correctly completed graph shows that the line

connecting the data points for days 25 and 35 crosses the 30-day line (horizontal axis) at an antibody level (vertical axis) very close to 110 arb. units.

WRONG CHOICES EXPLAINED

(1), (2), (4) The antibody level (in arb. units) of the individual on day 30 is *not* closest to *30*, *70*, or *160*. The graphed data do not support any of these values.

48. One credit is allowed for correctly identifying a procedure that could be used to make sure that the puppies all have long hair. Acceptable responses include but are not limited to: [1]

- *Breeding two dogs with long hair to produce long-haired puppies*
- *Selective breeding*
- *Cloning*
- *Genetic engineering*

49. **1** A possible explanation for the difference is that the Chihuahua puppy *was produced as a result of the recombination of genes during sexual reproduction*. The process of recombination produces new, unique genetic combinations that often result in phenotypes that are different from those of either parent. The genes that control the appearance of such traits are often recessive in nature. Traits controlled by recessive genes can be expressed only when they are paired with an allelic recessive gene as a result of fertilization.

WRONG CHOICES EXPLAINED

(2) A possible explanation for the difference is *not* that the Chihuahua puppy *was produced as a result of the process of asexual reproduction*. Dogs, like all mammalian species, reproduce sexually, not asexually.

(3) A possible explanation for the difference is *not* that the Chihuahua puppy *inherited a gene from one of its grandparents and not its parents*. The offspring of sexually reproducing species inherit all of their genes directly from their parents, but indirectly from their grandparents.

(4) A possible explanation for the difference is *not* that the Chihuahua puppy *had a mutation that occurred after it was born*. A mutation can affect the genetic inheritance of an individual organism only if it occurs in a gamete that participates in the fertilization of that individual. Mutations that occur in an individual organism after its birth are never expressed in that individual.

50. **2** Graph *2* best represents the DNA content found in each cell in each of the stages in the diagram. Graph 2 indicates that the amount of DNA in each cell in each stage is exactly the same. This is consistent with the fact that the diagram represents the stages of vegetative propagation of a carrot. Vegetative propagation is a type of asexual reproduction in which a

single diploid ($2n$) cell from a donor carrot is cultured to produce a cluster of genetically identical diploid cells. This cell cluster then undergoes mitotic cell division and differentiation to produce a new carrot with cells containing the diploid number of chromosomes. In all stages, each cell contains the same amount of DNA as represented by its diploid designation.

WRONG CHOICES EXPLAINED

(1), (3), (4) Graphs *1*, *3*, and *4* do *not* best represent the DNA content found in each cell in each of the stages in the diagram. These graphs do not correctly show the amount of DNA in each cell in this process.

51. One credit is allowed for correctly describing *one* specific reason why scientists would want to maintain the genetic makeup of a particular plant. Acceptable responses include but are not limited to: [1]

- *The plants have desirable traits that the growers want to appear in all offspring.*
- *This is done to maintain a plant variety that has increased nutritional and/or economic value.*
- *The technique maintains a plant that grows larger/matures faster.*
- *The parent plant might be resistant to diseases/herbicides, so the grower wants to maintain that trait in the offspring.*
- *This is a technique that might be used to save a plant species from extinction.*

52. One credit is allowed for correctly stating *one negative* effect of using miR-7 as the only treatment for breast cancer. Acceptable responses include but are not limited to: [1]

- *Certain levels of miR-7 expression can also stimulate the development of cancer cells.*
- *miR-7 can activate/turn on some cancer genes.*
- *miR-7 can cause the formation of cancer cells.*
- *miR-7 may prove to be a mutagenic agent that actually stimulates the formation of cancerous cells.*

53. One credit is allowed for stating *one* way cancer cells are different from normal body cells. Acceptable responses include but are not limited to: [1]

- *Cancer cells can metastasize/spread to other organs.*
- *Cancer cells keep dividing before they fully mature.*
- *Cancer cells can be larger/shaped differently than normal cells.*
- *Some cancer cells have more than one nucleus.*
- *Unlike normal cells, cancer cell division is uncontrolled.*
- *Cancer cells have more mutations.*

54. One credit is allowed for identifying *one* similarity, other than the genetic information, that these body cells have. Acceptable responses include but are not limited to: [1]

- *All of the cells contain organelles.*
- *All have a cell membrane/nucleus/cytoplasm/mitochondria.*
- *They carry out life processes/respiration/mitosis.*
- *They all use ATP/glucose energy.*

55. One credit is allowed for explaining why differences in these human body cells are a biological advantage. Acceptable responses include but are not limited to: [1]

- *This allows these cells to be specialized for a specific function.*
- *Differences in cells are related to the different functions in the body.*
- *Each cell type turns on genes that are needed but turns off genes that are not, conserving cell energy/metabolism/protein synthesis for essential functions.*

PART C

56. One credit is allowed for correctly describing how the barred owl population is having a *negative* effect on the spotted owl population. Acceptable responses include but are not limited to: [1]

- *They are eating the same food as the spotted owls, not leaving enough for the spotted owls.*
- *The barred owl is outcompeting the spotted owl for the same ecological niche.*
- *The two species compete for the same resources, and the barred owl is more successful.*
- *The barred owl is bigger and more aggressive and is taking over the habitat of the spotted owl.*
- *Barred owls now cover the spotted owls' range, outnumbering them in some cases.*
- *Interspecies competition for limited resources normally results in the elimination of one of the two species from the ecosystem.*

57. One credit is allowed for correctly explaining why it is important to protect the spotted owl from extinction. Acceptable responses include but are not limited to: [1]

- *Extinction represents the loss of genetic material and reduces the biodiversity of this ecosystem.*
- *Biodiversity helps to keep the ecosystem stable.*

- *Once the spotted owls are extinct, they cannot be brought back.*
- *Their prey could increase in number uncontrollably.*
- *Spotted owls perform an important role in the ecosystem food chain/ web.*

58. One credit is allowed for correctly describing the role that the habitat plays in the survival of an animal species such as the spotted owl. Acceptable responses include but are not limited to: [1]

- *Habitats provide the resources animals need to survive.*
- *Habitats provide food and shelter for birds and other species.*
- *The growth and survival of organisms depend on the physical conditions of their habitat.*
- *The resources available in the habitat limit the number of organisms it can support.*
- *The carrying capacity is dependent on the available resources of the habitat.*

59. One credit is allowed for correctly stating *one* advantage and *one* disadvantage of automobile emission testing. Acceptable responses include but are not limited to: [1]

Advantages:

- *The law helps to reduce the amount of air pollution in an area.*
- *The emissions from cars will contain fewer harmful chemicals.*
- *The law will result in cleaner air that will be better for humans and all living things to breathe.*
- *This law will force automobile owners to maintain their vehicles properly so the vehicles don't pollute as much.*

Disadvantages:

- *Because of the emissions control system, cars may be more expensive.*
- *The emissions system test is expensive.*
- *Repairing automobiles that fail to pass the test is expensive.*
- *The emissions control system may reduce gas mileage or have other negative effects on automobile efficiency.*

60. One credit is allowed for correctly identifying the end products of protein digestion that made up the contents of test tube *B* after the two hours. Acceptable responses include but are not limited to: [1]

- *Amino acids*
- *Dipeptides*

61. One credit is allowed for correctly explaining the importance of temperature in the functioning of enzymes. Acceptable responses include but are not limited to: [1]

- *Each enzyme works best within a specific range of temperatures.*
- *If the temperature is too high/low, the enzyme might not function.*
- *The enzyme could be denatured/change shape at extremely high temperatures.*
- *The enzymes produced by humans generally work best at body temperature.*
- *Temperature affects the rate of reaction/enzyme activity.*

62. One credit is allowed for correctly stating what the result would be if the same enzyme that was added to test tube *B* was added to a test tube containing starch and for supporting the answer. Acceptable responses include but are not limited to: [1]

- *Nothing would happen because enzymes are specific to particular substrates.*
- *There would be no reaction with the starch because this enzyme acts on only proteins.*
- *Enzymes act on only certain substances. So an enzyme that acts on protein would not act on starch.*

63. One credit is allowed for correctly stating what happens to these smaller molecules following digestion. Acceptable responses include but are not limited to: [1]

- *They diffuse into body cells/the bloodstream.*
- *They are absorbed from the small intestine into the blood.*
- *They are taken to the tissues by the circulatory system.*
- *The molecules are carried to the cells by the blood.*
- *The products of digestion are absorbed by the villi.*
- *They enter cells and are used there.*
- *They are used to build muscle or other compounds/proteins and to release energy.*

64–66. Three credits are allowed for correctly analyzing this experiment. In your answer, be sure to:

- Identify *one* error in the researcher's experimental design [1]
- Identify *one* way, other than affecting estrogen levels, that secondhand smoke could affect a developing embryo [1]
- Explain why the process of peer review is an important step in this research [1]

Acceptable responses include but are not limited to: [3]

This experiment was designed with too many variables, such as age of the women and the gender of the fetuses. [1] Variables should be limited to a single independent variable that the researcher controls, such as the amount of cigarette smoke to which the experimental group is exposed. Exposure to cigarette smoke can result in low-birth-weight babies that do not develop normally. [1] Peer review of scientific experiments is important because other scientists may discover errors in the research method that were missed by the original researchers. [1]

Too few experimental and control subjects were used in this experiment. [1] This can lead to statistical error that invalidates the experimental results. Smoking or exposure to secondhand smoke can result in miscarriage [1] or premature birth. Peer review of the experiment gives other researchers an opportunity to repeat the experiment to confirm/validate its results and conclusions. [1]

67. One credit is allowed for correctly identifying *one* population of organisms shown in the food web, other than rabbits or mice, that would likely be affected by the introduction of the snowy owls and for explaining why their population would be affected. Acceptable responses include but are not limited to: [1]

- *Snakes—The owl would compete with the snakes for the mouse population.*
- *Hawks—The owls feed on much of the same food as the hawks, so there would be fewer hawks.*
- *Mountain lions—The owls would compete for the rabbit population.*
- *Grasses—The owls would eat rabbits and mice, and there would be more grass.*

68. One credit is allowed for correctly identifying *one* condition that might cause snowy owls to leave their usual habitat and move to another area. Acceptable responses include but are not limited to: [1]

- *Lack of food*
- *Fewer lemmings*
- *Overpopulation of snowy owls*
- *Change in environmental temperature*
- *Loss of habitat/deforestation*
- *Increased competition*

69. One credit is allowed for correctly stating which level, *A*, *B*, or *C*, contains the *least* total available energy and for supporting your answer. Acceptable responses include but are not limited to: [1]

- *Level C—Energy is lost at each level going from the plants to the herbivores to the carnivores.*
- *Level C—Energy enters the ecosystem with green plants capturing energy from sunlight. There is less energy available as it is passed to the carnivores.*
- *C—Energy is lost at each feeding level from the plants to the carnivores.*

70. One credit is allowed for correctly stating *one* possible hypothesis that would explain the differences in the observed data between the two locations. Acceptable responses include but are not limited to: [1]

- *There are few predators present in New Mexico, so fur color does not influence survival.*
- *Predators can see the mice better if the fur color of the mice does not match the rock color in Arizona.*
- *In New Mexico, both fur colors are equally suited to the environment, unlike in Arizona.*
- *If there were no owls/hawks in New Mexico, the fur color would not matter.*

71. One credit is allowed for correctly explaining how it is possible for these two different populations to have identical gene sequences for dark fur color. Acceptable responses include but are not limited to: [1]

- *The mice in the two populations evolved from the same ancestral population.*
- *They evolved from a common ancestor.*
- *Some of the mice migrated.*
- *They are members of the same species.*
- *The populations interbred before they were geographically separated.*

72. One credit is allowed for correctly explaining what is meant by the statement: "While mutations are random, natural selection is not." Acceptable responses include but are not limited to: [1]

- *Many mutations occur by chance. Natural selection acts on traits.*
- *Depending on whether or not the mutation increases or decreases the chance of survival, the mutation for a trait might or might not be passed on to the next generation.*
- *Once a mutation occurs, natural selection acts on it methodically and either promotes it or eliminates it.*
- *Mutations happen by chance, but natural selection depends on the way mutations interact with environmental change.*

PART D

73. **3** Species *E and F* would be expected to have the most similar proteins. The diagram shows that species *F* branched off from species *E* relatively recently in the time period illustrated. This would indicate that they share many genetic traits (many genes) in common and therefore share the proteins coded by those genes.

WRONG CHOICES EXPLAINED

(1), (2) Species *A and C* and species *B and C* would *not* be expected to have the most similar proteins. These species split off from a common ancestor (species *J*) in the distant past. So they probably share some proteins in common but not many proteins.

(4) Species *H and M* would *not* be expected to have the most similar proteins. Although they may have shared many proteins in common in the distant past, these species are extinct.

74. **4** The new, more accepted classification is most probably based on an analysis of *the DNA present in the cells of squid*. Such an analysis would compare the base sequences of the DNA of representatives of different squid populations. If these sequences match in key genes, the populations would be considered to be the same species.

WRONG CHOICES EXPLAINED

(1) The new, more accepted classification is most probably *not* based on an analysis of *a greater number of squid carcasses*. This process was used in the past and did not yield good information. There is no reason to think that increasing the number of carcasses examined would yield any better information.

(2) The new, more accepted classification is most probably *not* based on an analysis of *the feeding habits of the three different species*. Animal behavior studies can yield some useful information on species relatedness. This information is not definitive and would not be the best method to use in this case.

(3) The new, more accepted classification is most probably *not* based on an analysis of *a number of newly found squid fossils*. Studies of fossils have provided a great deal of useful information concerning species relatedness over the past 300 years. This information is not definitive and would not be the best method to use in this case.

75. **2** *Water entered the bag due to the lower concentration of water inside. The extra water pushed the liquid up the tube as the bag filled* is the statement that correctly explains the rise of liquid in the tube at the end of the experiment. Diffusion is a process in which a substance moves with the concentration gradient from an area of higher relative concentration to an

area of lower relative concentration of that substance. This process does not require the expenditure of cellular energy. In the diagram, the movement of water molecules is shown to be with the concentration gradient. As the water molecules increased in number inside the bag, the water-sucrose solution was displaced up the tube.

WRONG CHOICES EXPLAINED

(1) *The concentration of sucrose molecules increased as water molecules entered the bag. This concentration increase pushed the liquid up the tube* is *not* the statement that correctly explains the rise of liquid in the tube at the end of the experiment. As water enters the bag, the concentration of sucrose would decrease, not increase.

(3) *Sucrose indicator entered the bag and reacted with the sucrose molecules. The reaction made the bag increase in size and pushed the liquid up the tube* is *not* the statement that correctly explains the rise of liquid in the tube at the end of the experiment. Sucrose indicator is not present in this setup, so it would have no means of entering the bag.

(4) *Sucrose molecules moved out of the bag and up the tube while water moved out, causing the rise of liquid in the tube* is *not* the statement that correctly explains the rise of liquid in the tube at the end of the experiment. Sucrose molecules are seen in both diagrams to be evenly distributed within the bag. Sucrose is not being concentrated and moved up the tube.

76. **4** Row *4* correctly pairs a finch species with its primary nutritional role and bill type. An examination of the diagram shows that the warbler finch has a probing bill and is a carnivore.

WRONG CHOICES EXPLAINED

(1) Row *1* does *not* correctly pair a finch species with its primary nutritional role and bill type. An examination of the diagram shows that the cactus finch is a herbivore, not a carnivore.

(2) Row *2* does *not* correctly pair a finch species with its primary nutritional role and bill type. An examination of the diagram shows that the medium ground finch has a crushing bill, not a grasping bill.

(3) Row *3* does *not* correctly pair a finch species with its primary nutritional role and bill type. An examination of the diagram shows that the large tree finch has a grasping bill, not a crushing bill. It is also a carnivore, not a herbivore.

77. One credit is allowed for correctly identifying the finch that is best adapted to feed on the nectar within tube-shaped flowers and for supporting your answer. Acceptable responses include but are not limited to: [1]

- *Cactus finch—Has a probing bill that can reach the nectar in the flowers*
- *The cactus finch is a plant eater/herbivore with a long, narrow beak.*

78. One credit is allowed for correctly stating *one* reason why the population of small ground finches has *not* been affected by the increasing number of small tree finches. Acceptable responses include but are not limited to: [1]

- *They aren't competing for the same food.*
- *Ground finches are mostly plant eaters, while tree finches are mostly animal eaters/carnivores.*
- *The two species occupy different niches.*
- *They live in different areas/nesting sites/feeding zones.*

79. One credit is allowed for correctly explaining why glucose molecules can cross a cell membrane and starch molecules can *not*. Acceptable responses include but are not limited to: [1]

- *Starch molecules are too large to move across the cell membrane, while glucose molecules are smaller and can easily move into and out of the cell.*
- *The pores in a cell membrane are big enough to transport tiny glucose molecules but are too small to transport large starch molecules.*

80. One credit is allowed for correctly explaining why allowing the student's pulse rate to return to normal might be a good indication that he had waited long enough for recovery. Acceptable responses include but are not limited to: [1]

- *If his pulse rate were normal, it would indicate that the muscles have removed wastes and have enough energy available to resume exercising.*
- *Normal pulse rate is the same as resting pulse rate.*
- *When his pulse returns to normal, it indicates that there is no longer a need for more oxygen/excretion of excess carbon dioxide/lactic acid.*
- *This shows that the body has returned to a homeostatic balance.*

81. **2** *Squeeze a clothespin until fatigued, rest, repeat. Over time, they should gradually increase the resistance of the clothespins they are squeezing* are the steps the students should take to follow the principles of PRE. These steps are listed in the paragraph as steps (1), (2), and (3).

WRONG CHOICES EXPLAINED

(1) *Measure their pulse rate after squeezing the clothespin until fatigued. Then increase the resistance of the clothespin for the next trial* are *not* the steps the students should take to follow the principles of PRE. These steps do not follow the principles of PRE listed in the paragraph.

(3) *Measure their pulse rate, squeeze a clothespin for one minute, rest, and measure their pulse rate* are *not* the steps the students should take to follow the principles of PRE. These steps do not follow the principles of PRE listed in the paragraph.

(4) *Squeeze a clothespin for as long as they can, measure their pulse rate, rest, eat some candy. Increase the resistance of the clothespin for the next trial* are *not* the steps the students should take to follow the principles of PRE. These steps do not follow the principles of PRE listed in the paragraph.

82. **1** *They could eventually lift heavier weights than when they started* is the observation that would indicate that their exercise program was successful. This observation is consistent with step (3) listed in the paragraph, which states that practitioners should increase the resistance (increasing the weight) as their ability to generate force increases.

WRONG CHOICES EXPLAINED

(2), (3), (4) *Their pulse rate increased more rapidly as they kept lifting weights, the number of weights their group could lift during competition decreased,* and *males and females could lift the same weight an equal number of times during competition* are *not* the observations that would indicate that their exercise program was successful. Each of these responses is inconsistent with step (3) listed in the paragraph, which states that practitioners should increase the resistance (increasing the weight) as their ability to generate force increases.

83. One credit is allowed for correctly sketching a line graph showing the changes in heart rate of a person who is walking slowly, then begins running, and then sits down to rest for a few minutes. Acceptable responses include but are not limited to: [1]

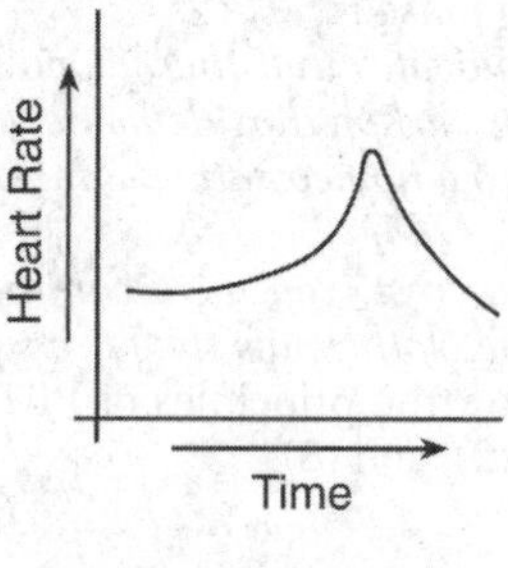

or

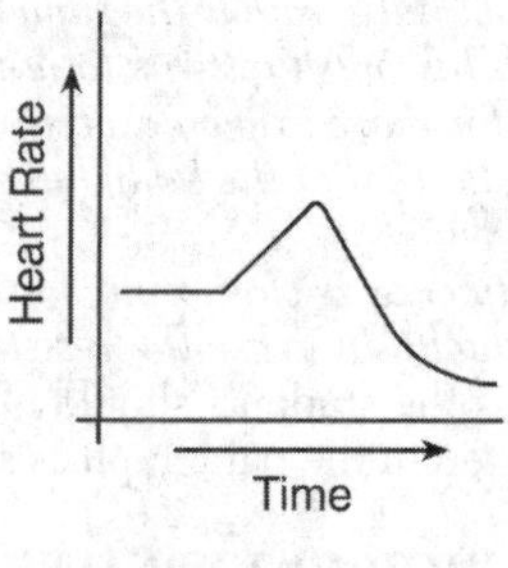

84. One credit is allowed for correctly identifying *one* waste product that is released during exercise and for explaining how this waste product leaves the body. Acceptable responses include but are not limited to: [1]

- *Carbon dioxide—It diffuses into the blood and is carried to the lungs, where it is exhaled.*
- *Water—It is absorbed by the blood and transported to the sweat glands, where it evaporates after being excreted.*

- *Heat—It is absorbed and carried to the skin by the blood, where it is radiated from the surface.*
- *Lactic acid—It is metabolized in the mitochondria to produce energy.*

85. One credit is allowed for correctly stating *one* way scientists could use the banding patterns produced by gel electrophoresis. Acceptable responses include but are not limited to: [1]

- *They could be used to determine evolutionary relationships between two similar species.*
- *Similarities between the banding patterns of an adult and a child can be used to determine parentage.*
- *The process could be used to screen for genetic disorders/diseases/mutations.*
- *Police might use the technique to identify a criminal from DNA evidence.*

STANDARDS/KEY IDEAS	AUGUST 2017 QUESTION NUMBERS	NUMBER OF CORRECT RESPONSES
STANDARD 1		
Key Idea 1: The central purpose of scientific inquiry is to develop explanations of natural phenomena in a continuing and creative process.	33, 52	
Key Idea 2: Beyond the use of reasoning and consensus, scientific inquiry involves the testing of proposed explanations involving the use of conventional techniques and procedures and usually requiring considerable ingenuity.	31, 37, 64, 70	
Key Idea 3: The observations made while testing proposed explanations, when analyzed using conventional and invented methods, provide new insights into natural phenomena.	36, 44, 45, 47, 66	
Laboratory Checklist	47	
STANDARD 4		
Key Idea 1: Living things are both similar to and different from each other and from nonliving things.	1, 5, 6, 9, 22, 24, 29, 54, 55, 60, 63	
Key Idea 2: Organisms inherit genetic information in a variety of ways that result in continuity of structure and function between parents and offspring.	4, 7, 16, 23, 25, 34, 48, 50	
Key Idea 3: Individual organisms and species change over time.	17, 26, 40, 41, 49, 71, 72	
Key Idea 4: The continuity of life is sustained through reproduction and development.	18, 27, 28, 32, 51, 65	
Key Idea 5: Organisms maintain a dynamic equilibrium that sustains life.	2, 3, 12, 35, 38, 39, 42, 43, 46, 53, 61, 62	
Key Idea 6: Plants and animals depend on each other and their physical environment.	10, 11, 56, 57, 58, 67, 68, 69	
Key Idea 7: Human decisions and activities have a profound impact on the physical and living environment.	8, 13, 14, 15, 19, 20, 21, 30, 59	

STANDARDS/KEY IDEAS	AUGUST 2017 QUESTION NUMBERS	NUMBER OF CORRECT RESPONSES
REQUIRED LABORATORIES		
Lab 1: "Relationships and Biodiversity"	73, 74, 85	
Lab 2: "Making Connections"	80, 81, 82, 83, 84	
Lab 3: "The Beaks of Finches"	76, 77, 78	
Lab 5: "Diffusion Through a Membrane"	75, 79	

Examination June 2018

Living Environment

PART A

Answer all questions in this part. [30]

Directions **(1–30): For *each* statement or question, record in the space provided the *number* of the word or expression that, of those given, best completes the statement or answers the question.**

1 Producers are generally found at the beginning of a food chain. Which statement best explains why this is true?

1 Producers are usually smaller in size than consumers.
2 Producers do not rely on other organisms for food.
3 There are always more consumers than producers in food chains.
4 Consumers are always more complex organisms than producers. 1 ____

2 A lion cub resembles its parents because it inherits genes that produce

1 DNA identical to all of the DNA found in both parents
2 proteins identical to all of the proteins found in both parents
3 ATP identical to some of the ATP found in each parent
4 enzymes identical to some of the enzymes found in each parent 2____

3 If body temperature is too high, some blood vessels increase in size and sweat glands will excrete sweat, resulting in a lower body temperature. These changes are an example of

1 a learned behavior
2 feedback mechanisms
3 an inherited disorder
4 genetic mutations 3____

4 A farmer grows beans that he sells to local markets. Over a period of 40 years, the farmer has identified the plants that produced the most beans and only used those beans to produce new plants. This procedure is part of the process of

1 selective breeding
2 genetic engineering
3 replication
4 cloning 4____

5 Although we rely on coal, oil, and natural gas to produce energy, some environmental scientists have proposed that we use less fossil fuel. One reason to support this proposal is to

1 enable us to preserve rain forests in tropical areas
2 help us to reduce the production of carbon dioxide gas
3 allow us to decrease the use of fertilizers on crops
4 encourage us to end research on wind and water power sources 5_____

6 The diagram below represents relationships in an ecosystem.

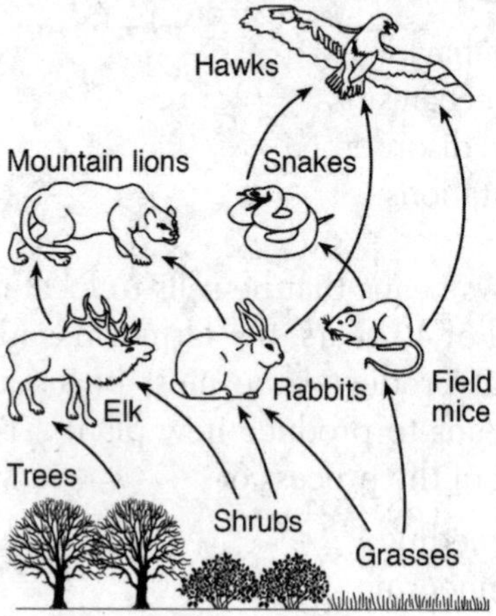

What is the primary source of energy in this environment?

1 cellular respiration in the plants
2 energy from minerals in the soil
3 fossil fuels
4 solar energy 6_____

7 Research has shown that treadmill training increases the number of certain energy-releasing structures in the brain cells of rats.

The cellular structures referred to in this study are most likely

1 mitochondria
2 nuclei
3 vacuoles
4 ribosomes

7 ____

8 Which process must first take place in order for the proteins in foods to be used by body cells?

1 digestion
2 storage
3 synthesis
4 excretion

8 ____

9 Which statement is characteristic of reproduction in humans?

1 The reproductive cells of males and females differ in chromosome number.
2 Males and females produce gametes in the ovaries.
3 Males and females produce the same number of gametes.
4 The reproductive cycles of males and females are regulated by hormones.

9 ____

10 Which row in the chart below represents the most likely changes in the atmosphere due to widespread deforestation?

Row	Oxygen Concentration	Carbon Dioxide Concentration
(1)	increases ↑	increases ↑
(2)	increases ↑	decreases ↓
(3)	decreases ↓	increases ↑
(4)	decreases ↓	decreases ↓

10 ____

11 The chart below represents some of the events that occur during the cycling of nutrients in an ecosystem.

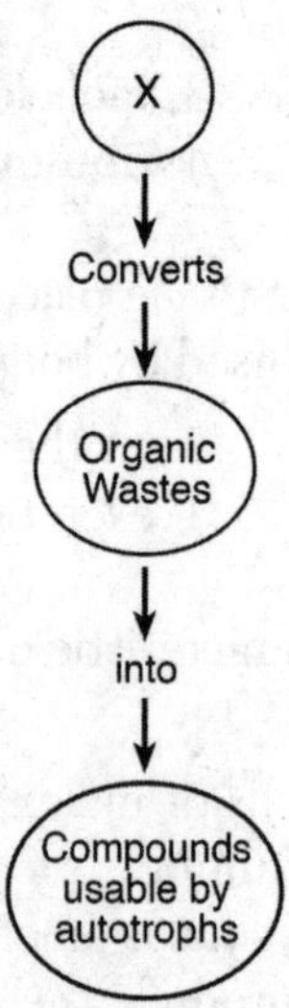

Which organisms would most appropriately complete the chart when written in the circle at *X*?

1 producers
2 herbivores
3 carnivores
4 decomposers

11 ____

12 The diagram below represents the formation of a cancerous growth.

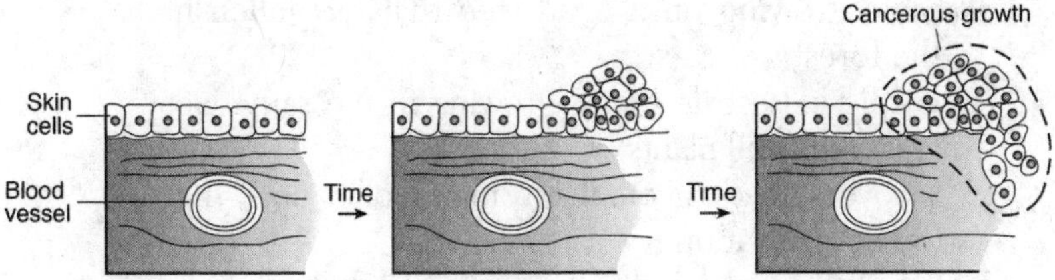

Which statement best explains the events represented in this diagram?

1 A gene mutation caused the cells to become muscle cells.
2 The growth resulted from the introduction of a vaccine.
3 A gene mutation caused abnormal mitotic cell division.
4 The growth resulted from uncontrolled meiotic cell division.

12 _____

13 A standard laboratory technique used to produce a new plant is represented in the diagram below.

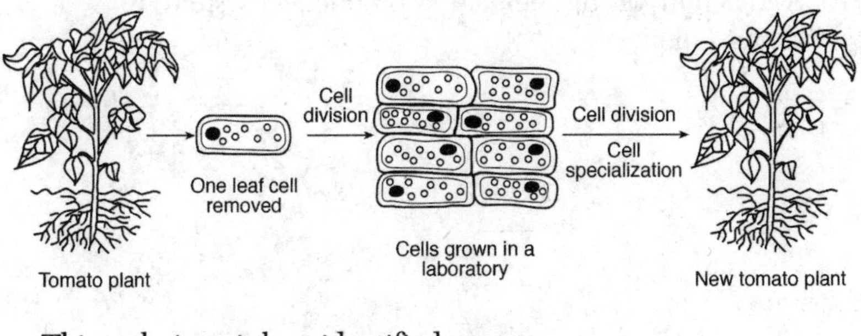

This technique is best identified as

1 gene alteration
2 selective breeding
3 replication
4 cloning

13 _____

14 An example of competition between members of two different species is

1 mold growing on a dead tree that has fallen in the forest
2 purple loosestrife plants growing in the same wet areas as cattail plants
3 a coyote feeding on the remains of a deer that died of starvation
4 two male turkeys displaying mating behaviors to attract a female turkey 14 ____

15 Which statement best explains why different body cells of the same individual look and function differently?

1 Each cell contains different genes.
2 Different genes are activated in different kinds of cells.
3 Cells are able to change to adapt to their surroundings.
4 Half of the genes in the cells came from the mother and half from the father. 15 ____

16 A diagram of the female reproductive system is shown below.

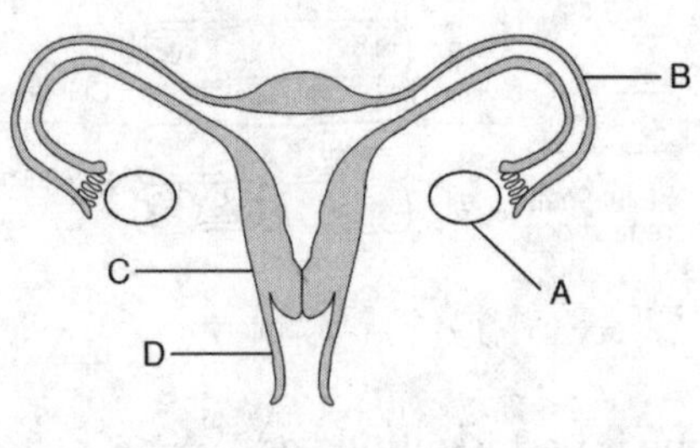

Identify the structure within which the egg cell is normally fertilized.

1 *A* 3 *C*
2 *B* 4 *D* 16 ____

17 The ameba, a single-celled organism, reproduces asexually. Variations in an ameba would most commonly occur through

1 differentiation during development
2 the fusion of gametes
3 random mutations
4 recombination during fertilization 17_____

18 The development of organs and tissues from a zygote includes

1 mitosis and differentiation
2 mitosis and gamete production
3 meiosis and gamete production
4 meiosis and fertilization 18_____

19 In the 1920s, over 25 million acres of the American southern plains were stripped of prairie grasses to provide more land for farmers to grow wheat. The prairie grasses had served to hold the soil in place and prevent erosion. In the early 1930s, a series of severe dust storms eroded topsoil from more than 13 million acres of the southern plains and dumped it as tons of dust particles over many cities in the Northeast. Farmland was destroyed and people were sickened from "dust pneumonia." This occurrence illustrates that

1 farmers should never clear land to grow crops as it always creates problems
2 once an ecosystem has been altered, it cannot be restored to normal
3 the farmers deliberately altered the equilibrium of the cities in the Northeast
4 when humans alter ecosystems, serious consequences may result 19_____

20 The Cornell University News Service reported, "The sugar maple is the most economically valuable tree in the eastern United States because of its high-priced lumber, syrup and tourist-attracting fall colors." The effects of acid rain now threaten the survival of these trees. This threat is the result of a human activity that has

1 introduced a foreign species by accident
2 stabilized a forest ecosystem through technology
3 weakened an ecosystem through pollution
4 weakened a species by direct harvesting 20_____

21 The human male reproductive system is adapted for the production of

1 sperm and the delivery of these cells for internal fertilization
2 gametes that transport food to the egg
3 zygotes and the development of these cells into a fetus
4 hormones that stimulate placenta formation in the male 21_____

22 The diagram below represents an important biological technique scientists rely on to produce replacement hormones.

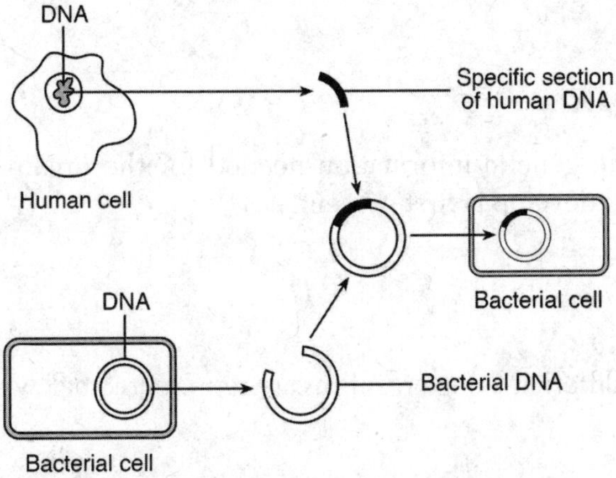

Which two processes are required for the technique to successfully produce hormones?

1 replication of DNA in bacterial cells and cell division
2 replication of DNA in bacterial cells and gamete formation
3 meiosis and development
4 mitosis and fertilization

22 _____

23 The diagram below summarizes some of the steps in the development of humans.

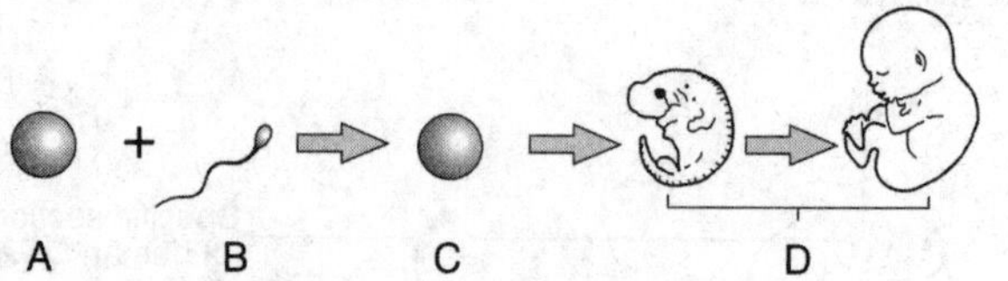

All the genetic information needed for the organism to develop is first present at

1 *A*
2 *B*
3 *C*
4 *D*

23 _____

24 Five different living organisms are represented below.

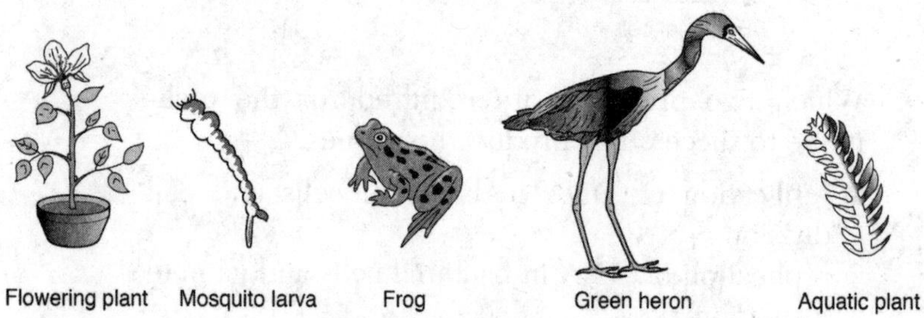

Which statement about the organisms represented above is correct?

1 All of the organisms are autotrophs.
2 Only the flowering plant, green heron, and aquatic plant carry out photosynthesis.
3 Only the frog and green heron can maintain homeostasis.
4 All of the organisms pass on traits through reproduction.

24 _____

25 Mistletoe is an evergreen shrub that can produce most of its own food. Often, mistletoe can be found living on trees and taking water and nutrients away from the tissues of the trees.

The relationship between mistletoe and trees is an example of

1 consumer/herbivore
2 predator/prey
3 scavenger/decomposer
4 parasite/host

25 _____

26 Rabbits are not native to Australia. They were imported by European settlers. In 1936, the myxoma virus was introduced into Australia as a means of biological control to infect and reduce the rabbit population. This method of controlling the rabbit population was an attempt to

1 stop the overpopulation of a native species
2 stop the overproduction of an introduced species
3 limit the food sources of the rabbit
4 limit the number of rabbits brought into the country

26 _____

27 The major role of carbohydrates in the human diet is to

1 form the membranes that surround mitochondria
2 act as a catalyst for cellular reactions
3 supply energy for the body
4 provide building blocks for amino acids 27____

28 Throughout New York State, some farmers have switched from growing a variety of vegetable crops to growing a single crop, such as corn. Other farmers are concerned that such a practice will make it more likely that an entire crop could be lost to disease or infestation by an insect pest. This is a valid concern because this practice

1 reduces the biodiversity of their fields
2 increases the number of decomposers in their fields
3 decreases the need to import food
4 increases the number of invasive species 28____

29 The breathing rate, heart rate, and blood hormone levels of an individual would directly provide information about that individual's

1 cellular organization
2 nutrition
3 inheritance
4 metabolic activity 29____

30 The diagram below represents an energy pyramid.

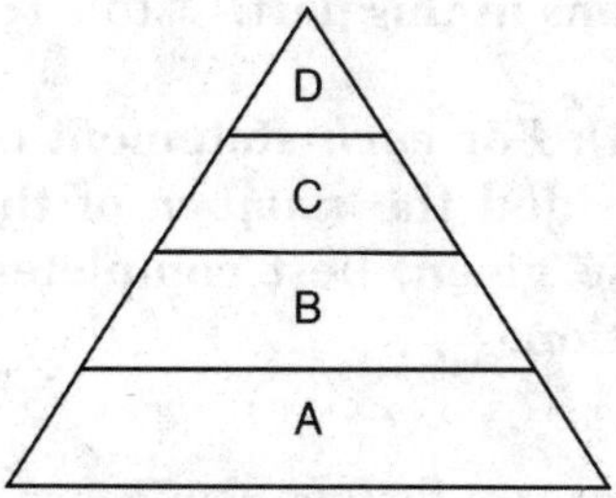

Which type of organism could occupy levels *B*, *C*, and *D* of this energy pyramid?

1 consumer
2 producer
3 autotroph
4 carnivore

30 ____

PART B–1

Answer all questions in this part. [13]

***Directions* (31–43): For *each* statement or question, record in the space provided the *number* of the word or expression that, of those given, best completes the statement or answers the question.**

31 In the 1920s, two conflicting newspaper headlines called attention to a mysterious new illness.

1921–"Don't breathe the air! Mysterious disease affecting thousands is caused by breathing the air in swamps."

1922–"Don't drink the water! Mysterious disease affecting thousands is caused by drinking the water in swamps."

Another series of headlines appeared in the 1940s and 50s.

1945–"New technology finds tiny worms on swamp vegetation."

1950–"Tiny worms found in lungs of patients suffering from mysterious swamp disease."

1952–"Mysterious disease known to be caused by worms given name Swamp Lung Disease."

Headlines such as these best illustrate the concept that

1 scientific explanations are tentative and subject to change
2 some newspapers are not honest and report incorrect information on purpose
3 worms can enter the body many different ways
4 worms found in swamps should not be used for fishing

31 ____

Base your answer to question 32 on the information and diagram below and on your knowledge of biology.

In the early 1600s, a scientist planted a willow tree that weighed 5 pounds in 200 pounds of dry soil. He placed it outside and watered it for 5 years. At the end of that time, he observed that the tree had gained 164 pounds 3 ounces, while the soil had lost just 2 ounces.

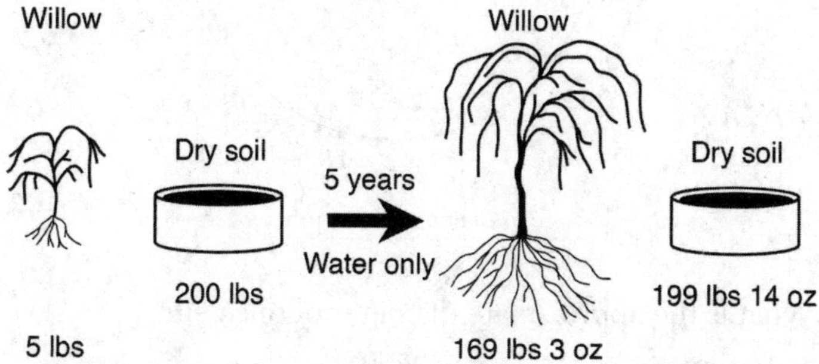

32 From this, he concluded that plants gain weight from the water they take in. His conclusion was based on.

1 the input of scientists from many countries doing similar studies
2 the application of advanced technologies to the study of a problem
3 careful observation, measurements, and inferences from his data
4 an extensive knowledge of the process of photosynthesis

32_____

33 A student observed five living cells in the field of view of a microscope as represented below.

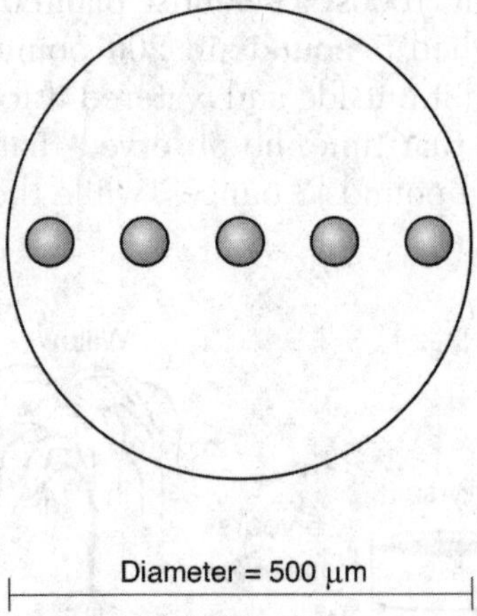

What is the approximate diameter of one cell?

1 10 µm
2 50 µm
3 250 µm
4 500 µm

33 ____

34 Ecologists are concerned that the golden-winged warbler population is at a dangerously low level. One reason this could lead to extinction of this warbler is that

1 after a species becomes extinct, it won't be able to carry out its role in the ecosystem
2 there may not be enough diversity among the birds for the species to be able to survive an environmental change
3 extinction always occurs when populations begin to decrease in number
4 an increase in biodiversity within a population often causes the population to be classified as threatened or endangered

34 ____

35 One primary function of the cell membrane is

1 regulating the flow of simple sugars into or out of the cell
2 synthesizing substances by breaking down cell organelles
3 storing carbohydrates, water, and starches for future use
4 digesting carbohydrates, fats, and protein

35____

36 For several years now, there has been discussion of constructing a large oil pipeline across the United States. Which statement expresses a major concern many people are likely to have about the proposed pipeline?

1 The pipeline will bring a large number of jobs to the area where it is being constructed.
2 The oil pipeline will increase the amount of finite resources.
3 If this pipeline were to leak, the oil could contaminate soil, water, and wildlife.
4 The pipeline is a technological fix for ozone depletion.

36____

37 The rings in the diagram below represent the annual growth of a tree approximately 20 years old.

Tree trunks grow wider each year by continuous growth in a thin layer of cells just beneath the bark. Since one new layer is added each year, the number of rings in a tree can be used to tell its age. The thickness of the rings provides information about the environmental conditions in past years.

By observing the annual rings in the diagram, one can infer that

1 environmental conditions did not change over the last 20 years
2 trees grow faster on the side that faces the Sun
3 some years provide better conditions for growth than other years
4 tree rings are not reliable because trees must be cut down to see them

37 _____

Base your answers to questions 38 through 40 on the information below and on your knowledge of biology.

Harmless Skin Virus Fights Acne

...Acne is caused when hair follicles become blocked with an oily substance called sebum, which the body makes to stop the hair and skin from drying out.

Normally harmless bacteria, such as *Propionibacterium acnes*, that live on the skin can then contaminate and infect the plugged follicles.

Phages [a type of virus] appear to help counteract this.

When the scientists sequenced the DNA coding of the phages, they discovered that, as well as sharing most of their genetic material, the viruses all had some key features in common.

All carry a gene that makes a protein called endolysin—an enzyme thought to destroy bacteria by breaking down their cell walls.

And unlike antibiotics, which kill many types of bacteria, including "good" ones that live in our gut, phages are programmed to target only specific bacteria...

Source: BBC News
September 25, 2012

38 This treatment for acne, using phages, is effective because phages

1 produce antibodies to clean out clogged pores and hair follicles
2 eliminate bacteria by attacking specific cell structures
3 carry genes and infect follicles
4 attack every known type of bacteria

38 _____

39 The protein endolysin belongs to which group of chemical substances?

1 hormones
2 receptors
3 biological catalysts
4 molecular bases

39 ____

40 The typical response of the human body to an infection by bacteria is to

1 stimulate the production of antigens
2 decrease the number of enzymes in the blood
3 ignore the organisms, unless they are pathogens
4 produce white blood cells and antibodies

40 ____

---

41 Two biological processes that occur in certain organelles are represented in the diagrams below.

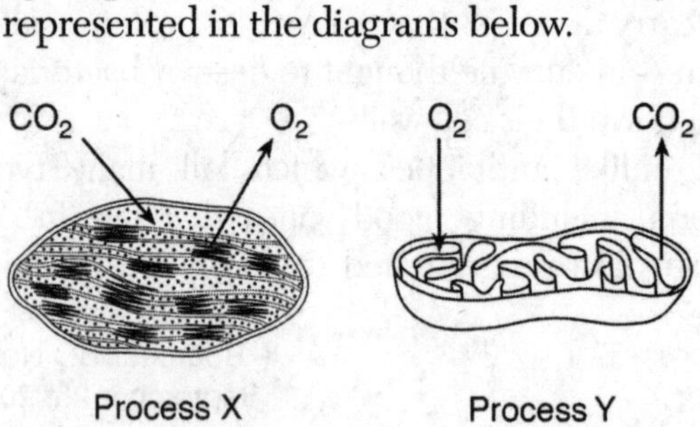

Which statement is correct regarding the types of organisms able to carry out these processes?

1 Process *X* occurs in heterotrophs, but not in autotrophs.
2 Process *Y* occurs in consumers, but not in producers.
3 Both processes *X* and *Y* occur in all living things.
4 Both processes *X* and *Y* occur in green plants.

41 ____

Base your answers to questions 42 and 43 on the diagram below and on your knowledge of biology. The diagram shows how ATP is used by some cell structures to perform various functions.

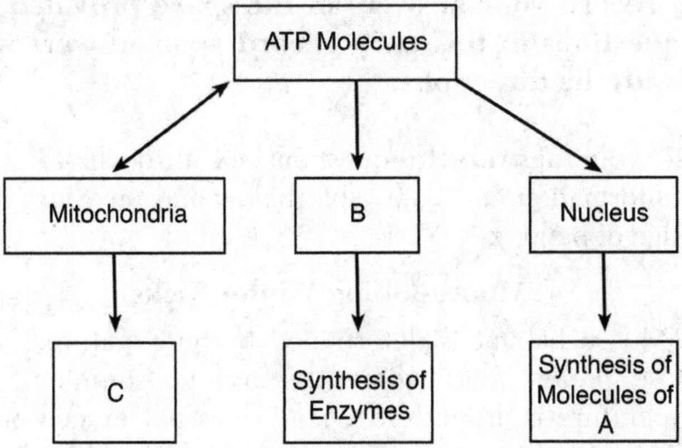

42 Which cell structure is represented by *B*?

1 vacuole
2 ribosome
3 cytoplasm
4 chloroplast

42 ____

43 The nucleus contains molecules of *A*, which

1 recycle waste products
2 remove water from the cell
3 store hereditary information
4 regulate the pH of cytoplasm

43 ____

PART B–2

Answer all questions in this part. [12]

***Directions* (44–55): For those questions that are multiple choice, record your answers in the space provided. For all other questions in this part, record your answers in accordance with the directions.**

Base your answers to questions 44 through 47 on the information and data table below and on your knowledge of biology.

Moose-killing Winter Ticks

Moose habitat is determined by temperature. Moose prefer areas where the average summer temperature is around 15°C and does not exceed 27°C for too long. The reason for this temperature dependency: Moose cannot sweat.

Besides the cooling effect of water, which moose are almost always near, aquatic environments provide them with a good supply of food, and in the past, have protected them against biting insects. However, the North American moose population is facing a new threat: a parasite called the winter tick. These ticks lodge themselves in the animal's fur and hold on through the winter, sucking the animal's blood. Many infected moose end up dying of exhaustion and weakness as a result of the large number of ticks feeding on them.

Ticks are most active during dry days in the fall. Adult ticks that drop off moose in the spring and land on snow cover have a poorer survival rate. Climate change can be predicted to improve conditions for winter ticks due to longer and warmer falls, and earlier snowmelt in the spring.

Surveys of the moose population in northeastern Minnesota have recorded the change shown below in the moose population between 2005 and 2013.

Estimated Moose Population
In Northeastern Minnesota

Survey Year	Estimated Moose Population
2005	8160
2006	8840
2007	6860
2008	7890
2009	7840
2010	5700
2011	4900
2012	4230
2013	2760

***Directions* (44–45): Using the information in the data table, construct a line graph on the grid below, following the directions below.**

44 Mark an appropriate scale, without any breaks in the data, on the axis labeled "Estimated Moose Population." [1]

45 Plot the data for the estimated moose population on the grid. Connect the points and surround each point with a small circle. [1]

Example: 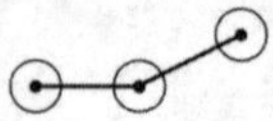

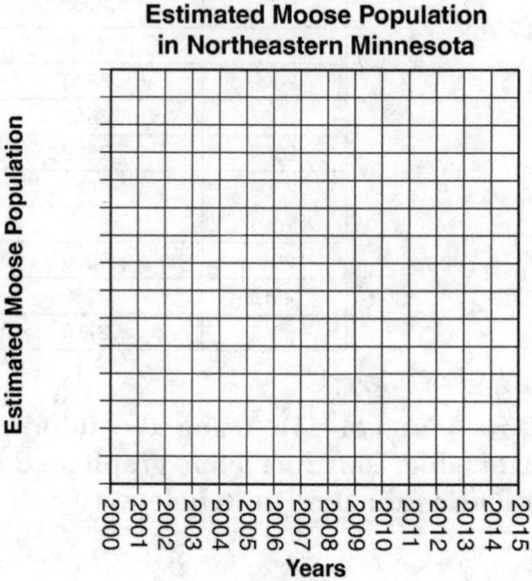

46 Explain how climate change could result in an increased number of moose infested with winter ticks. [1]

____

____

Note: The answer to question 47 should be recorded in the space provided.

47 Increased average yearly temperatures in regions presently inhabited by moose could result in a disruption in homeostasis in these animals because

1 a decrease in average temperatures will increase mutations in their skin cells
2 an increase in average temperatures will decrease the amount of blood ticks can consume
3 moose will not be able to maintain an appropriate body temperature, since they do not sweat
4 moose will sweat more and lose too much water from their bodies

47____

Base your answers to questions 48 and 49 on the information and diagram below and on your knowledge of biology. The diagram represents the evolutionary relationships among many organisms.

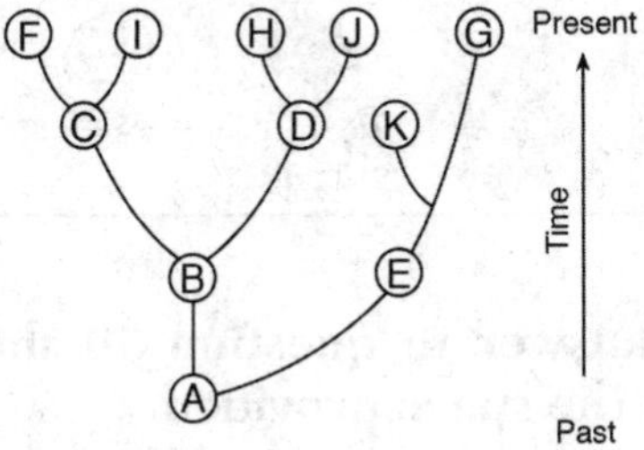

48 An environmental change severely affected the organism represented by species *K*. What was the result? Support your answer. [1]

____

____

____

Note: The answer to question 49 should be recorded in the space provided.

49 Three species with the most similar traits are most likely

1 *F, I, G*
2 *D, H, J*
3 *B, D, G*
4 *F, A, J*

49 _____

Base your answers to questions 50 and 51 on the diagram below and on your knowledge of biology. The diagram represents trophic levels in an ocean environment.

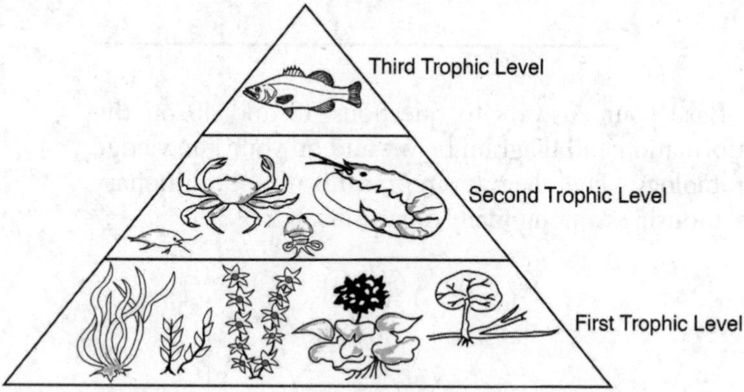

Note: The answer to question 50 should be recorded in the space provided.

50 The organisms found at the second trophic level of this pyramid would be

1 producers
2 decomposers
3 carnivores
4 herbivores

50 _____

51 State *one* reason why there is less energy available at each trophic level going from the first to the third trophic level. [1]

52 Stable predator-prey relationships are necessary to maintain a healthy ecosystem. The removal of a predator species from an area caused the deer population to sharply increase from 1910 to 1925. Changes in the deer population and carrying capacity of the area are represented in the graph below.

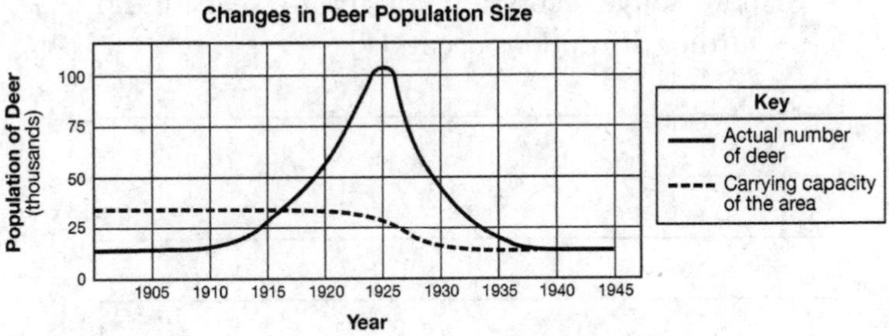

Based on the information provided, explain how the sharp population increase from 1910 to 1925 might have resulted in the decrease in the carrying capacity after 1925. [1]

Base your answers to question 53–54 on the diagram below and on your knowledge of biology. The diagram indicates a change in an ecosystem.

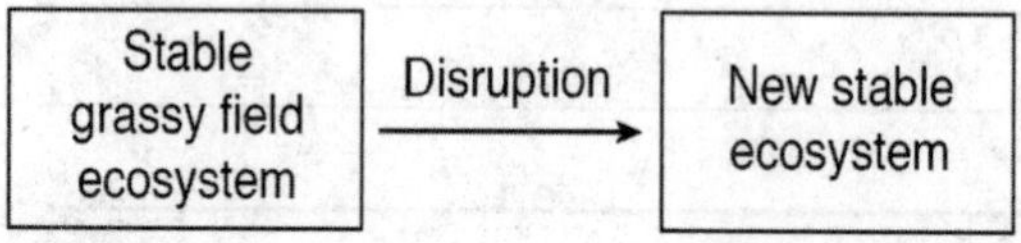

53–54 Identify some of the key events associated with the change. In your answer, be sure to:

- identify *one* natural event that could cause the disruption indicated in the diagram [1]
- state what would most likely happen to the new stable ecosystem in future years if no further disruptions occur [1]

55 Explain why a mutation that occurs in a body cell will *not* contribute to the evolution of a species. [1]

PART C

Answer all questions in this part. [17]

Directions **(56–72): Record your answers in the spaces provided.**

Base your answers to question 56 through 58 on the passage below and on your knowledge of biology.

Our [Nitrogen] Fertilized World

It is the engine of agriculture, the key to plenty in our crowded, hungry world. …

…Enter modern chemistry. Giant factories capture inert nitrogen gas from the vast stores in our atmosphere and force it into a chemical union with the hydrogen in natural gas, creating the reactive compounds that plants crave. That nitrogen fertilizer—more than a hundred million tons applied worldwide every year—fuels bountiful harvests. Without it, human civilization in its current form could not exist. Our planet's soil simply could not grow enough food to provide all seven billion of us our accustomed diet. In fact, almost half of the nitrogen found in our bodies' muscle and organ tissue started out in a fertilizer factory.

Source: National Geographic, May 2013

56 Nitrogen fertilizers are used by plants to synthesize amino acids. State *one* reason why a supply of amino acids is important for the survival of complex organisms. [1]

____

____

57 Identify *one* possible effect on the human population if nitrogen fertilizers were not available. [1]

____

____

58 Explain how the building of factories to produce fertilizer is an example of a trade-off. [1]

____

____

Base your answers to question 59–60 on the information and diagram below and on your knowledge of biology.

An experiment was carried out to determine the effect of exposure to ultraviolet (UV) light on the growth of bacteria. Equal quantities of bacterial cells were spread on Petri dishes that are used to grow colonies of bacteria. Half of each dish was shielded from the UV light with a UV screen. The other half was exposed to UV light for various amounts of time. After the UV treatment, the bacteria were grown in an incubator for 24 hours and the number of colonies was counted.

The diagram below represents the setup of the experiment.

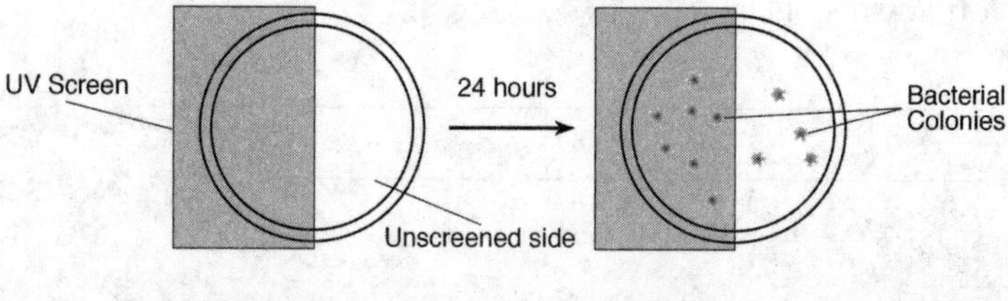

The table below contains the data collected at different exposure times by counting the number of bacterial colonies on both the screen-covered side and unscreened side.

Bacterial Growth

Exposure Time to UV Light (min)	**Colonies on Screened Side**	**Colonies on Unscreened Side**
0 (No exposure)	20	22
0.5	21	19
1.0	23	16
2.0	22	10
5.0	24	5
10.0	23	1

59–60 Analyze the experiment that produced the data in the table. In your answer, be sure to:

- state a hypothesis for the experiment [1]
- state whether the results of the experiment support or fail to support your hypothesis. Support your answer [1]

____

____

____

____

Base your answers to questions 61 and 62 on the information below and on your knowledge of biology.

Evolution leads to changes in how frequently certain traits appear in a population.

61 Explain the importance of the presence of variations within a population. [1]

62 Describe how the process of natural selection can result in an increase in frequencies of certain traits found in a population. [1]

Base your answers to questions 63 through 65 on the information below and on your knowledge of biology.

A typical human liver cell can have over 90,000 insulin receptors. Due to a genetic difference, some people have liver cells that contain only about 1000 insulin receptors.

63 Describe the importance of receptors in cellular communication. [1]

64 Describe the importance of the shape of receptor molecules for carrying out their function. [1]

---

65 Identify *one* effect a reduced number of insulin receptors might have on an individual. [1]

Base your answers to questions 66 through 68 on the passage below and on your knowledge of biology. Biologists have been studying the genes present in newborn twins.

Twins Don't Share Everything

...Chemicals called epigenetic markers can be attached to those [inherited] genes, like flags or balloons hanging off the sides of the DNA ladder. These don't just change the look of the genes. Like pieces of tape stuck over a light switch, these markers can force a gene to remain turned on or off. The type of marker scientists studied in the twins generally sticks the switch in the off position so that some proteins don't get made. And that means the proteins' jobs won't get done.

Every time a cell divides, new epigenetic markers may form. Foods, pollutants, and stress may all contribute to the development of new markers. So throughout our lives we tend to accumulate more and more. But a few are there from the day we're born.

...His [Jeffrey Craig's] team found that newborn twins have markers attached to different genes from the very start. It's true in identical twins, which come from the same fertilized egg. It's also true in fraternal twins, which come from different fertilized eggs. However, fraternal twins had more such differences than identical twins did.

Source: Science News for Students; July 31, 2012

66 Explain why the genetic material in an offspring produced by sexual reproduction contains genetic material that is *not* identical to the genetic material of either parent. [1]

____

____

67 State *one* reason why identical twins should have fewer genetic differences than fraternal twins. [1]

____

____

68 Identify *two* environmental factors that can lead to an increase in the number of epigenetic markers that modify gene expression. [1]

__________________ and __________________

Base your answers to questions 69 and 70 on the information below and on your knowledge of biology.

With the emotional roller coaster that pregnancy brings, it can be daunting [challenging] for pregnant women to take on the additional pressure of eating the "perfect" pregnancy diet. The good news: there is no single perfect diet for pregnancy. The best way for expectant mothers to meet their nutritional needs is to focus on consuming an overall healthy diet, with a variety of vegetables, fruits, whole grains, lean meats or meat substitutes, and low-fat dairy or dairy substitutes....

Source: US News Health 11/9/2012

69 Describe how nutrients move from the mother to the fetus. [1]

____

____

70 State *one* other way, in addition to consuming a balanced diet, pregnant women can help ensure proper development of the fetus. [1]

____

____

Base your answer to question 71 on the information below and on your knowledge of biology.

> ...Bacteria often evolve clever ways of evading chemical assaults, but they will always struggle to resist the old-fashioned way of killing them; heating them up. It takes only a relatively mild warming to kill bugs [bacteria] without discomfort or harm to tissues. So imagine if little electric heaters could be implanted into wounds and powered wirelessly to fry bacteria during healing before dissolving harmlessly into body fluids once their job is done....
>
> Source: BBC Future, May 24, 2013

71 State *one* way the use of these new "little electric heaters" might represent a long-term benefit over using antibiotics to treat bacterial infections. [1]

____

____

____

72 A child became ill with the measles. Measles is a disease that is highly contagious. The child's mother did not get sick, even though she and the child were close while the child was ill. State *one* reason why the mother did not get sick with the measles. [1]

____

PART D

Answer all questions in this part. [13]

Directions **(73–85): For those questions that are multiple choice, record your answer in the space provided. For all other questions in this part, record your answers in accordance with the directions.**

Note: The answer to question 73 should be recorded in the space provided.

73 The diagram below represents the major parts of a growing onion plant. Nutrients are represented in the soil around the onion.

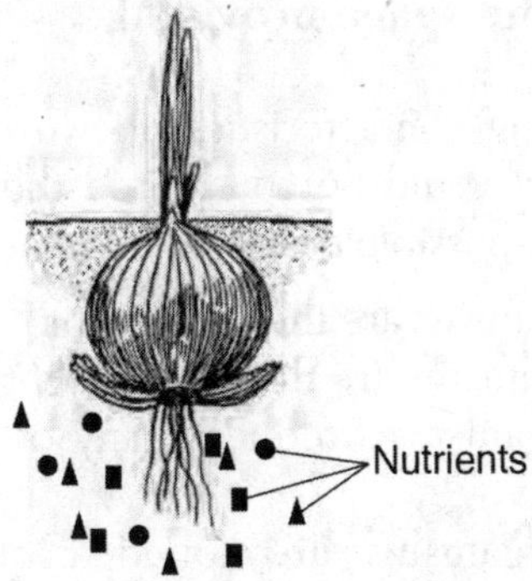

Which statement best describes how nutrients enter the root cells of the onion plant?

1 Only nutrients needed by the plant enter root cells.
2 The nutrients usually move from an area of high concentration in the soil to an area of low concentration in root cells.
3 Nutrients always move into the plant cells by active transport.
4 The nutrients always move from an area of low concentration in the soil to an area of high concentration in root cells.

73_____

Note: The answer to question 74 should be recorded in the space provided.

74 Which concept is correctly matched with an example from *The Beaks of Finches* lab?

1 Variation—different "beaks" were available.
2 Adaptation—different types of foods were available.
3 Selecting Agent—an insecticide was used to kill insects on one island.
4 Environment—"beaks" with similar qualities were used to gather seeds. 74____

Note: The answer to question 75 should be recorded in the space provided.

75 When comparing characteristics of two organisms, which evidence would be considered the strongest for supporting a possible evolutionary relationship?

1 The two organisms are the same color.
2 The two organisms are the same height.
3 The two organisms produce many of the same proteins.
4 The two organisms are found in the same locations. 75____

Note: The answer to question 76 should be recorded in the space provided.

76 *A* and *B* below represent two different slide preparations of elodea leaves. Elodea is a plant found in streams and ponds in New York State.

The water used on slide *A* contained 1% salt and 99% water.
The salt solution used on slide *B* contained 6% salt and 94% water.
Elodea cells normally contain 1% salt.

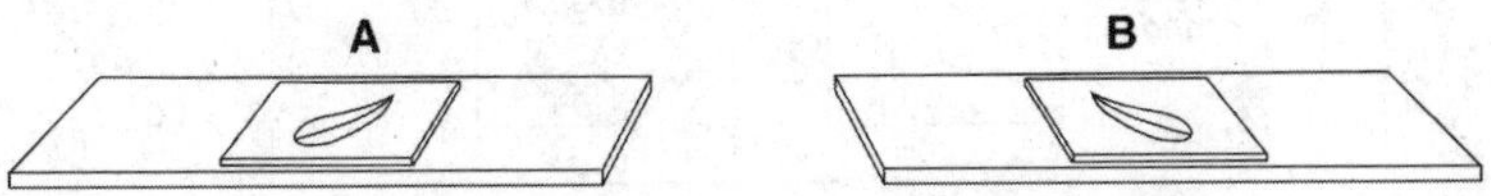

Elodea leaf mounted in 1% salt solution — Elodea leaf mounted in 6% salt solution

Five minutes after the slides were prepared, a student using a compound light microscope to observe the cells in leaves *A* and *B* would most likely see that

1 water had moved out of the cells of the leaf on slide *A*
2 salt had moved into the cells of the leaf on slide *A*
3 water had moved out of the cells of the leaf on slide *B*
4 salt had moved out of the cells of the leaf on slide *B*

76____

77 The table below shows the food sources for two different species of Galapagos finches on an island.

Two Galapagos Finches and Their Sources of Food

Name	Foods
Vegetarian finch *Platyspiza crassirostris* 	Buds, leaves, fruit of trees 
Cactus finch *Geospiza scandens* 	Cactus flowers and nectar 

State *one* reason why these two species probably do *not* live in the same area of this island. [1]

Base your answers to questions 78 and 79 on the information below and on your knowledge of biology.

During a lab activity, a 14-year-old student took his resting pulse rate. He counted 20 beats in 20 seconds. He calculated his pulse rate for a minute and compared the result to the data shown in the table below.

Normal Pulse Rate Ranges

Age Group	Resting Heart Rate (beats per minute)
Children (ages 6-15)	70 – 100
Adults (ages 18 and over)	60 – 100

78 According to the data table, does the student's pulse rate fall within the normal range? Circle yes or no and support your answer. [1]

yes no

____

____

79 Using a biological explanation, state *one* reason why a person's heart rate increases during exercise. [1]

____

____

80 The chart below shows the molecular comparison between several species.

Molecular Comparison Chart

Botana curus	DNA	GTG	GAC	TGA	GGA	CTC
	mRNA	CAC	CUG	ACU	CCU	GAG
	Amino acid	His	Leu	Thr	Pro	Glu

Species X	DNA	GTG	GAC	AGA	GGA	CAC
	mRNA	CAC	CUG	UCU	CCU	GUG
	Amino acid	His	Leu	Ser	Pro	Val

Species Y	DNA	GTG	GAC	AGA	GGA	CAC
	mRNA	CAC	CUG	UCU	CCU	GUG
	Amino acid	His	Leu	Ser	Pro	Val

Species Z	DNA	GTA	GAC	TGA	GGA	CTC
	mRNA	CAU	CUG	ACU	CCU	GAG
	Amino acid	His	Leu	Thr	Pro	Glu

Identify which species is likely to be more closely related to *Botana curus*. Support your answer. [1]

Species: ____________________

Support: ______________________________________

____

Note: The answer to question 81 should be recorded in the space provided.

81 A factor that contributed to the evolution of finches on the Galapagos Islands was most likely the

1 lack of variation in beak structure of the finches
2 isolation of the finches on separate islands
3 relatively constant atmospheric temperature
4 total lack of competition for food

81 ____

Base your answers to questions 82 and 83 on the diagram below and on your knowledge of biology. The diagram represents a laboratory setup.

A starch solution in a test tube was separated from the water in a beaker by a dialysis membrane. One hour later, it was observed that the liquid had risen in the test tube.

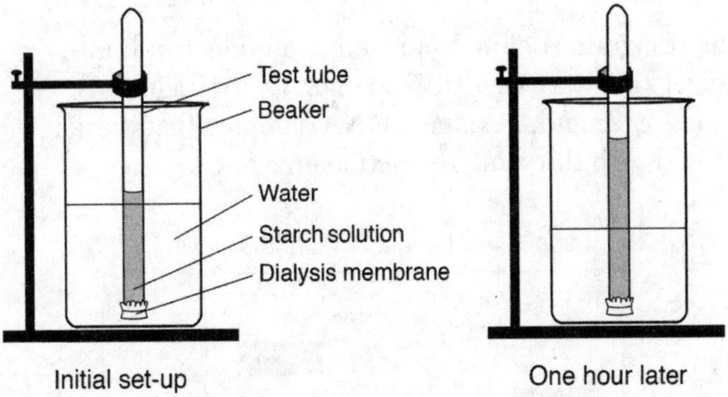

Note: The answer to question 82 should be recorded in the space provided.

82 The rise of the liquid in the test tube that was observed after one hour can be explained as a result of the

1 starch solution moving into the test tube and out of the beaker
2 water moving from the beaker into the test tube
3 large starch molecules blocking the dialysis membrane
4 dialysis membrane acting as a barrier to the water molecules

82 ____

83 If a starch indicator solution was initially added to the water in the beaker, describe *one* observation that would be made after one hour. [1]

84 The diagram below represents an electrophoresis gel that was used to separate DNA fragments. Lanes 1, 2, and 3 contain DNA samples that were treated with the same restriction enzyme.

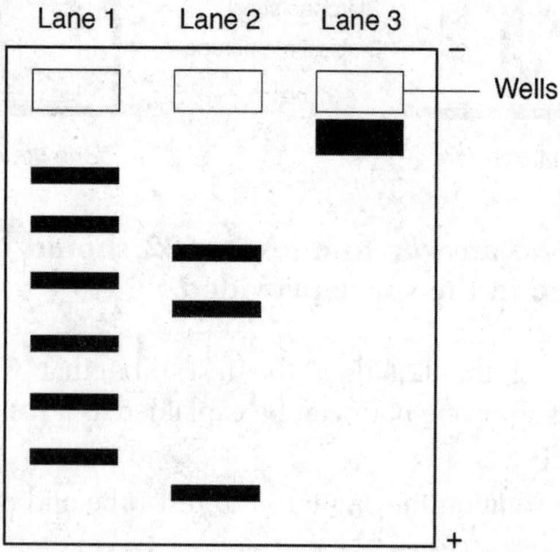

Explain why the DNA sample in lane 3 did *not* separate into fragments. [1]

85 An experiment is performed to determine the effect of watching basketball games on pulse rates. Ten students agreed to wear devices that monitor pulse rates while watching a basketball game between competitive opponents. Their pulse rates were measured every minute for five minutes in the first quarter of the game. The data collected indicated that pulse rates did not change significantly during the monitored period. State *one* way that this experiment could be improved to obtain a valid conclusion. [1]

____

____

Answers June 2018

Living Environment

Answer Key

PART A

1. 2	**6.** 4	**11.** 4	**16.** 2	**21.** 1	**26.** 2
2. 4	**7.** 1	**12.** 3	**17.** 3	**22.** 1	**27.** 3
3. 2	**8.** 1	**13.** 4	**18.** 1	**23.** 3	**28.** 1
4. 1	**9.** 4	**14.** 2	**19.** 4	**24.** 4	**29.** 4
5. 2	**10.** 3	**15.** 2	**20.** 3	**25.** 4	**30.** 1

PART B–1

31. 1	**34.** 2	**36.** 3	**38.** 2	**40.** 4	**42.** 2
32. 3	**35.** 1	**37.** 3	**39.** 3	**41.** 4	**43.** 3
33. 2					

PART B–2

44. *See* Answers Explained.
45. *See* Answers Explained.
46. *See* Answers Explained.
47. 3
48. *See* Answers Explained.
49. 2
50. 4
51. *See* Answers Explained.
52. *See* Answers Explained.
53. *See* Answers Explained.
54. *See* Answers Explained.
55. *See* Answers Explained.

PART C. *See* **Answers Explained**.

PART D

73. 2
74. 1
75. 3
76. 3
77. *See* Answers Explained.
78. *See* Answers Explained.
79. *See* Answers Explained.
80. *See* Answers Explained.
81. 2
82. 2
83. *See* Answers Explained.
84. *See* Answers Explained.
85. *See* Answers Explained.

Answers Explained

PART A

1. **2** *Producers do not rely on other organisms for food* is the statement that best explains why producers are generally found at the beginning of a food chain. Food chains are graphical representations of simple nutritional relationships commonly found in nature. Food chains illustrate such nutritional relationships in terms of the transfer of energy from producers (green plants), which manufacture their own food via photosynthesis, to sequential levels of consumer organisms. For this reason, producers are generally represented at the beginning of all food chains.

WRONG CHOICES EXPLAINED

(1) *Producers are usually smaller in size than consumers* is *not* the statement that best explains why producers are generally found at the beginning of a food chain. Many producers are far larger in size (e.g., redwood trees, oak trees) than the consumers (e.g., elk, whitetail deer) that inhabit the same environment. Relative body size is not a criterion for placement in a food chain representation.

(3) *There are always more consumers than producers in food chains* is *not* the statement that best explains why producers are generally found at the beginning of a food chain. In fact in any stable environment, producers always far outnumber consumers by a factor of 10:1 or greater. The relative number of organisms is not a criterion for placement in a food chain representation.

(4) *Consumers are always more complex organisms than producers* is *not* the statement that best explains why producers are generally found at the beginning of a food chain. Many producers are far more complex than the consumers with which they interact (e.g., white oak tree (producer) vs. oak wilt fungus (consumer)). Relative body complexity is not a criterion for placement in a food chain representation.

2. **4** A lion cub resembles its parents because it inherits genes that produce *enzymes identical to some of the enzymes found in each parent.* According to the one gene–one enzyme hypothesis, each functional gene in the genome of an organism is responsible for the synthesis of a single unique enzyme. The role of each of these enzymes is the catalysis of a single biochemical reaction, resulting in a single genetic phenotype. Since a lion cub receives half of its genetic information from each of its parents, the enzymes produced in that cub should represent half of the enzymes produced by each parent.

WRONG CHOICES EXPLAINED

(1), (2) It is *not* true that a lion cub resembles its parents because it inherits genes that produce *DNA identical to all of the DNA found in both parents* or *proteins identical to all of the proteins found in both parents*. Sexual reproduction ensures that offspring receive exactly half, not all, of the genes available in each parent's genome. By extension, this principle means that half of the proteins found in an offspring may be traced back to one of its parents and the other half of the proteins may be traced back to the other parent.

(3) It is *not* true that a lion cub resembles its parents because it inherits genes that produce *ATP identical to some of the ATP found in each parent*. ATP (adenosine triphosphate) is a biochemical found in most, if not all, living organisms. It has the function of transferring cellular energy from one biochemical reaction to another. The chemical form of ATP is identical in all living things.

3. **2** These changes are an example of *feedback mechanisms*. In this example, the human body senses its own temperature and initiates responses to modify its temperature. When the body's temperature is reduced to an appropriate level, the body senses this also and the responses are reversed. This stimulus-response (feedback) mechanism helps the body to stabilize its temperature within a relatively narrow range that promotes the survival of the individual.

WRONG CHOICES EXPLAINED

(1) These changes are *not* an example of *a learned behavior*. Feedback mechanisms are reflex reactions that are beneficial inherited traits, not learned behaviors.

(3) These changes are *not* an example of *an inherited disorder*. Feedback mechanisms are reflex reactions that are beneficial inherited traits, not disorders.

(4) These changes are *not* an example of *genetic mutations*. Feedback mechanisms are reflex reactions that are beneficial inherited traits, not mutations.

4. **1** This procedure is part of the process of *selective breeding*. Selective breeding is a technique used by animal and plant breeders in which organisms with desirable traits are cross-bred with the hope of producing offspring that also display those traits. The resulting offspring would theoretically be capable of passing on these desirable traits to future generations.

WRONG CHOICES EXPLAINED

(2) This procedure is *not* part of the process of *genetic engineering*. Genetic engineering is a laboratory technique in which a gene for a desired trait is snipped from the DNA of a donor cell and inserted into the genome of

a recipient cell. This technique is not used by traditional farmers to select the best beans to plant to produce better bean crops.

(3) This procedure is *not* part of the process of *replication*. Replication is the exact self-duplication of DNA that occurs in the processes of mitotic and meiotic cell division. This technique is not used by traditional farmers to select the best beans to plant to produce better bean crops.

(4) This procedure is *not* part of the process of *cloning*. Cloning is a laboratory technique that involves the removal of a diploid nucleus from a donor organism and its insertion into a denucleated fertilized egg cell of the same or related species. This nucleus contains the DNA of the two parents of the donor. This technique is not used by traditional farmers to select the best beans to plant to produce better bean crops.

5. **2** One reason to support this proposal is to *help us to reduce the production of carbon dioxide gas*. Fossil fuels, such as coal, oil, and natural gas, are naturally occurring hydrocarbon substances that contain an abundance of stored energy. When burned, this stored energy is released. The burning of fossil fuels has been used for hundreds or thousands of years to power human industrial processes. It also produces polluting by-products, such as carbon dioxide, particulate matter, and sulfur and nitrogen oxides. All of these by-products pose a threat to the worldwide environment and to human health. Proposals aimed at reducing our use of fossil fuels help to reduce these threats.

WRONG CHOICES EXPLAINED

(1) One reason to support this proposal is *not* to *enable us to preserve rain forests in tropical areas*. In fact, the burning of fossil fuels and its resulting pollution have been scientifically linked to global climate change. Pollution that occurs in the United States can easily diffuse into the atmosphere and cause unanticipated changes in the global climate. We can no longer afford to assume that pollution produced locally affects only the local environment. Instead, we should understand that it can and does adversely affect the global environment, including tropical rain forests.

[NOTE: This is a potential correct answer and may even be the best answer to the question as written.]

(3) One reason to support this proposal is *not* to *allow us to decrease the use of fertilizers on crops*. Crops grown on commercial farms require the judicious application of nutrient fertilizer in order to maximize production potential. The production of such fertilizers does not directly depend on the burning of fossil fuels. This proposal would probably not negatively affect the use of agricultural fertilizers.

(4) One reason to support this proposal is *not* to *encourage us to end research on wind and water power sources*. Wind and water power sources provide economically viable alternatives to the burning of fossil fuels. Research and development of alternative energy sources should be sup-

ported, not discouraged, by proposals such as this one. This proposal would probably not negatively affect the future of research on wind and water power sources.

6. **4** The primary source of energy in this environment is *solar energy*. Food webs are graphical representations of interactive nutritional relationships found commonly in nature. Food webs illustrate such relationships in terms of the transfer of energy from producers (green plants), which manufacture their own food via photosynthesis, to sequential levels of consumer organisms. In turn, the photosynthetic reactions depend on the energy of the sun (solar energy) to supply the energy needed to forge the chemical bonds in glucose. In this way, solar energy is known to be the primary source of energy in all natural environments.

WRONG CHOICES EXPLAINED

(1) It is *not* true that the primary source of energy in this environment is *cellular respiration in the plants*. Cellular respiration in plants is a biochemical process by which the energy stored in the chemical bonds of glucose molecules is released in a controlled manner and is transferred to molecules of ATP for use in other biochemical processes. Although essential to the survival of individual plants, cellular respiration does not provide primary energy to green plants or to the environment in any circumstance.

(2) It is *not* true that the primary source of energy in this environment is *energy from minerals in the soil*. Soil minerals supply nitrogen, potassium, and phosphorus compounds to plants and are essential to their healthy growth. Soil minerals do not provide primary energy to green plants or to the environment in any circumstance.

(3) It is *not* true that the primary source of energy in this environment is *fossil fuels*. Fossil fuels, such as coal, oil, and natural gas, are naturally occurring hydrocarbon substances that contain an abundance of stored energy but not in a form that can be accessed by green plants. Fossil fuels do not provide primary energy to green plants or to the environment in any circumstance.

7. **1** The cellular structures referred to in this study are most likely *mitochondria*. Mitochondria are cell organelles that contain the enzymes that catalyze the respiratory reactions. These reactions are responsible for releasing the stored energy in glucose for use in other cellular reactions.

WRONG CHOICES EXPLAINED

(2) The cellular structures referred to in this study are *not* most likely *nuclei*. Nuclei are cell organelles that contain the genetic material in the form of chromosomes. Nuclei and their contents are not directly responsible for releasing the stored energy in glucose for use in other cellular reactions.

(3) The cellular structures referred to in this study are *not* most likely *vacuoles*. Vacuoles are cell organelles that store nutrients or cellular wastes. Vacuoles are not directly responsible for releasing the stored energy in glucose for use in other cellular reactions.

(4) The cellular structures referred to in this study are *not* most likely *ribosomes*. Ribosomes are cell organelles that serve as platforms for protein synthesis in cells. Ribosomes are not directly responsible for releasing the stored energy in glucose for use in other cellular reactions.

8. **1** *Digestion* is the process that must first take place in order for the proteins in foods to be used by body cells. The digestive reactions convert complex proteins to amino acids. These amino acids can be absorbed into the bloodstream, transported throughout the body, and delivered to cells for the process of protein synthesis.

WRONG CHOICES EXPLAINED

(2) *Storage* is *not* the process that must first take place in order for the proteins in foods to be used by body cells. Storage is a general term that relates to the temporary or permanent placement of materials in a defined space.

(3) *Synthesis* is *not* the process that must first take place in order for the proteins in foods to be used by body cells. Synthesis is a term that describes the formation of complex molecules from simple subcomponents.

(4) *Excretion* is *not* the process that must first take place in order for the proteins in foods to be used by body cells. Excretion is a term that describes the elimination of toxic metabolic waste from the body.

9. **4** *The reproductive cycles of males and females are regulated by hormones* is the statement that is characteristic of reproduction in humans. In human females, the hormones estrogen, progesterone, follicle-stimulating hormone (FSH), and luteinizing hormone (LH), among others, are integral to the formation, fertilization, and maintenance of egg cells. In human males, the hormone testosterone is integral to the formation of sperm cells.

WRONG CHOICES EXPLAINED

(1) *The reproductive cells of males and females differ in chromosome number* is *not* the statement that is characteristic of reproduction in humans. The reproductive cells of both male and female humans carry the haploid (*n*) number of chromosomes. In humans, the haploid number of chromosomes is 23.

(2) *Males and females produce gametes in the ovaries* is *not* the statement that is characteristic of reproduction in humans. Female humans produce egg cells in the ovaries. Male humans produce sperm cells in the testes.

(3) *Males and females produce the same number of gametes* is *not* the statement that is characteristic of reproduction in humans. Males produce functional sperm cells at a rate approximately four times greater than the rate at which females produce functional egg cells.

10. **3** Row *3* in the chart represents the most likely changes in the atmosphere due to widespread deforestation. Widespread deforestation reduces the populations of producer organisms that are known to consume carbon dioxide gas and produce oxygen gas in the process of photosynthesis. The most likely result of such deforestation is a decrease in atmospheric oxygen content and an increase in atmospheric carbon dioxide content.

WRONG CHOICES EXPLAINED

(1), (2), (4) Rows *1*, *2*, and *4* in the chart do *not* represent the most likely changes in the atmosphere due to widespread deforestation. Each of these rows contains information inconsistent with the correct answer shown above.

11. **4** *Decomposers* are the organisms that would most appropriately complete the chart when written in the circle at *X*. In every healthy natural environment, decomposers such as soil bacteria and fungi are active in reducing complex organic matter to simple components that can be absorbed by green plants as fertilizing nutrients.

WRONG CHOICES EXPLAINED

(1) *Producers* are *not* the organisms that would most appropriately complete the chart when written in the circle at *X*. Producers (green plants) are autotrophic organisms that manufacture their own food by means of photosynthesis. Producers do not convert organic wastes into compounds usable by autotrophs.

(2) *Herbivores* are *not* the organisms that would most appropriately complete the chart when written in the circle at *X*. Herbivores (plant eaters) consume green plants growing in their environment, not dead and decomposing plant matter such as dead leaves and trees. Herbivores do not convert organic wastes into compounds usable by autotrophs.

(3) *Carnivores* are *not* the organisms that would most appropriately complete the chart when written in the circle at *X*. Carnivores (flesh eaters) consume prey animals in their environment, not dead and decomposing plant matter such as dead leaves and trees. Carnivores do not convert organic wastes into compounds usable by autotrophs.

12. **3** *A gene mutation caused abnormal mitotic cell division* is the statement that best explains the events represented in the diagram. Skin cells are known to reproduce themselves via mitotic cell division. In the presence of mutagenic agents, this cell division may become abnormal, resulting in the

formation of a cancerous growth. Abnormal skin growths, such as moles and scabs, should be checked by a doctor to ensure that they are not cancerous growths.

WRONG CHOICES EXPLAINED

(1) *A gene mutation caused the cells to become muscle cells* is *not* the statement that best explains the events represented in the diagram. Gene mutations are random events that may introduce small changes to the genome of an affected cell. Gene mutations are not capable of converting one cell type (skin) to a different cell type (muscle).

(2) *The growth resulted from the introduction of a vaccine* is *not* the statement that best explains the events represented in the diagram. Vaccines are substances created in laboratories and are targeted to provide immunity to specific microbial diseases. Vaccines are not known to promote cancer.

(4) *The growth resulted from uncontrolled meiotic cell division* is *not* the statement that best explains the events represented in the diagram. The diagram illustrates an occurrence in the skin. Meiotic cell division occurs only in the ovaries or the testes, not in the skin.

13. **4** This technique is best identified as *cloning*. Plant cloning is a laboratory technique that involves the removal of a viable diploid cell from a donor organism and its growth in laboratory culture to produce a genetically identical "offspring" organism.

WRONG CHOICES EXPLAINED

(1) This technique is *not* best identified as *gene alteration*. Gene alterations (mutations) are random events in which genetic material is altered through the effects of mutagenic agents including radiation. Gene alterations are not represented in this diagram as a causative agent.

(2) This technique is *not* best identified as *selective breeding*. Selective breeding is a technique used by animal and plant breeders in which organisms with desirable traits are cross-bred with the hope of producing offspring that also display those traits. Selective breeding is not represented in this diagram as a causative agent.

(3) This technique is *not* best identified as *replication*. Replication is the exact self-duplication of DNA that occurs in the processes of mitotic and meiotic cell division. Except for the fact that replication occurs each and every time a cell divides by mitotic cell division, replication is not represented in this diagram as a causative agent.

14. **2** An example of competition between members of two different species is *purple loosestrife plants growing in the same wet areas as cattail plants*. In this example, the invasive purple loosestrife plant is in direct interspecies (between different species) competition with the native cattail plant because the two species have very similar environmental requirements. In

this competition, purple loosestrife often has the advantage as it has few natural enemies in its adopted North American environment.

WRONG CHOICES EXPLAINED

(1) It is *not* true that an example of competition between members of two different species is *mold growing on a dead tree that has fallen in the forest*. This is an example of the actions of a decomposer.

(3) It is *not* true that an example of competition between members of two different species is *a coyote feeding on the remains of a deer that died of starvation*. This is an example of scavenging behavior.

(4) It is *not* true that an example of competition between members of two different species is *two male turkeys displaying mating behaviors to attract a female turkey*. This is an example of intraspecies (within a species) competition.

15. **2** *Different genes are activated in different kinds of cells* is the statement that best explains why different body cells of the same individual look and function differently. Differentiation is the process by which embryonic cells undergo specialization to become specific body tissues. Differentiation is accomplished when genes relating to a particular tissue type are switched on while those not related to that particular tissue type are switched off. This process allows each tissue type to function at maximum efficiency within the body.

WRONG CHOICES EXPLAINED

(1) *Each cell contains different genes* is *not* the statement that best explains why different body cells of the same individual look and function differently. All somatic (body) cells of an individual organism contain exactly the same genetic complement.

(3) *Cells are able to change to adapt to their surroundings* is *not* the statement that best explains why different body cells of the same individual look and function differently. Although cells are able to monitor and respond to stimuli in their surroundings, they do not adapt their specialized functions as a result.

(4) *Half of the genes in the cells came from the mother and half from the father* is *not* the statement that best explains why different body cells of the same individual look and function differently. Although this is a true statement, it does not explain the process of differentiation as defined above.

16. **2** Structure *B* is the structure within which the egg cell is normally fertilized. Structure *B* is the oviduct (Fallopian tube). It transports a mature egg from the ovary (structure *A*) to the uterus (structure *C*) during the human female reproductive cycle. Sperm cells implanted in the vagina (structure *D*) travel upward through the female reproductive tract and normally encounter and fertilize the egg cell within the oviduct.

WRONG CHOICES EXPLAINED

(1) Structure *A* is *not* the structure within which the egg cell is normally fertilized. Structure *A* is the ovary, within which the egg cell matures and from which it is released into the oviduct (structure *B*) following ovulation.

(3) Structure *C* is *not* the structure within which the egg cell is normally fertilized. Structure *C* is the uterus, within which the fertilized egg cell develops into a fetus prior to birth via the vagina (structure *D*).

(4) Structure *D* is *not* the structure within which the egg cell is normally fertilized. Structure *D* is the vagina (birth canal), within which the sperm cells are implanted prior to fertilization and through which the fetus passes during birth.

17. **3** Variations in an ameba would most commonly occur through *random mutations*. Because the ameba is a single-celled organism that reproduces via mitotic cell division, the only way its genetic information can be altered is through the action of mutagenic agents (e.g., radiation, chemicals). Any changes that occur within the genome of any individual ameba will be passed on to the offspring cells that result from future mitotic cell divisions.

WRONG CHOICES EXPLAINED

(1) Variations in an ameba would *not* most commonly occur through *differentiation during development*. The concept of differentiation refers to the specialization of tissues in multicellular organisms. However, the ameba is a single-celled, not a multicellular, organism.

(2), (4) Variations in an ameba would *not* most commonly occur through *the fusion of gametes* or through *recombination during fertilization*. The concepts of gametic fusion and genetic recombination relate to sexually reproducing organisms. However, the ameba is an asexually reproducing, not a sexually reproducing, organism.

18. **1** The development of organs and tissues from a zygote includes *mitosis and differentiation*. A zygote (fertilized egg cell) reproduces asexually via mitotic cell division until a critical mass of embryonic cells is achieved. Once the critical cell mass has formed, these generic embryonic cells undergo the process of differentiation to convert themselves into the specialized tissue types needed to produce a viable fetus.

WRONG CHOICES EXPLAINED

(2), (3), (4) The development of organs and tissues from a zygote does *not* include *mitosis and gamete production*, *meiosis and gamete production*, or *meiosis and fertilization*. Gamete (sex cell) production via meiotic cell division is a process that occurs only during maturation of the offspring. This maturation occurs significantly after (often years after) the basic development of organs and tissues via differentiation.

19. **4** This occurrence illustrates that *when humans alter ecosystems, serious consequences may result*. The human species has often altered, and continues to alter, the natural environment in multiple ways to serve its economic and survival needs. Before such an alteration occurs, the potential consequences to the natural environment should be taken into consideration. As a result of this consideration, humans should attempt to minimize any negative impact the alterations will have on the natural environment.

WRONG CHOICES EXPLAINED

(1) This occurrence does *not* illustrate that *farmers should never clear land to grow crops as it always creates problems*. Farming is an example of an economic activity that significantly alters the natural environment. Farming practices can be adjusted to minimize their impact on the natural environment while still providing the food resources necessary for human survival.

(2) This occurrence does *not* illustrate that *once an ecosystem has been altered, it can not be restored to normal*. Any environment, no matter how extensively altered by humans, can recover and restore itself, if given time and the absence of human activity, via the process of ecological succession.

(3) This occurrence does *not* illustrate that *the farmers deliberately altered the equilibrium of the cities in the Northeast*. Midwestern farmers in the 1920s had little understanding of the effects that their farming practices were having on the national and global climates. Today, after continuing extensive scientific research, we should have a much better understanding of the ways that human activities can negatively affect the natural environment.

20. **3** This threat is the result of a human activity that has *weakened an ecosystem through pollution*. The decades-long assault of acid rain on formerly unpolluted forest environments has altered soil chemistry to the point that some long-established native plant communities, such as the sugar maple, can no longer thrive. As a result, the New York State woodland ecosystem and the human economy it supports are in danger of collapsing.

WRONG CHOICES EXPLAINED

(1) It is *not* true that this threat is the result of a human activity that has *introduced a foreign species by accident*. The sugar maple tree is a species native to New York State, not a foreign invasive species. Therefore, the sugar maple tree should be protected by any means necessary.

(2) It is *not* true that this threat is the result of a human activity that has *stabilized a forest ecosystem through technology*. Acid rain is a national and global environmental problem that needs to be corrected immediately to save our treasured natural environment and the human industries that depend on it. Industries that create acid rain are not stabilizing but, instead, are destabilizing technologies.

(4) It is *not* true that this threat is the result of a human activity that has *weakened a species by direct harvesting*. Direct harvesting is a human practice in which plants or animals are removed (harvested) from a natural environment for their economic value. Although sugar maple trees are often directly harvested for their wood and sap, this practice is not under question in this case.

21. **1** The human male reproductive system is adapted for the production of *sperm and the delivery of these cells for internal fertilization*. Sperm cells are gametes that contain the haploid (n) number of chromosomes. Sperm cells are produced in the male testes and are transported via the vasa deferentia and urethra to the female reproductive tract where fertilization of an egg cell may occur. These structures and processes enable the human species to survive from generation to generation.

WRONG CHOICES EXPLAINED

(2) It is *not* true that the human male reproductive system is adapted for the production of *gametes that transport food to the egg*. An egg cell produced in the ovary in females contains sufficient food to power multiple cell divisions prior to the formation of the placenta during development. Male gametes do not transport food. This is a nonsense distracter.

(3) It is *not* true that the human male reproductive system is adapted for the production of *zygotes and the development of these cells into a fetus*. A zygote (fertilized egg) is formed following fertilization of an egg cell by a sperm cell in the oviduct of the female. Fetal development occurs in the uterus of the female. The oviduct and uterus are not part of the male reproductive system.

(4) It is *not* true that the human male reproductive system is adapted for the production of *hormones that stimulate placenta formation in the male*. The placenta forms in the uterus of the female following fertilization and embedding of the zygote in the uterine lining. The placenta is not part of the male reproductive system.

22. **1** *Replication of DNA in bacterial cells and cell division* are the two processes required for the technique to successfully produce hormones. The technique illustrated is genetic engineering. Genetic engineering is a laboratory technique in which a gene for a desired trait is snipped from the DNA of a donor cell and inserted into the genome of a recipient cell (in this case, a bacterial cell). The recipient bacterial cell is then cultured and allowed to reproduce rapidly by mitotic cell division such that a bacterial colony is established. All the cells within this colony will be capable of producing the desired hormone, which can be drawn off and purified for use in treating a human hormone deficiency.

WRONG CHOICES EXPLAINED

(2), (3), (4) *Replication of DNA in bacterial cells and gamete formation, meiosis and development*, and *mitosis and fertilization* are *not* the two processes required for the technique to successfully produce hormones. Bacterial cells reproduce asexually by the process of mitosis followed by binary fission. Gametes (formed by meiotic cell division) and fertilization (fusion of male and female gametes) are sexual reproductive processes that are not involved in this technique.

23. **3** All the genetic information needed for the organism to develop is first present at step *C*. At this stage, the structure illustrated is the zygote, which results from the fusion of a haploid (n) egg cell (step *A*) with a haploid (n) sperm cell (step *B*). The zygote contains the diploid ($2n$) number of chromosomes and all the genetic information needed to form a new human offspring.

WRONG CHOICES EXPLAINED

(1) It is *not* true that all the genetic information needed for the organism to develop is first present at step *A*. Step *A* represents the unfertilized egg cell, which contains only half of the genetic information needed to form a new human offspring.

(2) It is *not* true that all the genetic information needed for the organism to develop is first present at step *B*. Step *B* represents the sperm cell, which contains only half of the genetic information needed to form a new human offspring.

(4) It is *not* true that all the genetic information needed for the organism to develop is first present at step *D*. Step *D* represents the embryo/fetus, whose cells contain all of the genetic information needed to form a new human offspring. However, step *D* is not the first stage in which this full genetic complement is found.

24. **4** *All of the organisms pass on traits through reproduction* is the statement about the organisms represented that is correct. Every viable species, whether plant, animal, or single-celled organism, must reproduce in order to pass on the genetic traits that describe the species.

WRONG CHOICES EXPLAINED

(1) *All of the organisms are autotrophs* is *not* the statement about the organisms represented that is correct. Autotrophs are organisms that manufacture their own food by means of photosynthesis. Of the organisms represented, only the flowering plant and the aquatic plant are autotrophs.

(2) *Only the flowering plant, green heron, and aquatic plant carry out photosynthesis* is *not* the statement about the organisms represented that is correct. The green heron is an animal and a heterotroph. So the green heron cannot carry out photosynthesis.

(3) *Only the frog and green heron can maintain homeostasis* is *not* the statement about the organisms represented that is correct. All organisms of all types in all environments must maintain homeostasis (steady state) in order to survive.

25. **4** The relationship between mistletoe and trees is an example of *parasite/host*. A parasite is defined as an organism that lives on or in another organism (host) and derives benefit from that organism while causing harm to it. The description given of the mistletoe plant and trees clearly meets this definition.

WRONG CHOICES EXPLAINED

(1) The relationship between mistletoe and trees is *not* an example of *consumer/herbivore*. A consumer is an organism that derives its nutrition from the tissues of other plant or animal organisms. An herbivore is an animal consumer that derives its nutrition solely from plant matter. The description given of the mistletoe plant and trees does not meet this definition.

(2) The relationship between mistletoe and trees is *not* an example of *predator/prey*. A predator is an animal that hunts, kills, and consumes the bodies of other animals. Prey organisms are animals that are hunted and consumed by predators. The description given of the mistletoe plant and trees does not meet this definition.

(3) The relationship between mistletoe and trees is *not* an example of *scavenger/decomposer*. A scavenger is a consumer animal that consumes the dead bodies of other animals for nutrition. A decomposer is a bacterium or fungus that breaks down dead plants and animals for their nutrition. The description given of the mistletoe plant and trees does not meet this definition.

26. **2** This method of controlling the rabbit population was an attempt to *stop the overproduction of an introduced species*. Although the exact effect of the myxoma virus is not stated in the paragraph, it is mentioned that the virus was introduced to infect the rabbits for the purpose of reducing the population of this introduced (invasive) species.

WRONG CHOICES EXPLAINED

(1) It is *not* true that his method of controlling the rabbit population was an attempt to *stop the overproduction of a native species*. The paragraph states clearly that rabbits are not native to Australia.

(3) It is *not* true that his method of controlling the rabbit population was an attempt to *limit the food sources of the rabbit*. No information is presented in the paragraph that would support this inference.

(4) It is *not* true that his method of controlling the rabbit population was an attempt to *limit the number of rabbits brought into the country*. No information is presented in the paragraph that would support this inference.

27. **3** The major role of carbohydrates in the human diet is to *supply energy for the body*. The digestive process functions to reduce complex carbohydrates to simple sugars such as glucose. This glucose enters the body's cells where it is broken down chemically. This chemical breakdown releases energy that is transferred to molecules of ATP. ATP, in turn, releases this energy to fuel cellular reactions.

WRONG CHOICES EXPLAINED

(1) It is *not* true that the major role of carbohydrates in the human diet is to *form the membranes that surround the mitochondria*. The membranes surrounding mitochondria are composed of proteins and lipids, not carbohydrates.

(2) It is *not* true that the major role of carbohydrates in the human diet is to *act as a catalyst for cellular reactions*. Cellular reactions are catalyzed by enzymes (specialized proteins), not by carbohydrates.

(4) It is *not* true that the major role of carbohydrates in the human diet is to *provide building blocks for amino acids*. Amino acids are the simple building blocks of proteins and are not composed of carbohydrates.

28. **1** This is a valid concern because this practice *reduces the biodiversity of their fields*. Biodiversity is a measure of the total number of different species that inhabit and interact in a given environment. When forests or fallow (resting) fields are cleared for the purpose of planting crops, biodiversity is significantly reduced. The planting of a single species in such areas (known as monocropping) maximizes the lack of biodiversity. When the environment changes, monocrop farm fields are more susceptible to drought, disease, and soil nutrient depletion than if the fields were planted in alternating rows of compatible crops (e.g., corn, beans, and squash).

WRONG CHOICES EXPLAINED

(2) It is *not* true that this is a valid concern because this practice *increases the number of decomposers in their fields*. Decomposers are most abundant in soils that contain an abundance of organic matter and that are kept shaded and moist. Monocrop fields may lack these conditions and so may be less likely to support decomposers than fields planted in multiple crop species.

(3) It is *not* true that this is a valid concern because this practice *decreases the need to import food*. Planting only one crop may reduce the availability of vegetables and fruits within the state and so would increase the need to import these foods from other states or countries.

(4) It is *not* true that this is a valid concern because this practice *increases the number of invasive species*. Invasive species enter the environment in multiple ways, including through purposeful introduction by humans.

29. **4** The breathing rate, heart rate, and blood hormone levels of an individual would directly provide information about that individual's *metabolic activity*. The term *metabolism* refers to the sum total of all the chemical and physical activities that occur in the body. When a doctor takes measurements such as those listed, he or she can make informed judgments about a patient's overall metabolic health.

WRONG CHOICES EXPLAINED

(1) It is *not* true that the breathing rate, heart rate, and blood hormone levels of an individual would directly provide information about that individual's *cellular organization*. Cellular organization is determined by the processes of growth and differentiation and cannot be determined through measures such as those listed.

(2) It is *not* true that the breathing rate, heart rate, and blood hormone levels of an individual would directly provide information about that individual's *nutrition*. A doctor could best determine a patient's nutrition by asking the patient to answer questions concerning his or her dietary habits, not through measures such as those listed.

(3) It is *not* true that the breathing rate, heart rate, and blood hormone levels of an individual would directly provide information about that individual's *inheritance*. Inheritance of genetic traits is determined at the moment of fertilization and might best be determined through DNA testing, not through measures such as those listed.

30. **1** *Consumer* is the type of organism that could occupy levels *B*, *C*, and *D* of this energy pyramid. In any traditional energy pyramid, the base of the pyramid is reserved for producer organisms whose photosynthetic activity supports the upper levels. The upper levels of the energy pyramid are reserved for the consumer organisms, including herbivores (level *B*), carnivores (level *C*), and top predators (level *D*).

WRONG CHOICES EXPLAINED

(2), (3) *Producer* and *autotroph* are *not* the types of organisms that could occupy levels *B*, *C*, and *D* of this energy pyramid. Producers, also known as autotrophs, are listed in a traditional energy pyramid at level *A*.

(4) *Carnivore* is *not* the type of organism that could occupy levels *B*, *C*, and *D* of this energy pyramid. Carnivores, including top predators, are listed in a traditional energy pyramid at levels *C* and *D* but not *B*. Level *B* traditionally contains only herbivores.

PART B–1

31. **1** Headlines such as these best illustrate the concept that *scientific explanations are tentative and subject to change*. Humans' understanding of natural phenomena is constantly improving as more and better scientific information derived from professional research is published in scientific journals by competent scientists. The newspaper headlines indicate that this understanding improved significantly over an approximately 30-year period during which competent scientific research was carried out. Newspapers (and, in more modern times, the Internet) are not necessarily the best sources of scientific information. Rather, the best sources of scientific information are the scientific journals in which actual research results are published and vetted by other scientists.

WRONG CHOICES EXPLAINED

(2), (3), (4) It is *not* true that headlines such as these best illustrate the concept that *some newspapers are not honest and report incorrect information on purpose*, *worms can enter the body many different ways*, or *worms found in swamps should not be used for fishing*. No information is presented in the question that supports any of these inferences.

32. **3** The scientist's conclusion was based on *careful observation, measurements, and inferences from his data*. Although his inferences were only partially correct, the scientist applied what he understood to be careful scientific methodologies (observation and measurement) to the problem. He followed through on a long-term experiment and drew inferences from his results. We know now, after 400 years of additional scientific experimentation, that the plant's weight gain was actually the result of biochemical activity that chemically combined water, carbon dioxide, and soil minerals to produce molecules of carbohydrates, proteins, and other essential organic components.

WRONG CHOICES EXPLAINED

(1), (2), (4) It is *not* true that the scientist's conclusion was based on *the input of scientists from many countries doing similar studies*, *the application of advanced technologies to the study of a problem*, or *an extensive knowledge of the process of photosynthesis*. Such peer collaboration, advanced technologies, and biochemical knowledge were not available to scientists working in the early 1600s.

33. **2** The approximate diameter of one cell is *50 μm*. This value can be estimated by placing a piece of paper against one of the cells and marking its diameter along the paper edge, transferring that dimension to the 500 μm diameter line in the diagram, and estimating that approximately 10 of these

cells can fit across the diameter of the field of view. Dividing 500 µm by 10 cell diameters yields a cell diameter of 50 µm.

WRONG CHOICES EXPLAINED

(1), (3), (4) It is *not* true that the approximate diameter of one cell is *10 µm, 250 µm*, or *500 µm*. None of these values can be supported by application of the method described above or by any similar method.

34. **2** One reason this could lead to extinction of this warbler is that *there may not be enough diversity among the birds for the species to be able to survive an environmental change*. In general, a species with many members and with a large and diverse gene pool contains sufficient variation to allow a large proportion of its members to survive any change in environmental conditions that might threaten its existence. By contrast, a species with few members and with a shrinking gene pool contains fewer variations and is less likely to survive such changes. This latter condition increases the likelihood of extinction.

WRONG CHOICES EXPLAINED

(1) One reason this could lead to extinction of this warbler is *not* that *after a species becomes extinct, it won't be able to carry out its role in the ecosystem*. Although a species that has become extinct can no longer carry out its role in the environment, this phenomenon cannot lead to extinction of that species. This is a nonsense distracter.

(3) One reason this could lead to extinction of this warbler is *not* that *extinction always occurs when populations begin to reduce in number*. All natural populations experience periods of population decline, especially under changing environmental conditions. The majority of such population declines do not result in extinction. Instead, these declines are usually reversed as surviving members reproduce and pass on their favorable traits to ensure the continuing survival of future generations.

(4) One reason this could lead to extinction of this warbler is *not* that *an increase in biodiversity within a population often causes the population to be classified as threatened or endangered*. Biodiversity is defined as the number of different species that inhabit an environment. Using the term *biodiversity* to describe genetic variation within a species is not scientifically correct. This is a nonsense distracter.

35. **1** One primary function of the cell membrane is *regulating the flow of simple sugars into or out of the cell*. In addition to bounding the cell and its contents, the cell membrane implements several methods to control the movement of substances into and out of the cell. Simple sugars are normally brought into a cell by diffusion though the cell membrane.

WRONG CHOICES EXPLAINED

(2) It is *not* true that one primary function of the cell membrane is *synthesizing substances by breaking down cell organelles*. This is not a function of the cell membrane or any other cell component. This is a nonsense distracter.

(3) It is *not* true that one primary function of the cell membrane is *storing carbohydrates, water, and starches for future use*. This is a function of a food vacuole in a cell, not the cell membrane.

(4) It is *not* true that one primary function of the cell membrane is *digesting carbohydrates, fats, and protein*. The process of digestion of complex organic molecules normally occurs in the cell's external environment or within the digestive tracts of complex animals but not in the cell membrane.

36. **3** *If this pipeline were to leak, the oil could contaminate soil, water, and wildlife* is the statement that expresses a major concern many people are likely to have about the proposed pipeline. Most people understand the complex and fragile relationships that exist in nature. They do not want ecosystems to be disrupted or destroyed indiscriminately by human commercial and industrial activities.

WRONG CHOICES EXPLAINED

(1) *The pipeline will bring a large number of jobs to the area where it is being constructed* is *not* the statement that expresses a major concern many people are likely to have about the proposed pipeline. Oil pipelines provide a short-term advantage in the form of construction jobs. However, they also provide long-term risks to the environments they cross. Job creation is not usually the subject of environmental concerns about such projects.

(2) *The oil pipeline will increase the amount of finite resources* is *not* the statement that expresses a major concern many people are likely to have about the proposed pipeline. Oil and other fossil fuels are finite resources that are dwindling and will eventually become exhausted. Although this oil pipeline would increase access to these finite resources, it would not increase the overall amount of oil.

(4) *The pipeline is a technological fix for ozone depletion* is *not* the statement that expresses a major concern many people are likely to have about the proposed pipeline. Ozone is known to be depleted by the release of CFC (chlorofluorocarbons) and other aerosol propellants. No known relationship exists between oil pipelines and ozone depletion.

37. **3** By observing the annual rings in the diagram, one can infer that *some years provide better conditions for growth than other years*. An examination of the diagram reveals that some annual rings in this tree are wider or narrower than others. Narrow rings indicate relatively slow growth, while wide rings indicate relatively rapid growth. It can reasonably be inferred from this examination that this tree encountered environmental conditions that limited its growth in years 10–12. It may also be inferred that from year 13

to the present, the tree had relatively rapid growth, indicating more favorable environmental conditions in those years.

WRONG CHOICES EXPLAINED

(1) It is *not* true that by observing the annual rings in the diagram, one can infer that *environmental conditions did not change over the last 20 years*. If one accepts the relationship between a tree's annual rings and its growth patterns, then one should infer that environmental conditions have varied as described above.

(2) It is *not* true that by observing the annual rings in the diagram, one can infer that *trees grow faster on the side that faces the Sun*. No information provided in the question would lead one to make this inference.

(4) It is *not* true that by observing the annual rings in the diagram, one can infer that *tree rings are not reliable because trees must be cut down to see them*. In fact, it is possible to examine a tree's annual ring pattern without cutting down the tree. This technique involves taking a core sample through the tree's bark to its center and then sealing the hole to prevent entry by fungi or insects.

38. **2** This treatment for acne, using phages, is effective because phages *eliminate bacteria by attacking specific cell structures*. According to the paragraph, phages produce a protein, endolysin, that breaks down the cell walls of acne bacteria.

WRONG CHOICES EXPLAINED

(1), (3) It is *not* true that this treatment for acne, using phages, is effective because phages *produce antibodies to clean out clogged pores and hair follicles* or *carry genes and infect follicles*. No information is presented in the paragraph concerning the production of antibodies or the infection of follicles.

(4) It is *not* true that this treatment for acne, using phages, is effective because phages *attack every known type of bacteria*. The paragraph specifically states that phages attack only acne bacteria, not all types of bacteria.

39. **3** The protein endolysin belongs to the group of chemical substances known as *biological catalysts*. A catalyst is a substance that speeds up or slows down the rate of a chemical reaction. In this case, the biological catalyst endolysin targets and breaks apart the specific materials that make up the cell walls of acne bacteria.

WRONG CHOICES EXPLAINED

(1) The protein endolysin does *not* belong to the group of chemical substances known as *hormones*. Hormones are specific proteins that help to regulate metabolic activities of plants and animals. Hormones are not known to break apart the cell walls of bacteria.

(2) The protein endolysin does *not* belong to the group of chemical substances known as *receptors*. Receptors are specific proteins embedded in the membranes of cells and that regulate cell-to-cell communication. Receptors are not known to break apart the cell walls of bacteria.

(4) The protein endolysin does *not* belong to the group of chemical substances known as *molecular bases*. Molecular bases, such as adenine, thymine, cytosine, and guanine, are components of large nucleic acids, such as DNA and RNA. Molecular bases are not known to break apart the cell walls of bacteria.

40. **4** The typical response of the human body to an infection by bacteria is to *produce white blood cells and antibodies*. The immune response in humans detects the antigens of foreign invaders. It responds to these antigens by manufacturing antibodies that specifically target and neutralize the infectious agent. Likewise, the body programs white blood cells to engulf and destroy these invaders.

WRONG CHOICES EXPLAINED

(1) The typical response of the human body to an infection by bacteria is *not* to *stimulate the production of antigens*. Antigens are proteins produced by the cells of all living things and are specific to the species that produces them. Antigens are not produced to protect the body from infection.

(2) The typical response of the human body to an infection by bacteria is *not* to *decrease the number of enzymes in the blood*. Enzymes are biological catalysts produced by the cells of all living things and that help to regulate the body's metabolism. The number of enzymes in the blood is unaffected by the immune response.

(3) The typical response of the human body to an infection by bacteria is *not* to *ignore the organisms, unless they are pathogens*. The human immune response does not discriminate between infectious and noninfectious agents in the exercise of its protective function.

41. **4** *Both processes X and Y occur in green plants* is the correct statement regarding the types of organisms able to carry out these processes. Process *X* is photosynthesis, in which green plants take in CO_2 (and water and solar energy) and release O_2 (and sugar). Process *Y* is respiration, in which both green plants and animals take in O_2 (and sugar) and release CO_2 (and water and ATP energy).

WRONG CHOICES EXPLAINED

(1) *Process X occurs in heterotrophs, but not in autotrophs* is *not* the correct statement regarding the types of organisms able to carry out these processes. Process *X* (photosynthesis) does not occur in heterotrophs (animals) but does occur in autotrophs (green plants).

(2) *Process Y occurs in consumers, but not in producers* is *not* the correct statement regarding the types of organisms able to carry out these processes. Process *Y* (respiration) is carried out by both consumers (animals) and producers (green plants).

(3) *Both processes X and Y occur in all living things* is *not* the correct statement regarding the types of organisms able to carry out these processes. Process *X* (photosynthesis) is carried out by green plants (producers, autotrophs) but not by animals (consumers, heterotrophs).

42. **2** *Ribosome* is the structure represented by *B*. In the process of protein (enzyme) synthesis, the DNA code is read by molecules of mRNA and transferred to ribosomes for translation into specific protein molecules.

WRONG CHOICES EXPLAINED

(1) *Vacuole* is *not* the structure represented by *B*. Vacuoles in cells serve to store materials that are too large to readily pass through the cell membrane.

(3) *Cytoplasm* is *not* the structure represented by *B*. Cytoplasm is the watery medium of the cell interior that supports the organelles and assists in the transport of materials within the cell.

(4) *Chloroplast* is *not* the structure represented by *B*. Chloroplasts in green plant cells function to convert solar energy to the chemical bonds of glucose molecules.

43. **3** The nucleus contains molecules of *A*, which *store hereditary information*. Molecule *A* is DNA, which carries the genetic code that provides information for the production of proteins in the cell.

WRONG CHOICES EXPLAINED

(1) It is *not* true that the nucleus contains molecules of *A*, which *recycle waste products*. DNA is not known to recycle waste products.

(2) It is *not* true that the nucleus contains molecules of *A*, which *remove water from the cell*. This is a function of organelles known as contractile vacuoles, not DNA.

(4) It is *not* true that the nucleus contains molecules of *A*, which *regulate the pH of cytoplasm*. DNA is not known to regulate the pH of the cytoplasm.

PART B–2

44. One credit is allowed for correctly marking an appropriate scale, without any breaks in the data, on the axis labeled "Estimated Moose Population." [1]

45. One credit is allowed for correctly plotting the data for the estimated moose population on the grid, connecting the points, and surrounding each point with a small circle. [1]

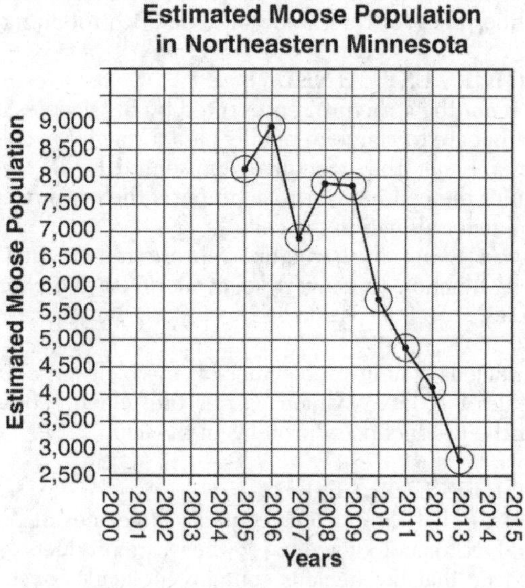

46. One credit is allowed for correctly explaining how climate change could result in an increased number of moose infected with winter ticks. Acceptable responses include but are not limited to: [1]

- *The fall seasons are longer and warmer, increasing the chances of ticks surviving the winters and infesting moose.*
- *Climate change can be predicted to improve conditions for the survival of winter ticks through earlier snowmelt in the spring.*
- *Ticks that fall off moose onto soil instead of snow are more likely to survive and then reinfect moose in the spring.*
- *In the past, moose in northern climates were protected from winter ticks by long periods of cold weather that killed the ticks. However, this is no longer the case due to global climate change that has warmed average temperatures worldwide.*

47. **3** Increased average yearly temperatures in regions presently inhabited by moose could result in a disruption in homeostasis in these animals because *moose will not be able to maintain an appropriate body temperature, since they do not sweat*. The sweating mechanism present in humans enables us to regulate our body temperature as sweat evaporates from the skin surface and cools the blood. Since they lack this mechanism, moose can cool their bodies only by exposure to cool air or cool water. Failure to regulate body temperature within an acceptable range can significantly disrupt the body's homeostatic balance.

WRONG CHOICES EXPLAINED

(1), (2), (4) It is *not* true that increased average yearly temperatures in regions presently inhabited by moose could result in a disruption in homeostasis in these animals because *a decrease in average temperatures will increase mutations in their skin cells*, *an increase in average temperatures will decrease the amount of blood ticks can consume*, or *moose will sweat more and lose too much water from their bodies*. No information is presented in the paragraph that would support any of these inferences.

48. One credit is allowed for correctly describing the probable result of an environmental change that severely affected the organism represented by species *K*. Acceptable responses include but are not limited to: [1]

- *Species K did not survive because they were unable to adapt to their new environment.*
- *Species K became extinct because they could not survive the changed environment.*
- *They became extinct. The diagram shown that they did not survive to the present.*

49. **2** Three species with the most similar traits are most likely *D, H, J*. The diagram indicates that species *D* is the most direct common ancestor of species *H* and *J*. These three species likely share a similar gene pool and, therefore, likely share similar traits.

WRONG CHOICES EXPLAINED

(1), (3), (4) Three species with the most similar traits are *not* most likely *F, I, G*; *B, D, G*; or *F, A, J*. The diagram indicates that these species are widely separated in the evolutionary history of this genetic grouping. The only common ancestor linking these species is long-extinct species *A*. So it is unlikely that *F, A, J* are the three species with the most similar traits.

50. **4** The organisms found at the second trophic level of this pyramid would be *herbivores*. The diagram represents a food (energy) pyramid. Traditionally, organisms shown in the second trophic level of a food pyramid

are herbivores, which derive their nutritional energy primarily from the producers found at the first trophic level.

WRONG CHOICES EXPLAINED

(1) The organisms found at the second trophic level of this pyramid would *not* be *producers*. Producers (green plants) are traditionally represented as inhabiting the first trophic level in a food pyramid.

(2) The organisms found at the second trophic level of this pyramid would *not* be *decomposers*. Despite their crucial role in the ecological food web, decomposers are often not included in representations of a food pyramid, including this one.

(3) The organisms found at the second trophic level of this pyramid would *not* be *carnivores*. Carnivores are traditionally represented as inhabiting the third trophic level in a food pyramid.

51. One credit is allowed for correctly stating *one* reason why there is less energy available at each trophic level going from the first to the third trophic level. Acceptable responses include but are not limited to: [1]

- *Energy is lost at each level as heat that dissipates into the environment.*
- *Some energy is used by the organisms at each level and is not available to the organisms at the next level.*
- *An amount of energy is used at each level to support metabolic activities. This energy is lost and cannot be used by organisms at the next level of the food/energy pyramid.*

52. One credit is allowed for correctly explaining how the sharp population increase from 1910 to 1925 might have resulted in the decrease in the carrying capacity after 1925. Acceptable responses include but are not limited to: [1]

- *The overpopulation of deer in 1925 caused their habitat to be overgrazed. This did not allow the food sources in the area to regrow for the future deer population.*
- *The interference of humans in the predator-prey relationship was the cause of this problem. By removing natural predators of the deer, the deer population exploded. There was not enough vegetation to support the increased population of deer after 1925.*
- *In 1925, the deer ate too much food, reducing the amount of food available in future years.*
- *By removing the predator species that had kept the deer population in check, the deer population increased geometrically. This increase exceeded the environment's carrying capacity because the additional deer overgrazed the habitat from 1920 through 1925. From 1925 to 1937, the deer population dropped at a rapid rate, stabilizing after 1937.*
- *Overpopulation of deer in 1925 destroyed some of the resources deer needed. So fewer deer could live there.*

53–54. Two credits are allowed for identifying some of the key events associated with the change. In your answer, be sure to:

- Identify *one* natural event that could cause the disruption indicated in the diagram. [1]
- State what would most likely happen to the new stable ecosystem in future years if no further disruptions occur. [1]

Acceptable responses include but are not limited to: [2]

- *A possible disruption in this ecosystem is forest fire. [1] The new ecosystem will undergo ecological succession until a new climax community compatible with the larger environment is achieved. [1]*
- *Volcanic eruption might have disrupted this ecosystem. [1] A long period of change will precede the formation of any new stable ecosystem. Once established, the new stable ecosystem will last until the next disruption. [1]*

55. One credit is allowed for correctly explaining why a mutation that occurs in a body cell will not contribute to the evolution of a species. Acceptable responses include but are not limited to: [1]

- *Genetic information in body cells isn't passed on to offspring.*
- *The mutation isn't in sperm or egg cells.*
- *Mutations must be in gametes to be passed on to offspring.*

PART C

56. One credit is allowed for correctly stating *one* reason why a supply of amino acids is important for the survival of complex organisms. Acceptable responses include but are not limited to: [1]

- *Amino acids are the building blocks of protein molecules.*
- *Amino acids are found in enzymes, which regulate chemical activity in complex organisms.*
- *Amino acids are the chemical components of the proteins found in our bodies' muscle and organ tissues.*

57. One credit is allowed for correctly identifying *one* possible effect on the human population if nitrogen fertilizers were not available. Acceptable responses include but are not limited to: [1]

- *People would starve.*
- *Farmers could not grow enough crops to feed the population.*
- *The human population would decrease.*
- *The food supply would decrease.*

- *Human civilization as we know it could not exist.*
- *Humans would have deficiencies in their diets.*
- *Humans would become malnourished.*

58. One credit is allowed for correctly explaining how the building of factories to produce fertilizer is an example of a trade-off. Acceptable responses include but are not limited to: [1]

- *The factories produce useful products to help increase the food supply but they also produce pollutants.*
- *The factories provide fertilizers for the production of more food but they also disrupt the habitats of some organisms.*
- *Fertilizers are useful for the growth of healthy plants but can also be harmful if they wash off into lakes/waterways.*
- *Fertilizers help plants grow but could harm the environment.*

59–60. Two credits are allowed for correctly analyzing the experiment that produced the data in the table. In your answer, be sure to:

- State a hypothesis for the experiment. [1]
- State whether the results of the experiment support or fail to support your hypothesis; support your answer. [1]

Acceptable responses include but are not limited to: [2]

- *Hypothesis: If bacteria are exposed to UV light, they will form fewer colonies than those not exposed. [1] All UV-screened areas of exposed Petri dishes formed an average of 22.2 colonies, whereas unscreened (UV-exposed) areas formed an average of only 12.2 colonies. Therefore, my hypothesis appears to be supported by the experimental data.[1]*
- *My hypothesis for this experiment is that longer exposure time to UV light will limit bacterial growth more than shorter exposure time. [1] My analysis of the data in the table is that the data support my hypothesis. The data show that 0.0 min. of UV exposure allowed 22 colonies to grow, while 10.0 min. of UV exposure allowed only 1 colony to grow. Each increase in exposure time had more and more impact on the growth of the bacteria. [1]*
- *I think that if bacteria are exposed to UV light, they will not form colonies (hypothesis). [1] This hypothesis isn't supported by the data, unfortunately. The data indicate that bacterial colonies grew in each UV-exposed Petri dish, even the one that received 10 minutes of exposure time! [1]*

61. One credit is allowed for correctly explaining the importance of the presence of variations within a population. Acceptable responses include but are not limited to: [1]

- *More variation increases the chances that some members of the population will survive in a changing environment.*
- *Differences among offspring increase the likelihood that some individuals will have adaptations to increase their chances of survival in a new environment.*
- *Variations provide raw material for natural selection when environments change.*
- *More variation increases the stability of the population.*
- *Without variations, the population may not survive in a changing environment.*

62. One credit is allowed for correctly describing how the process of natural selection can result in an increase in frequencies of certain traits found in a population. Acceptable responses include but are not limited to: [1]

- *More individuals with favorable traits usually survive, passing on those traits to the next generation.*
- *As a result of natural selection, the number of individuals with favorable traits increases.*
- *The fittest individuals will survive, causing an increase in the frequency of favorable traits.*
- *Individuals with unfavorable traits will succumb to environmental pressures and so will not be able to pass on those unfavorable traits to the next generation.*

63. One credit is allowed for correctly describing the importance of receptors in cellular communication. Acceptable responses include but are not limited to: [1]

- *Receptors receive messages sent by other cells.*
- *They enable cells to respond appropriately.*
- *They assist cells to recognize and admit hormones and neurotransmitters.*
- *In white blood cells, receptors assist in the recognition of antigens on the surfaces of foreign invaders.*

64. One credit is allowed for correctly describing the importance of the shape of receptor molecules for carrying out their function. Acceptable responses include but are not limited to: [1]

- *Receptor molecules have specific shapes that influence how they interact with other molecules.*
- *The shape of the receptor molecule determines with which molecules they can interact.*
- *The shape of the receptor molecule and messenger molecule must match.*
- *The shapes must match/fit together at the active site.*

65. One credit is allowed for correctly identifying *one* effect a reduced number of insulin receptors might have on an individual. Acceptable responses include but are not limited to: [1]

- *The individual might experience high levels of glucose in the blood/urine.*
- *The person might be/become diabetic as a result.*
- *If insulin receptors are reduced, less glucose will go to the cells.*
- *The blood glucose level will not be appropriately controlled.*

66. One credit is allowed for correctly explaining why the genetic material in an offspring produced by sexual reproduction contains genetic material that is *not* identical to the genetic material of either parent. Acceptable responses include but are not limited to: [1]

- *Offspring receive genetic information from each of the two parents.*
- *Each parent contributes half of the offspring's genetic material.*
- *Recombination of genes occurs at fertilization.*
- *The processes of meiosis, segregation, fertilization, and recombination ensure that each offspring receives a unique genetic profile.*
- *In rare cases, mutations such as deletion, addition, and crossing-over may occur to change the offspring's genetic complement.*

67. One credit is allowed for correctly stating *one* reason why identical twins should have fewer genetic differences than fraternal twins. Acceptable responses include but are not limited to: [1]

- *Fraternal twins develop from two separately fertilized egg cells.*
- *Fraternal twins are no more closely related than siblings born at different times because fraternal twins arise from separate fertilizations of two different egg cells by two different sperm cells.*
- *Identical twins are produced from a single zygote/fertilized egg, so all of their cells are identical.*
- *Identical twins come from a single fertilized egg, so they are genetically identical.*

68. One credit is allowed for correctly identifying *two* environmental factors that can lead to an increase in the number of epigenetic markers that modify gene expression. Acceptable responses include but are not limited to: [1]

- *Stress*
- *Pollutants*
- *Food*
- *Smoke*
- *Pesticides*
- *Alcohol*

- *Drugs*
- *Temperature*

69. One credit is allowed for correctly describing how nutrients move from the mother to the fetus. Acceptable responses include but are not limited to: [1]

- *Nutrients move across the placenta from mother to fetus.*
- *Most soluble substances pass readily across the placental membranes via diffusion.*
- *In the placenta, a series of maternal and fetal membranes separate the mother's blood from the fetus's blood but allow the diffusion of soluble materials between maternal and fetal bloodstreams.*

70. One credit is allowed for correctly stating *one* other way, in addition to consuming a balanced diet, pregnant women can help ensure proper development of the fetus. Acceptable responses include but are not limited to: [1]

- *Pregnant women should have regular doctor visits to ensure the health of mother and fetus.*
- *Pregnant women should continue to exercise appropriately during pregnancy.*
- *Pregnant women should avoid alcohol/drugs/tobacco products.*
- *Pregnant women should maintain an adequate waking/sleeping cycle.*
- *Pregnant women should avoid contact with infectious agents and their vectors, such as deer ticks carrying Lyme disease bacteria and mosquitos carrying Zika virus.*
- *Pregnant women should avoid unprotected sex and its associated risks, such as AIDS/gonorrhea/syphilis, that could affect the mother's health and the baby's development.*

71. One credit is allowed for correctly stating *one* way the use of these new "little electric heaters" might represent a long-term benefit over using antibiotics to treat bacterial infections. Acceptable responses include but are not limited to: [1]

- *The bacteria may not become resistant to the heaters.*
- *Antibiotics may have negative side effects.*
- *Heaters will work faster than the antibiotics.*

72. One credit is allowed for correctly stating *one* reason why the mother did not get sick with the measles. Acceptable responses include but are not limited to: [1]

- *The mother had antibodies against the measles.*
- *The mother had been vaccinated against the measles.*

- *The mother had measles when younger and still had a natural immunity to measles.*
- *The mother is immune to the measles.*
- *The mother had a mutation that made her resistant to the measles.*

PART D

73. **2** *The nutrients usually move from an area of high concentration in the soil to an area of low concentration in root cells* is the statement that best describes how nutrients enter the root cells of the onion plant. Dissolved nutrients, such as nitrates and phosphates, are absorbed from the soil into the root hair cells via the process of diffusion. Diffusion in the direction of the concentration gradient is accomplished without the expenditure of cell energy.

WRONG CHOICES EXPLAINED

(1) *Only nutrients needed by the plant enter root cells* is *not* the statement that best describes how nutrients enter the root cells of the onion plant. Any dissolved substance of proper chemical composition may diffuse into the root cells of a plant.

(3), (4) *Nutrients always move into the plant cells by active transport* and *the nutrients always move from an area of low concentration in the soil to an area of high concentration in root cells* are *not* the statements that best describe how nutrients enter the root cells of the onion plant. Active transport is a process by which substances are transported into root cells from an area of low relative concentration to an area of high relative concentration. Active transport against the concentration gradient requires the expenditure of cell energy.

74. **1** *Variation—different "beaks" were available* is the concept correctly matched with an example from *The Beaks of Finches* lab. The various tools of different shapes and sizes used by students to gather seeds simulated the natural variation of beak shapes and sizes found among the Galapagos finches.

WRONG CHOICES EXPLAINED

(2) *Adaptation—different types of food were available* is *not* the concept correctly matched with an example from *The Beaks of Finches* lab. The adaptations in this lab were the various tools of different shapes and sizes used by students to gather seeds, not the seeds themselves.

(3) *Selecting Agent—an insecticide was used to kill insects on one island* is *not* the concept correctly matched with an example from *The Beaks of Finches* lab. The selecting agent in this lab was the quantity of seeds available to be gathered by the various tools of different shapes and sizes.

(4) *Environment—"beaks" with similar qualities were used to gather seeds* is *not* the concept correctly matched with an example from *The Beaks of Finches* lab. In this lab, students used various tools of different shapes and sizes ("beaks"), not tools with similar qualities.

75. **3** *The two organisms produce many of the same proteins* is the evidence that would be considered the strongest for supporting a possible evolutionary relationship when comparing characteristics of two organisms. If two organisms share many proteins in common, this would indicate that they may share many genetic traits (many genes) in common and therefore may be related evolutionarily.

WRONG CHOICES EXPLAINED

(1), (2) *The two organisms are the same color* or *the two organisms are the same height* are *not* the evidences that would be considered the strongest for supporting a possible evolutionary relationship when comparing characteristics of two organisms. These are superficial characteristics that may be shared among many organisms with widely different evolutionary histories.

(4) *The two organisms are found in the same locations* is *not* the evidence that would be considered the strongest for supporting a possible evolutionary relationship when comparing characteristics of two organisms. Every stable ecosystem is populated by hundreds or thousands of different species with widely varying evolutionary histories.

76. **3** Five minutes after the slides were prepared, a student using a compound light microscope to observe the cells in leaves *A* and *B* would most likely see that *water had moved out of the cells of the leaf on slide B*. Actually, the student would more likely see that the contents of the cells in leaf *B* had shrunk and pulled away from their cell walls. The student might then correctly infer from this observation and his/her knowledge of biology that water had moved out of those cells via osmosis to the surrounding 6% salt and 94% water on the slide.

WRONG CHOICES EXPLAINED

(1), (2) Five minutes after the slides were prepared, a student using a compound light microscope to observe the cells in leaves *A* and *B* would *not* most likely see that *water had moved out of the cells of the leaf on slide* A or that *salt had moved into the cells of the leaf on slide* A. The leaf on slide *A* is mounted in a 1% salt and 99% water solution that is compatible with the Elodea's cell contents. The student would likely observe that the shape and size of these cells remained unchanged and correctly infer that their salt to water ratio was in equilibrium with the water on the slide.

(4) Five minutes after the slides were prepared, a student using a compound light microscope to observe the cells in leaves *A* and *B* would *not* most likely see that *salt had moved out of the cells of the leaf on slide B*. The stu-

dent would likely make the observation noted in the correct answer above and use his/her knowledge of biology to correctly infer that water, not salt, had moved out of the cells of this leaf.

77. One credit is allowed for correctly stating *one* reason why these two species probably do *not* live in the same area of this island. Acceptable responses include but are not limited to: [1]

- *Trees and cacti do not usually grow in the same areas/environmental conditions.*
- *Their food may be available only in certain areas on the island and the foods are different.*
- *Their foods do not grow in the same place since these foods require different growing conditions.*
- *There is another resource one of the species needs that is not available where the other species lives.*

78. One credit is allowed for circling "no" and for correctly supporting the answer. Acceptable responses include but are not limited to: [1]

- *The student's pulse rate, at 60 bpm, is below the range for his age group.*
- *It falls in the range for people older than 14.*
- *His rate of 60 beats per minute puts him in the adult range.*

79. One credit is allowed for correctly stating *one* reason why a person's heart rate increases during exercise. Acceptable responses include but are not limited to: [1]

- *During exercise, there is an increase in carbon dioxide in the blood, which causes the heart to beat faster.*
- *There is a feedback mechanism that causes the heart to beat faster in response to an increase in physical activity.*
- *When the heart beats faster, it increases the flow of oxygen-rich blood to the muscles.*
- *The heart pumping faster removes wastes faster.*

80. One credit is allowed for correctly identifying which species is likely to be more closely related to *Botana curus* and for supporting the answer. Acceptable responses include but are not limited to: [1]

- *Species Z—This species has the same amino acid sequence as* Botana curus.
- *Species Z—*Botana curus *has more DNA bases in common with this species than it does with species X or species Y.*

81. **2** A factor that contributed to the evolution of finches on the Galapagos Islands was most likely the *isolation of the finches on separate islands.* Any species that experiences the geographic isolation of its population into several smaller groups will undergo the phenomenon known as speciation. During speciation, genetic variations that exist among the parent species' populations will be subjected to natural selection in response to different environmental conditions. If separated long enough, these populations will continue to vary to the point that they can no longer interbreed. At this point, they may be considered different related species.

WRONG CHOICES EXPLAINED

(1) It is *not* true that a factor that contributed to the evolution of finches on the Galapagos Islands was most likely the *lack of variation in beak structure of the finches*. In fact, the beak shape among the several Galapagos finch species exhibits considerable variation.

(3) It is *not* true that a factor that contributed to the evolution of finches on the Galapagos Islands was most likely the *relatively constant atmospheric temperature*. The more similar the temperatures of the environments into which the finch populations separated, the more slowly evolution would occur.

(4) It is *not* true that a factor that contributed to the evolution of finches on the Galapagos Islands was most likely the *total lack of competition for food*. In fact, the Galapagos Islands do not contain abundant food supplies. For this reason, competition has always been a factor in the evolution of the finches.

82. **2** The rise of the liquid in the test tube that was observed after one hour can be explained as a result of the *water moving from the beaker into the test tube*. Diffusion is a process in which a substance moves with the concentration gradient from an area of higher relative concentration to an area of lower relative concentration of that substance. This process does not require the expenditure of cellular energy. In the diagram, water molecules are more concentrated in the beaker and less concentrated in the tube. As a result, the net movement of water is from the breaker, through the membrane, and into the tube. As the water molecules increase in number inside the tube, the water-starch solution is displaced up the tube.

WRONG CHOICES EXPLAINED

(1) It is *not* true that the rise of the liquid in the test tube that was observed after one hour can be explained as a result of the *starch solution moving into the test tube and out of the beaker*. Starch molecules are too large and complex to move through the extremely small pores in the dialysis membrane.

(3), (4) It is *not* true that the rise of the liquid in the test tube that was observed after one hour can be explained as a result of the *large starch molecules blocking the dialysis membrane* or the *dialysis membrane acting as a barrier to the water molecules*. If the dialysis membrane was being blocked or was actively blocking to prevent the movement of substances through it, the fluid level in the test tube would not change.

83. One credit is allowed for correctly describing *one* observation that would be made after one hour if a starch indicator solution was initially added to the water in the beaker. Acceptable responses include but are not limited to: [1]

- *The starch solution in the tube would turn blue-black.*
- *The water in the beaker would be amber/tan colored.*
- *The fluid in the test tube would be blue-black, indicating the presence of starch.*
- *A color change would occur in the test tube.*

84. One credit is allowed for correctly explaining why the DNA sample in lane 3 did not separate into fragments. Acceptable responses include but are not limited to: [1]

- *The restriction enzyme did not have a long enough time to cut the DNA sample.*
- *The DNA sample was not cut by this restriction enzyme.*
- *The restriction enzyme was denatured/did not work in the sample in lane 3.*

85. One credit is allowed for correctly stating *one* way that this experiment could be improved to obtain a valid conclusion. Acceptable responses include but are not limited to: [1]

- *The experimental group should have been larger.*
- *The data should have been collected for a longer time period.*
- *A control group should have been included in the study.*
- *Data should have been collected during the last quarter of the game.*
- *The experiment should be repeated by other competent scientists.*

STANDARDS/KEY IDEAS	JUNE 2018 QUESTION NUMBERS	NUMBER OF CORRECT RESPONSES
STANDARD 1		
Key Idea 1: The central purpose of scientific inquiry is to develop explanations of natural phenomena in a continuing and creative process.	31, 32	
Key Idea 2: Beyond the use of reasoning and consensus, scientific inquiry involves the testing of proposed explanations involving the use of conventional techniques and procedures and usually requiring considerable ingenuity.	33, 59	
Key Idea 3: The observations made while testing proposed explanations, when analyzed using conventional and invented methods, provide new insights into natural phenomena.	37, 44, 45, 60	
Laboratory Checklist	33, 44, 45, 59, 60	
STANDARD 4		
Key Idea 1: Living things are both similar to and different from each other and from nonliving things.	7, 8, 11, 14, 24, 29, 35, 42, 43, 63	
Key Idea 2: Organisms inherit genetic information in a variety of ways that result in continuity of structure and function between parents and offspring.	2, 4, 15, 17, 22, 66, 67, 68	
Key Idea 3: Individual organisms and species change over time.	34, 48, 49, 55, 61, 62	
Key Idea 4: The continuity of life is sustained through reproduction and development.	9, 13, 16, 18, 21, 23, 69, 70	
Key Idea 5: Organisms maintain a dynamic equilibrium that sustains life.	3, 12, 27, 38, 39, 40, 41, 47, 56, 64, 65, 71, 72	
Key Idea 6: Plants and animals depend on each other and their physical environment.	1, 6, 25, 30, 50, 51, 52, 53, 54	
Key Idea 7: Human decisions and activities have a profound impact on the physical and living environment.	5, 10, 19, 20, 26, 28, 36, 46, 57, 58	

STANDARDS/KEY IDEAS	JUNE 2018 QUESTION NUMBERS	NUMBER OF CORRECT RESPONSES
REQUIRED LABORATORIES		
Lab 1: "Relationships and Biodiversity"	75, 80, 84	
Lab 2: "Making Connections"	78, 79, 85	
Lab 3: "The Beaks of Finches"	74, 77, 81	
Lab 5: "Diffusion Through a Membrane"	73, 76, 82, 83	